Roswitha B. Grannell

IN MEMORY OF

Alfred Hall Drummond, Jr.
from his wife, Naomi

This volume is part of the Drummond Collection,
a gift to the University.

KEITH AND SHIRLEY CAMPBELL LIBRARY
THE MUSIC LIBRARY AT WILSON HALL
FRANK H. STEWART ROOM AND UNIVERSITY ARCHIVES

Introduction to Theoretical Geophysics

Introduction to Theoretical Geophysics

Charles B. Officer

Springer-Verlag New York · Heidelberg · Berlin
1974

Charles B. Officer
Partner, Marine Environmental Services
Hanover, New Hampshire

Library of Congress Cataloging in Publication Data

Officer, Charles B
    Introduction to theoretical geophysics.

    1. Geophysics. I. Title.
QE501.O33   551   73-15605
ISBN 0-387-06485-0

All rights reserved.

No part of this book may be translated or reproduced in
any form without written permission from Springer-Verlag.

© 1974 by Springer-Verlag New York Inc.

Printed in the United States of America.

ISBN 0-387-06485-0 Springer-Verlag New York Heidelberg Berlin
ISBN 3-540-06485-0 Springer-Verlag Berlin Heidelberg New York

Preface

It has been my intention in this book to give a coordinated treatment of the whole of theoretical geophysics. The book assumes a mathematical background through calculus and differential equations. It also assumes a reasonable background in physics and in elementary vector analysis. The level of the book is commensurate with that of a senior undergraduate or first year graduate course. Its aim is to provide the reader with a survey of the whole of theoretical geophysics.

The emphasis has been on the basic and the elementary. The expert in any one of the several disciplines covered here will find much lacking from his particular area of investigation; no apology is made for that. In order to treat all aspects in a coordinated manner, the simplest type of mathematical notation for the various physical problems has been used, namely, that of scalars, three-dimensional vectors, and the vector operators, gradient, curl, divergence, etc. It is appreciated that this elementary notation often may not be the most conducive to the solution of some of the more complex geophysical problems.

The derivations are, in almost every case, carried through in considerable detail. Sometimes the particulars of the algebra and calculus have been omitted and relegated to one of the problems following the section. The emphasis has been on the physics of the derivations and on explaining the various physical principles important in geophysics, such as continuity, mixing, diffusion, conduction, convection, precession, wobble, rays, waves, dispersion, and potential theory.

The problems are considered an important part of the text. They include filling in mathematical details in the derivations and applications and extensions of the derivations. They do vary considerably in complexity.

The order of the chapters has not been entirely arbitrary. Part I is necessarily a hodgepodge. It includes discussions of the solutions of some of the basic partial differential equations of physics and of the properties of some of the functions particularly pertinent to geophysics. Part II is on the thermodynamics of the earth and the hydrodynamics of the earth's outer surface, the ocean. Part III—seismology, gravity, and magnetism—is largely on the applications of various physical principles to a definition and investigation of the earth and its properties. Part IV is on the dynamics of the earth itself and necessarily includes results from various of the preceding chapters. In addition, there is some increase in complexity as the chapters move along.

I should like to express my thanks to Dartmouth College for the use of the facilities at the Baker Library and at the Department of Earth Sciences.

Hanover, New Hampshire CHARLES B. OFFICER
January, 1974

Contents

Part I Introduction   1

CHAPTER 1. MATHEMATICAL CONSIDERATIONS   3
- 1.1 Vector Analysis 3
- 1.2 Curvilinear Coordinates 13
- 1.3 Vector Relations in Curvilinear Coordinates . 15
- 1.4 Spherical Coordinates 18
- 1.5 Cylindrical Coordinates 20
- 1.6 Legendre Polynomials 21
- 1.7 Laplace's Equation 27
- 1.8 Fourier Series 34
- 1.9 Fourier Integrals 37
- 1.10 Wave Equation 42
- 1.11 Associated Legendre Polynomials . . . 49
- 1.12 Bessel Functions 52
- 1.13 Velocity and Acceleration Referred to Moving Axes 56
- References 59

Part II Thermodynamics and Hydrodynamics   61

CHAPTER 2. THERMODYNAMICS OF THE EARTH   63
- 2.1 Thermodynamics 63
- 2.2 Implicit Functions in Thermodynamics . . 65
- 2.3 Gravitational Adiabatic Equilibrium . . . 70
- 2.4 Heat Conduction Equation 72
- 2.5 Periodic Flow of Heat 74
- 2.6 Heat Flow 76
- 2.7 Internal Heat of the Earth 82
- References 87

CHAPTER 3. HYDRODYNAMICS — 89
- 3.1 Kinematic Preliminaries 89
- 3.2 Conjugate Functions 92
- 3.3 Equation of Continuity 94
- 3.4 Equation of Motion 96
- 3.5 Kelvin's Circulation Theorem . . . 99
- 3.6 Equation of Motion of a Viscous Fluid . . 101
- References 103

CHAPTER 4. PHYSICAL OCEANOGRAPHY—CIRCULATION 104
- 4.1 Thermodynamic Circulation 104
- 4.2 Continuity, Mixing, and Diffusion . . . 107
- 4.3 Ocean Mixing 114
- 4.4 Estuary Mixing 117
- 4.5 Equation of Motion and Circulation Theorem . 123
- 4.6 Pressure Gradient and Geostrophic Effects . . 125
- 4.7 Inertia Effects 130
- 4.8 Equation of Motion with Internal Friction . . 131
- 4.9 Friction and Geostrophic Effects . . . 133
- 4.10 Friction, Geostrophic, and Pressure Gradient Effects 136
- 4.11 Wind-Driven Ocean Circulation . . . 139
- References 143

CHAPTER 5. PHYSICAL OCEANOGRAPHY—WAVES AND TIDES 144
- 5.1 Tidal Waves 144
- 5.2 Driven Tidal Waves 150
- 5.3 Seiches 152
- 5.4 Geostrophic Effects on Tidal Waves . . . 154
- 5.5 Internal Tidal Waves 159
- 5.6 Surface Waves 163
- 5.7 Permanent Waves 166
- 5.8 Waves Due to a Local Disturbance . . . 169
- 5.9 Equilibrium Theory of the Tides . . . 173
- 5.10 Dynamical Theory of the Tides . . . 176
- References 178

Part III Seismology, Gravity, and Magnetism — 179

CHAPTER 6. SEISMOLOGY—RAY THEORY — 181
- 6.1 Dynamics — 181
- 6.2 Bodily Elastic Waves — 186
- 6.3 Reflection and Refraction of Elastic Waves — 190
- 6.4 Development of Solution in Terms of Rays — 202
- 6.5 Ray Characteristics for a Flat Earth — 210
- 6.6 Ray Characteristics for a Spherically Stratified Earth — 225
- References — 233

CHAPTER 7. SEISMOLOGY—WAVE THEORY — 234
- 7.1 Normal Modes — 234
- 7.2 Dispersion — 241
- 7.3 Dispersive Normal Modes — 248
- 7.4 Rayleigh Waves — 253
- 7.5 Love Waves — 256
- References — 258

CHAPTER 8. GRAVITY — 260
- 8.1 Fundamental Relations — 260
- 8.2 Gauss' Law and Green's Equivalent Layer — 262
- 8.3 Gravitational Potential and Attraction for an Ellipsoidal Earth — 269
- 8.4 Geocentric and Geographic Latitude — 274
- 8.5 Deviations of the Geoid from a Reference Ellipsoid — 275
- 8.6 Deflection of the Vertical — 280
- 8.7 Gravity Corrections and Gravity Anomalies — 283
- 8.8 Internal Mass Distribution of the Earth — 292
- References — 302

CHAPTER 9. GEOMAGNETISM — 303
- 9.1 Electromagnetism — 303
- 9.2 Earth's Main Magnetic Field — 307
- 9.3 Electromagnet Theory — 311
- 9.4 Internal and Atmospheric Dynamos — 316
- 9.5 Electromagnetic Induction in the Earth — 321
- References — 325

Part IV Dynamics of the Earth 327

CHAPTER 10. EARTH MOTION, ROTATION, AND DEFORMATION 329

10.1 Motion under an Inverse Square Force of Attraction 329
10.2 Precession of the Equinoxes 330
10.3 Gyroscopic Compass 335
10.4 Eulerian Free Motion 337
10.5 Marine Tides 339
10.6 Earth Tides 343
10.7 Chandler Wobble 347
10.8 Tidal Friction 350
10.9 Free Oscillations of the Earth . . . 354
References 358

CHAPTER 11. EARTH CRUSTAL AND MANTLE DEFORMATION 359

11.1 Introduction 359
11.2 Loading of an Elastic Crust . . . 360
11.3 Loading of a Floating Lithosphere . . 364
11.4 Postglacial Uplift 369
11.5 Thermal Contraction of the Earth . . 370
11.6 Thermal Convection Currents in the Mantle . 373
References 378

INDEX 379

Part One

Introduction

CHAPTER 1

MATHEMATICAL CONSIDERATIONS

1.1 Vector Analysis

It is assumed that the reader has a working knowledge of elementary, three-dimensional vector analysis. We shall review, or rather restate, the more important relations, which will be needed in the derivations in the later sections.

A vector will be denoted by a letter in **bold face** type, the scalar magnitude being indicated by the same **bold face** letter between vertical bars, or simply by the same letter in *italics*. Thus, the scalar magnitude of the vector **P** is designated by either $|\mathbf{P}|$ or P.

A *unit vector* is one whose magnitude is unity, and the three unit vectors parallel, respectively, to the X, Y, Z rectangular axes will be designated by **i**, **j**, **k**. Rectangular axes are always taken to constitute a right-handed set; that is, the positive direction of the Z axis is determined by the sense of advance of a right-handed screw rotated in the XY plane from the positive direction of the X axis to that of the Y axis through the right angle between them.

Occasionally, it is convenient to specify the direction of a vector by its three-dimensional cosines l, m, n with the axes X, Y, Z, and the reader is reminded of the geometrical relations

$$l^2 + m^2 + n^2 = 1 \tag{1.1}$$

$$\cos\theta = l_1 l_2 + m_1 m_2 + n_1 n_2 \tag{1.2}$$

In the second of these, θ is the angle between two lines whose direction cosines are l_1, m_1, n_1 and l_2, m_2, n_2, respectively.

Two vectors add as shown in Fig. 1.1 so that magnitude of their *sum* or *resultant*, **R**, is simply

$$R^2 = P^2 + Q^2 + 2PQ\cos\alpha \tag{1.3}$$

The *components* of a vector **P** are simply any vectors whose sum is **P**. The components most frequently used are those parallel to the axes X, Y, Z,

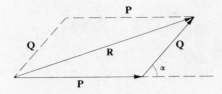

Fig. 1.1

designated by P_x, P_y, P_z and referred to as the *rectangular components*, for which the relation

$$P^2 = P_x^2 + P_y^2 + P_z^2 \tag{1.4}$$

holds.

A surface, such as ABC in Fig. 1.2, may be represented by a vector **σ**, whose magnitude is equal to the area of the surface and whose direction is normal to the surface. For a closed surface, the outward drawn normal is taken as positive; for a surface that is not closed, the positive sense of describing the periphery is connected with the positive sense of the normal, as shown in Fig. 1.2.

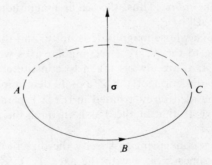

Fig. 1.2

Transformation of the components of a vector **P** from one set of rectangular axes to another is the same as the transformation of the coordinates themselves. We then have from the diagram of Fig. 1.3, where l_1, l_2, l_3 are the direction cosines of X' referred to X, Y, Z, and so on, that

	X	Y	Z
X'	l_1	l_2	l_3
Y'	m_1	m_2	m_3
Z'	n_1	n_2	n_3

Fig. 1.3

$$x' = l_1 x + l_2 y + l_3 z$$
$$y' = m_1 x + m_2 y + m_3 z$$
$$z' = n_1 x + n_2 y + n_3 z$$

or (1.5)

$$x = l_1 x' + m_1 y' + n_1 z'$$
$$y = l_2 x' + m_2 y' + n_2 z'$$
$$z = l_3 x' + m_3 y' + n_3 z'$$

or the same relations for P_x' in terms of P_x, P_y, P_z, and so on, and $\mathbf{i}'$ in terms of $\mathbf{i}, \mathbf{j}, \mathbf{k}$, and so on, and vice versa.

The *scalar* or *dot product* of two vectors is defined as a scalar equal in magnitude to the product of the magnitudes of the two vectors by the cosine of the angle between them. From this definition, we then have the relations

$$\mathbf{Q} \cdot \mathbf{P} = \mathbf{P} \cdot \mathbf{Q} \tag{1.6}$$

and
$$(\mathbf{P} + \mathbf{Q}) \cdot \mathbf{R} = \mathbf{P} \cdot \mathbf{R} + \mathbf{Q} \cdot \mathbf{R} \tag{1.7}$$

$$\mathbf{i} \cdot \mathbf{i} = \mathbf{j} \cdot \mathbf{j} = \mathbf{k} \cdot \mathbf{k} = 1 \tag{1.8}$$
$$\mathbf{i} \cdot \mathbf{j} = \mathbf{j} \cdot \mathbf{k} = \mathbf{k} \cdot \mathbf{i} = 0$$

and
$$\mathbf{P} \cdot \mathbf{Q} = P_x Q_x + P_y Q_y + P_z Q_z \tag{1.9}$$
$$P^2 = \mathbf{P} \cdot \mathbf{P} = P_x^2 + P_y^2 + P_z^2$$

The *vector* or *cross product* of two vectors is a vector perpendicular to their plane in the sense of a right-handed screw rotated from the first to the second through the smaller angle between their positive directions as shown in Fig. 1.4, and with a magnitude equal to the product of the magnitudes of

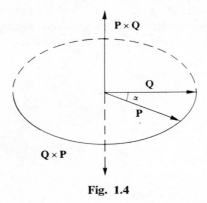

Fig. 1.4

the two vectors by the sine of the angle between them. From this definition, we then have the relations

6 ¶ Introduction

and
$$Q \times P = -P \times Q \qquad P \times P = 0 \tag{1.10}$$
and
$$(P+Q) \times R = P \times R + Q \times R \tag{1.11}$$
and
$$i \times j = k \qquad j \times k = i \qquad k \times i = j$$
$$i \times i = j \times j = k \times k = 0 \tag{1.12}$$

$$P \times Q = \begin{vmatrix} i & j & k \\ P_x & P_y & P_z \\ Q_x & Q_y & Q_z \end{vmatrix} \tag{1.13}$$
$$= i(P_y Q_z - P_z Q_y) + j(P_z Q_x - P_x Q_z) + k(P_x Q_y - P_y Q_x)$$

where the expression of the right-hand side of Eq. (1.13) is the determinant.

For products involving three vectors, we can have two distinct combinations. The *triple scalar product* is the scalar

$$(P \times Q) \cdot R = R \cdot (P \times Q) = -R \cdot (Q \times P) = -(Q \times P) \cdot R \tag{1.14a}$$

where the vector product $P \times Q$ is formed first, and then the scalar product of this vector is taken with R. We can then obtain the further relations

$$(P \times Q) \cdot R = P \cdot (Q \times R) \tag{1.14b}$$

and

$$P \times Q \cdot R = \begin{vmatrix} P_x & P_y & P_z \\ Q_x & Q_y & Q_z \\ R_x & R_y & R_z \end{vmatrix} \tag{1.15}$$

and

$$i \times j \cdot k = i \cdot j \times k = -i \cdot k \times j = 1 \tag{1.16}$$

The *triple vector product* is the vector

$$(P \times Q) \times R = -R \times (P \times Q) = R \times (Q \times P) \tag{1.17}$$

where the vector product $P \times Q$ is formed first, and then the cross product of this vector is taken with R. It is apparent from Fig. 1.5 that $(P \times Q) \times R$ will lie in the plane of P and Q. We can then obtain the relations

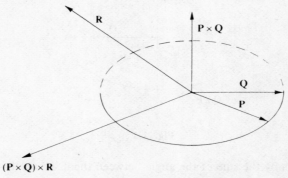

Fig. 1.5

$$(\mathbf{P} \times \mathbf{Q}) \times \mathbf{R} = \mathbf{R} \cdot \mathbf{P}\mathbf{Q} - \mathbf{R} \cdot \mathbf{Q}\mathbf{P}$$
$$\mathbf{R} \times (\mathbf{P} \times \mathbf{Q}) = \mathbf{R} \cdot \mathbf{Q}\mathbf{P} - \mathbf{R} \cdot \mathbf{P}\mathbf{Q} \tag{1.18}$$

For the derivative of a vector function $\mathbf{r}$ of the scalar t, defined by

$$\mathbf{r}(t) = \mathbf{i}x(t) + \mathbf{j}y(t) + \mathbf{k}z(t)$$

and where X, Y, Z are fixed rectangular coordinate axes, we have

$$\frac{d\mathbf{r}}{dt} = \mathbf{i}\frac{dx}{dt} + \mathbf{j}\frac{dy}{dt} + \mathbf{k}\frac{dz}{dt}$$

and

$$\frac{d^2\mathbf{r}}{dt^2} = \mathbf{i}\frac{d^2 x}{dt^2} + \mathbf{j}\frac{d^2 y}{dt^2} + \mathbf{k}\frac{d^2 z}{dt^2} \tag{1.19}$$

For the derivatives of the scalar and vector products, we shall also have

$$\frac{d}{dt}(\mathbf{P} \cdot \mathbf{Q}) = \frac{d\mathbf{P}}{dt} \cdot \mathbf{Q} + \mathbf{P} \cdot \frac{d\mathbf{Q}}{dt}$$

and

$$\frac{d}{dt}(\mathbf{P} \times \mathbf{Q}) = \frac{d\mathbf{P}}{dt} \times \mathbf{Q} + \mathbf{P} \times \frac{d\mathbf{Q}}{dt} \tag{1.20}$$

The vector differential operator ∇ is defined as the quantity

$$\nabla = \mathbf{i}\frac{\partial}{\partial x} + \mathbf{j}\frac{\partial}{\partial y} + \mathbf{k}\frac{\partial}{\partial z} \tag{1.21}$$

If we wish to use spherical coordinates r, θ, φ—where r is the radius vector, θ the polar angle, and φ the azimuth and where $\mathbf{r}_1, \boldsymbol{\theta}_1,$ and $\boldsymbol{\varphi}_1$, are the respective unit vectors—∇ will be given by

$$\nabla = \mathbf{r}_1 \frac{\partial}{\partial r} + \boldsymbol{\theta}_1 \frac{\partial}{r \partial \theta} + \boldsymbol{\varphi}_1 \frac{\partial}{r \sin \theta \, \delta \varphi} \tag{1.22}$$

If we wish to use cylinderical coordinates ρ, φ, z—where ρ is the distance coordinate normal to the cylindrical axis, φ the azimuth, and z the axial distance coordinate and where $\boldsymbol{\rho}_1, \boldsymbol{\varphi}_1$, and $\mathbf{k}$ are the respective unit vectors—∇ will be given by

$$\nabla = \boldsymbol{\rho}_1 \frac{\partial}{\partial \rho} + \boldsymbol{\varphi}_1 \frac{\partial}{\rho \partial \varphi} + \mathbf{k}\frac{\partial}{\partial z} \tag{1.23}$$

If $\Phi(x, y, z)$ is a proper scalar function of the coordinates, then the vector

$$\nabla \Phi = \mathbf{i}\frac{\partial \Phi}{\partial x} + \mathbf{j}\frac{\partial \Phi}{\partial y} + \mathbf{k}\frac{\partial \Phi}{\partial z} \tag{1.24}$$

is called the *gradient* of Φ. Considering $\Phi(x, y, z) = C$, a constant, to define

8 ¶ *Introduction*

a surface in space, the geometrical significance of $\nabla\Phi$ is that it is a vector having both the magnitude and the direction of the greatest space rate of increase of Φ. It is normal to the surface, $\Phi = C$, and its component in any direction is equal to the space rate of increase of Φ in that direction so that, as shown in Fig. 1.6, the incremental increase in Φ from A to B is simply

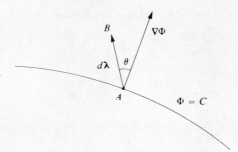

Fig. 1.6

$$d\Phi = \nabla\Phi d\lambda \cos\theta = \nabla\Phi \cdot d\mathbf{k} \qquad (1.25)$$

If $\mathbf{V}(x, y, z)$ is a proper vector function of the coordinates, then the scalar

$$\nabla \cdot \mathbf{V} = \frac{\partial V_x}{\partial x} + \frac{\partial V_y}{\partial y} + \frac{\partial V_z}{\partial z} \qquad (1.26)$$

is called the *divergence* of $\mathbf{V}$. Considering $\mathbf{V} = \rho\mathbf{v}$ to represent the mass of fluid flowing through a unit cross section in a unit time, where ρ is the fluid density and $\mathbf{v}$ its velocity, the quantity $-\nabla \cdot \mathbf{V}$ represents the total increase in mass per unit volume per unit time due to the flow and leads directly to the *equation of continuity*:

$$\frac{\partial \rho}{\partial t} = -\nabla \cdot \mathbf{V} \qquad (1.27)$$

or

$$\frac{\partial \rho}{\partial t} + \nabla \cdot (\rho\mathbf{v}) = 0 \qquad (1.28)$$

If the fluid is incompressible, that is, ρ a constant, these equations reduce simply to

$$\nabla \cdot \mathbf{V} = \nabla \cdot \mathbf{v} = 0 \qquad (1.29)$$

Again if $\mathbf{V}(x, y, z)$ is a proper vector function of the coordinate, then the vector

$$\nabla \times \mathbf{V} = \begin{vmatrix} \mathbf{i} & \mathbf{j} & \mathbf{k} \\ \frac{\partial}{\partial x} & \frac{\partial}{\partial y} & \frac{\partial}{\partial z} \\ V_x & V_y & V_z \end{vmatrix} \qquad (1.30)$$

is called the *curl* or *rotation* of **V**. Considering a rigid body, which has a constant angular velocity **ω** about an origin O fixed in the body, O itself having the constant linear velocity $\mathbf{v}_0$, as shown in Fig. 1.7, the total linear

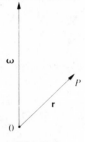

Fig. 1.7

velocity of a point P will be

$$\mathbf{v} = \mathbf{v}_0 + \boldsymbol{\omega} \times \mathbf{r} \tag{1.31}$$

and we can obtain the relation

$$\nabla \times \mathbf{v} = \boldsymbol{\omega} \tag{1.32}$$

so that the curl of the linear velocity is equal to twice the angular velocity.

Taking successive application of ∇, we have that the divergence of the gradient is

$$\nabla \cdot \nabla \Phi = \frac{\partial^2 \Phi}{\partial x^2} + \frac{\partial^2 \Phi}{\partial y^2} + \frac{\partial^2 \Phi}{\partial z^2} \tag{1.33}$$

The operator

$$\nabla \cdot \nabla = \frac{\partial^2}{\partial x^2} + \frac{\partial^2}{\partial y^2} + \frac{\partial^2}{\partial z^2} \tag{1.34}$$

is referred to as the *Laplacian* and may be equally well applied to a vector function **V**. We also have that the curl of the gradient is uniquely zero, or

$$\nabla \times \nabla \Phi = 0 \tag{1.35}$$

and the divergence of curl is uniquely zero, or

$$\nabla \cdot \nabla \times \mathbf{V} = 0 \tag{1.36}$$

If the curl of a vector function of position in space vanishes everywhere in a region τ, **W** is said to be *irrotational*; from Eq. (1.35) it is seen that if **W** is the gradient of a scalar function Φ, then **W** is irrotational. Further, if the divergence of a vector function **W** vanishes everywhere, **W** is said to be *solenoidal*; from Eq. (1.36) we see that if **W** is the curl of a vector function **V**, then **W** is solenoidal. The vector product, $\nabla \times \nabla \times \mathbf{V}$, of the curl of the vector **V** is also important and is given by

$$\nabla \times \nabla \times \mathbf{V} = \nabla \times (\nabla \times \mathbf{V}) = \nabla(\nabla \cdot \mathbf{V}) - \nabla \cdot \nabla \mathbf{V} \tag{1.37}$$

From the relations above, we can also obtain the useful result that any vector **W**, which can be expressed in terms of the Laplacian of another vector, can be considered to be composed of two vector functions, one of which is irrotational and the other of which is solenoidal, or

$$\mathbf{W} = \nabla\Phi + \nabla \times \mathbf{A} \tag{1.38}$$

$$\nabla \cdot \mathbf{A} = 0$$

where Φ is referred to as the *scalar potential* of **W** and **A** as its *vector potential*.

If **V** (x, y, z) is again a proper vector function, then the integral

$$\int_A^B \mathbf{V} \cdot d\boldsymbol{\lambda}$$

as defined by Fig. 1.8, is referred to as the *line integral* of **V** along the path AB.

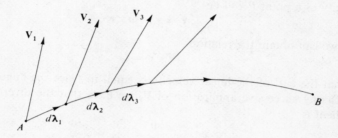

Fig. 1.8

Considering **V** to be the force on a moving particle, then the line integral of **V** over the path described by the particle is simply the work done by the force. If **V** is the gradient $\nabla\Phi$ of a scalar function of position, it follows that the line integral is independent of the path from A to B and that the line integral around a closed curve, denoted by

$$\oint \mathbf{V} \cdot d\boldsymbol{\lambda}$$

is uniquely zero. It also follows that if the line integral of **V** vanishes about every closed path, **V** must be the gradient of some scalar function Φ and therefore irrotational.

Considering a surface σ, as shown in Fig. 1.9, the integral

$$\int_\sigma \mathbf{V} \cdot d\boldsymbol{\sigma}$$

is referred to as the *surface integral* of **V** over the surface σ. The surface integral of a vector **V** is called the *flux* of **V** through the surface. If again **V** is the mass of fluid flowing through a unit cross section per unit time, then the

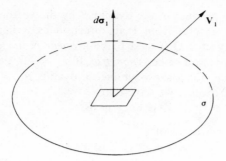

Fig. 1.9

surface integral is the mass of fluid passing through the entire surface per unit time.

We next have the integral relation, known as *Gauss' theorem*, which states that the volume integral of the divergence of a vector function **V** taken over any volume τ is equal to the surface integral of **V** taken over the closed surface surrounding the volume τ, or

$$\int_\tau \nabla \cdot \mathbf{V}\, d\tau = \int_\sigma \mathbf{V} \cdot d\boldsymbol{\sigma} \tag{1.39}$$

and its corollary, *Green's theorem*, which states that

$$\int_\tau (u\nabla \cdot \nabla v - v\nabla \cdot \nabla u) d\tau = \int_\sigma (u\nabla v - v\nabla u) \cdot d\boldsymbol{\sigma} \tag{1.40}$$

where u and v are two scalar functions of the coordinates.

We also have the integral relation, known as *Stokes' theorem*, which states that the surface integral of the curl of a vector function **V** taken over any surface σ is equal to the line integral of **V** around the periphery λ of the surface, or

$$\int_\sigma \nabla \times \mathbf{V} \cdot d\boldsymbol{\sigma} = \oint \mathbf{V} \cdot d\boldsymbol{\lambda} \tag{1.41}$$

If the surface to which Stokes' theorem is applied is a closed surface, the length of the periphery is zero, and hence the left-hand side of Eq. (1.41) is equal to zero.

We shall also have brief need of the simplest of vector operators, Ψ known as a *dyadic* or *tensor of the second rank*, to transform one vector into another vector and defined by

$$\begin{aligned}\Psi = &\ a_{11}\mathbf{ii} + a_{12}\mathbf{ij} + a_{13}\mathbf{ik} \\ &+ a_{21}\mathbf{ji} + a_{22}\mathbf{jj} + a_{23}\mathbf{jk} \\ &+ a_{31}\mathbf{ki} + a_{32}\mathbf{kj} + a_{33}\mathbf{kk}\end{aligned} \tag{1.42}$$

By taking the scalar product of the dyadic Ψ by the vector $\mathbf{P}$, a new vector $\mathbf{Q}$, which differs in general from $\mathbf{P}$ in direction as well as magnitude, is obtained. The new vector $\mathbf{Q}$ is said to be a *linear vector function* of $\mathbf{P}$. If $a_{ij} = a_{ji}$, the dyadic is said to be *symmetric*. If $a_{ij} = -a_{ji}$, the dyadic is said to be *skew-symmetric*; for a skew-symmetric dyadic, $a_{11} = a_{22} = a_{33} = 0$.

For a symmetric dyadic, we have that

$$\mathbf{P}\cdot\Psi = \Psi\cdot\mathbf{P} \tag{1.43}$$

and for a skew-symmetric dyadic,

$$\mathbf{P}\cdot\Psi = -\Psi\cdot\mathbf{P}$$

The *conjugate* Ψ_c of any dyadic Ψ is obtained by interchanging the order of the unit vectors in each of the members of Eq. (1.42) so that

$$\mathbf{P}\cdot\Psi = \Psi_c\cdot\mathbf{P} \tag{1.44}$$

The symmetric dyadic

$$I = \mathbf{ii} + \mathbf{jj} + \mathbf{kk} \tag{1.45}$$

is known as the *idemfactor* and has the simple property that

$$\mathbf{P}\cdot I = I\cdot\mathbf{P} = \mathbf{P} \tag{1.46}$$

It can be shown that any dyadic may be expressed as the sum of a symmetric and a skew-symmetric dyadic and that any symmetric dyadic may be reduced to

$$\Psi = a_x\mathbf{ii} + a_y\mathbf{jj} + a_z\mathbf{kk} \tag{1.47}$$

by a suitable orientation of coordinate axes.

Problem 1.1(a) By vector algebra, prove that the diagonals of a parallelogram bisect each other.

Problem 1.1(b) By vector algebra, prove that the line that joins one corner of a parallelogram to the middle point of an opposite side trisects the diagonal and is trisected by it.

Problem 1.1(c) A vector has components $x^2A_x + xyA_y + xzA_z$, $xyA_x + y^2A_y + yzA_z$, $xzA_x + yzA_y + z^2A_z$ relative to X, Y, Z where $\mathbf{A}$ is a constant vector. Show that its components relative to X', Y', Z' are given by the same function of $x', y', z', A'_x, A'_y, A'_z$. Such a vector is known as a *proper vector function*.

Problem 1.1(d) Show by use of the expansion of the scalar product that if two vectors have direction cosines l_1, m_1, n_1 and l_2, m_2, n_2, respectively, and θ is the angle between them, then $\cos\theta = l_1l_2 + m_1m_2 + n_1n_2$.

Problem 1.1(e) If $\mathbf{A}$ and $\mathbf{B}$ are the sides of a parallelogram, $\mathbf{C}$ and $\mathbf{D}$ the diagonals, and θ the angle between $\mathbf{A}$ and $\mathbf{B}$, show by vector algebra that $C^2 + D^2 = 2(A^2 + B^2)$ and that $C^2 - D^2 = 4AB\cos\theta$.

Problem 1.1(f) If $l_1, l_2, l_3; m_1, m_2, m_3; n_1, n_2, n_3$ are the direction cosines of the rectangular axes X', Y', Z' relative to X, Y, Z, deduce by vector methods the relations $l_1 = m_2n_3 - n_2m_3; l_2 = m_3n_1 - m_1n_3; l_3 = m_1n_2 - m_2n_1$, and so on.

Problem 1.1(g) Verify the relations

$$(\mathbf{A} \times \mathbf{B}) \cdot (\mathbf{C} \times \mathbf{D}) = (\mathbf{A} \cdot \mathbf{C})(\mathbf{B} \cdot \mathbf{D}) - (\mathbf{A} \cdot \mathbf{D})(\mathbf{B} \cdot \mathbf{C})$$

$$(\mathbf{A} \times \mathbf{B}) \times (\mathbf{C} \times \mathbf{D}) = (\mathbf{A} \cdot \mathbf{C} \times \mathbf{D})\mathbf{B} - (\mathbf{B} \cdot \mathbf{C} \times \mathbf{D})\mathbf{A}$$

$$= (\mathbf{A} \cdot \mathbf{B} \times \mathbf{D})\mathbf{C} - (\mathbf{A} \cdot \mathbf{B} \times \mathbf{C})\mathbf{D}$$

Problem 1.1(h) If $\boldsymbol{\alpha}$ is a vector constant in magnitude but variable in direction, show that $\boldsymbol{\alpha} \cdot d\boldsymbol{\alpha} = 0$.

Problem 1.1(i) If $\mathbf{r}_1$ is a unit vector of variable direction along the radius vector, the position vector of a moving point may be written: $\mathbf{r} = \mathbf{r}_1 r$. Find, by vector methods, the components of the acceleration parallel and perpendicular to the radius vector of a particle moving in the XY plane.

Problem 1.1(j) Show that $\nabla \cdot \mathbf{r} = 3$ and $\nabla \times \mathbf{r} = 0$ where $\mathbf{r} = \mathbf{i}x + \mathbf{j}y + \mathbf{k}z$.

Problem 1.1(k) Show by expanding the vectors in terms of their components that

$$\nabla \cdot (\mathbf{U} \times \mathbf{V}) = \mathbf{V} \cdot \nabla \times \mathbf{U} - \mathbf{U} \cdot \nabla \times \mathbf{V}$$

and that

$$\nabla \times (\mathbf{U} \times \mathbf{V}) = (\mathbf{V} \cdot \nabla)\mathbf{U} - \mathbf{V}(\nabla \cdot \mathbf{U}) - (\mathbf{U} \cdot \nabla)\mathbf{V} + \mathbf{U}(\nabla \cdot \mathbf{V})$$

Problem 1.1(l) Show that $\nabla \cdot \nabla(1/r) = 0$ where $\mathbf{r} = \mathbf{i}x + \mathbf{j}y + \mathbf{k}z$.

Problem 1.1(m) Derive the equation of continuity by the use of Gauss' theorem.

Problem 1.1(n) By the use of Gauss' theorem and Stokes' theorem prove that $\nabla \cdot \nabla \times \mathbf{V} = 0$ and that $\nabla \times \nabla \Phi = 0$.

1.2 Curvilinear Coordinates

Frequently it is necessary, or convenient, to derive various physical laws or solve problems in coordinates other than rectangular (Cartesian) coordinates. It is the purpose of the next four sections to show how vectors and vector operators may be formulated in more general, *curvilinear* coordinates.

In Cartesian coordinates, the position of a point is determined by the three coordinates x, y, z and is the intersection of the three mutually perpendicular planes determined by each of these coordinates, respectively. If we designate by q_1, q_2, q_3 our three generalized coordinates, we can express x, y, and z in terms of them by

$$x = x(q_1, q_2, q_3)$$
$$y = y(q_1, q_2, q_3) \qquad (1.48)$$
$$z = z(q_1, q_2, q_3)$$

and conversely

$$q_1 = q_1(x, y, z)$$
$$q_2 = q_2(x, y, z) \qquad (1.49)$$
$$q_3 = q_3(x, y, z)$$

Any point may then be specified by either x, y, z or q_1, q_2, q_3.

From Eqs. (1.48) we may write directly,

$$dx = \frac{\partial x}{\partial q_1} dq_1 + \frac{\partial x}{\partial q_2} dq_2 + \frac{\partial x}{\partial q_3} dq_3$$
$$dy = \frac{\partial y}{\partial q_1} dq_1 + \frac{\partial y}{\partial q_2} dq_2 + \frac{\partial y}{\partial q_3} dq_3 \qquad (1.50)$$
$$dz = \frac{\partial z}{\partial q_1} dq_1 + \frac{\partial z}{\partial q_2} dq_2 + \frac{\partial z}{\partial q_3} dq_3$$

We shall designate by ds the incremental distance between two adjacent points. Clearly, in Cartesian coordinates $ds^2 = dx^2 + dy^2 + dz^2$ or in terms of our curvilinear coordinates from Eqs. (1.50)

$$\begin{aligned} ds^2 &= dx^2 + dy^2 + dz^2 \\ &= Q_1^2 dq_1^2 + Q_2^2 dq_2^2 + Q_3^2 dq_3^2 \\ &\quad + 2Q_{12} dq_1 dq_2 + 2Q_{13} dq_1 dq_3 + 2Q_{23} dq_2 dq_3 \end{aligned} \qquad (1.51)$$

where

$$Q_i^2 = \left(\frac{\partial x}{\partial q_i}\right)^2 + \left(\frac{\partial y}{\partial q_i}\right)^2 + \left(\frac{\partial z}{\partial q_i}\right)^2 \qquad (i = 1, 2, 3)$$

and

$$Q_{ij} = \frac{\partial x}{\partial q_i} \frac{\partial x}{\partial q_j} + \frac{\partial y}{\partial q_i} \frac{\partial y}{\partial q_j} + \frac{\partial z}{\partial q_i} \frac{\partial z}{\partial q_j} \qquad (i, j = 1, 2, 3, i \neq j)$$

Now the distance along any curvilinear coordinate direction we shall designate for convenience by s_1, s_2, s_3 corresponding to q_1, q_2, q_3. In general, and unlike Cartesian coordinates, the distance and coordinate elements in curvilinear coordinates are not the same. For example, in two-dimensional polar coordinates, the coordinate is the polar angle and the distance the arc length. From Eq. (1.51) we may write for the incremental coordinate distances, when variation is limited to only one of the q_i,

$$ds_i = Q_i dq_i \qquad (i = 1, 2, 3) \tag{1.52}$$

In general, we are only interested in curvilinear coordinate systems that are *orthogonal*, that is, coordinate systems in which the three coordinate surfaces always intersect at right angles. For orthogonal coordinates, the incremental distance ds between two points is simply in terms of the coordinate incremental distances

$$ds^2 = ds_1^2 + ds_2^2 + ds_3^2 \tag{1.53}$$

From Eqs. (1.51) and (1.52) we see that Eq. (1.53) will only be valid for all points if $Q_{ij} = 0$. For orthogonal curvilinear coordinates, Eq. (1.51) reduces to

$$ds^2 = Q_1^2 dq_1^2 + Q_2^2 dq_2^2 + Q_3^2 dq_3^2 \tag{1.54}$$

The incremental volume element $d\tau$ and the three incremental surface elements normal to the three coordinate directions become

$$d\tau = ds_1 ds_2 ds_3 = Q_1 Q_2 Q_3 dq_1 dq_2 dq_3 \tag{1.55}$$

and

$$d\sigma_{ij} = ds_i ds_j = Q_i Q_j dq_i dq_j \qquad (i_1 j = 1, 2, 3, i \neq j) \tag{1.56}$$

1.3 Vector Relations in Curvilinear Coordinates

For convenience, we shall designate by $\mathbf{u}_1$, $\mathbf{u}_2$, $\mathbf{u}_3$ the three unit vectors along the curvilinear coordinate axes, corresponding to $\mathbf{i}$, $\mathbf{j}$, $\mathbf{k}$ along the rectangular coordinate axes. Then we may write, for any proper vector function $\mathbf{V}$ in terms of its components V_1, V_2, V_3 along the curvilinear coordinate axes,

$$\mathbf{V} = \mathbf{u}_1 V_1 + \mathbf{u}_2 V_2 + \mathbf{u}_3 V_3 \tag{1.57}$$

From Section 1.1 we know that $\nabla \Phi$ is a vector whose magnitude and direction give the maximum space rate of change of Φ and that a component of $\nabla \Phi$ is its directional derivative along a given coordinate direction. From Eq. (1.52) we may write, for these space derivatives,

$$\frac{\partial}{\partial s_i} = \frac{1}{Q_i} \frac{\partial}{\partial q_i} \qquad (i = 1, 2, 3) \tag{1.58}$$

Since ∇ is a proper vector operator, we may then write for ∇ in orthogonal curvilinear coordinates,

$$\nabla = \frac{\mathbf{u}_1}{Q_1} \frac{\partial}{\partial q_1} + \frac{\mathbf{u}_2}{Q_2} \frac{\partial}{\partial q_2} + \frac{\mathbf{u}_3}{Q_3} \frac{\partial}{\partial q_3} \tag{1.59}$$

and for the gradient

$$\nabla \Phi = \frac{\mathbf{u}_1}{Q_1} \frac{\delta \Phi}{\delta q_1} + \frac{\mathbf{u}_2}{Q_2} \frac{\delta \Phi}{\delta q_2} + \frac{\mathbf{u}_3}{Q_3} \frac{\delta \Phi}{\delta q_3} \tag{1.60}$$

16 ¶ Introduction

We now wish to find what $\nabla \cdot \mathbf{V}$ and $\nabla \times \mathbf{V}$ are in terms of the orthogonal curvilinear coordinates q_1, q_2, q_3 and the curvilinear unit vectors $\mathbf{u}_1, \mathbf{u}_2, \mathbf{u}_3$. To do this, we must first proceed by evaluating $\nabla \times \mathbf{u}_i$ and $\nabla \cdot \mathbf{u}_i$. From Eq. (1.60) we can write that

$$\nabla q_1 = \frac{\mathbf{u}_1}{Q_1}$$

so that

$$\nabla \times \nabla q_1 = \nabla \times \frac{\mathbf{u}_1}{Q_1} = \nabla\left(\frac{1}{Q_1}\right) \times \mathbf{u}_1 + \frac{1}{Q_1} \nabla \times \mathbf{u}_1$$

$$= -\mathbf{u}_1 \times \nabla\left(\frac{1}{Q_1}\right) + \frac{1}{Q_1} \nabla \times \mathbf{u}_1 \qquad (1.61)$$

using Eqs. (1.10). From Eq. (1.35) we have that the curl of the gradient of a scalar function is uniquely zero so that Eq. (1.61) becomes

$$\mathbf{u}_1 \times \nabla\left(\frac{1}{Q_1}\right) = \frac{1}{Q_1} \nabla \times \mathbf{u}_1 \qquad (1.62)$$

Performing the differentiation on the second member of the left-hand side of Eq. (1.62) with the use of Eq. (1.59), we obtain

$$\nabla\left(\frac{1}{Q_1}\right) = -\frac{\mathbf{u}_1}{Q_1^3}\frac{\partial Q_1}{\partial q_1} - \frac{\mathbf{u}_2}{Q_1^2 Q_2}\frac{\partial Q_1}{\partial q_2} - \frac{\mathbf{u}_3}{Q_1^2 Q_3}\frac{\partial Q_1}{\partial q_3} \qquad (1.63)$$

Since we are using orthogonal curvilinear coordinates, the same relations for the vector products of the unit vectors will pertain as we had in Eqs. (1.12), or

$$\mathbf{u}_1 \times \mathbf{u}_2 = \mathbf{u}_3 \qquad \mathbf{u}_2 \times \mathbf{u}_3 = \mathbf{u}_1 \qquad \mathbf{u}_3 \times \mathbf{u}_1 = \mathbf{u}_2 \qquad (1.64)$$
$$\mathbf{u}_1 \times \mathbf{u}_1 = \mathbf{u}_2 \times \mathbf{u}_2 = \mathbf{u}_3 \times \mathbf{u}_3 = 0$$

Remembering these relations in the evaluation of the left-hand side of Eq. (1.62), we obtain for $\nabla \times \mathbf{u}_1$,

$$\nabla \times \mathbf{u}_1 = \frac{\mathbf{u}_2}{Q_1 Q_3}\frac{\partial Q_1}{\partial q_3} - \frac{\mathbf{u}_3}{Q_1 Q_2}\frac{\partial Q_1}{\partial q_2}$$

and similarly

$$\nabla \times \mathbf{u}_2 = \frac{\mathbf{u}_3}{Q_1 Q_2}\frac{\partial Q_2}{\partial q_1} - \frac{\mathbf{u}_1}{Q_2 Q_3}\frac{\partial Q_2}{\partial q_3} \qquad (1.65)$$

$$\nabla \times \mathbf{u}_3 = \frac{\mathbf{u}_1}{Q_2 Q_3}\frac{\partial Q_3}{\partial q_2} - \frac{\mathbf{u}_2}{Q_1 Q_3}\frac{\partial Q_3}{\partial q_1}$$

We shall have the same relations for the scalar products of the unit vectors $\mathbf{u}_1, \mathbf{u}_2, \mathbf{u}_3$ as we had in Eqs. (1.8) or

$$\mathbf{u}_1 \cdot \mathbf{u}_1 = \mathbf{u}_2 \cdot \mathbf{u}_2 = \mathbf{u}_3 \cdot \mathbf{u}_3 = 1 \qquad (1.66)$$
$$\mathbf{u}_1 \cdot \mathbf{u}_2 = \mathbf{u}_2 \cdot \mathbf{u}_3 = \mathbf{u}_3 \cdot \mathbf{u}_1 = 0$$

We may then write for $\nabla \cdot \mathbf{u}_1$, using Eqs. (1.64) and the results of Problem 1.1(k),

$$\nabla \cdot \mathbf{u}_1 = \nabla \cdot (\mathbf{u}_2 \times \mathbf{u}_3) = \mathbf{u}_3 \cdot (\nabla \times \mathbf{u}_2) - \mathbf{u}_2 \cdot (\nabla \times \mathbf{u}_3) \qquad (1.67)$$

Evaluating the right-hand side of Eq. (1.67) with the substitution of Eqs. (1.65) and the use of Eqs. (1.66), we obtain

$$\nabla \cdot \mathbf{u}_1 = \frac{1}{Q_1 Q_2} \frac{\partial Q_2}{\partial q_1} + \frac{1}{Q_1 Q_3} \frac{\partial Q_3}{\partial q_1} = \frac{1}{Q_1 Q_2 Q_3} \frac{\partial (Q_2 Q_3)}{\partial q_1}$$

and similarly

$$\nabla \cdot \mathbf{u}_2 = \frac{1}{Q_1 Q_2 Q_3} \frac{\partial (Q_1 Q_3)}{\partial q_2} \qquad (1.68)$$

$$\nabla \cdot \mathbf{u}_3 = \frac{1}{Q_1 Q_3 Q_2} \frac{\partial (Q_1 Q_2)}{\partial q_3}$$

For $\nabla \cdot \mathbf{V}$ in orthogonal curvilinear coordinates, we may then write, using Eq. (1.6) and remembering that $\mathbf{u}_i$ is a vector and V_i a scalar,

$$\begin{aligned} \nabla \cdot \mathbf{V} &= \nabla \cdot (\mathbf{u}_1 V_1) + \nabla \cdot (\mathbf{u}_2 V_2) + \nabla \cdot (\mathbf{u}_3 V_3) \\ &= V_1 \nabla \cdot \mathbf{u}_1 + \mathbf{u}_1 \cdot \nabla V_1 + V_2 \nabla \cdot \mathbf{u}_2 \\ &\quad + \mathbf{u}_2 \cdot \nabla V_2 + V_3 \nabla \cdot \mathbf{u}_3 + \mathbf{u}_3 \cdot \nabla V_3 \end{aligned} \qquad (1.69)$$

Evaluating the first and second terms of the right-hand side, using, respectively, Eqs. (1.68) and (1.59) and (1.66), we obtain

$$V_1 \nabla \cdot \mathbf{u}_1 = \frac{V_1}{Q_1 Q_2 Q_3} \frac{\partial (Q_2 Q_3)}{\partial q_1} \qquad (1.70)$$

and

$$\mathbf{u}_1 \cdot \nabla V_1 = \frac{1}{Q_1} \frac{\partial V_1}{\partial q_1} \qquad (1.71)$$

Substituting Eqs. (1.70) and (1.71) and the similar expressions for the other terms on the right-hand side of Eq. (1.69) into Eq. (1.69) and combining terms, we obtain finally

$$\nabla \cdot \mathbf{V} = \frac{1}{Q_1 Q_2 Q_3} \left[\frac{\partial}{\partial q_1} (V_1 Q_2 Q_3) + \frac{\partial}{\partial q_2} (V_2 Q_1 Q_3) + \frac{\partial}{\partial q_3} (V_3 Q_1 Q_2) \right] \qquad (1.72)$$

For $\nabla \times \mathbf{V}$ in orthogonal curvilinear coordinates, we may proceed in the same manner, using Eqs. (1.10) and again remembering that $\mathbf{u}_i$ is a vector and V_i a scalar,

$$\begin{aligned} \nabla \times \mathbf{V} &= \nabla \times (\mathbf{u}_1 V_1) + \nabla \times (\mathbf{u}_2 V_2) + \nabla \times (\mathbf{u}_3 V_3) \\ &= V_1 \nabla \times \mathbf{u}_1 - \mathbf{u}_1 \times \nabla V_1 + V_2 \nabla \times \mathbf{u}_2 \\ &\quad - \mathbf{u}_2 \times \nabla V_2 + V_3 \nabla \times \mathbf{u}_3 - \mathbf{u}_3 \times \nabla V_3 \end{aligned} \qquad (1.73)$$

Evaluating, as before, the first and second terms of the right-hand side, using, respectively, Eqs. (1.65) and (1.59) and (1.64), we obtain

$$V_1 \nabla \times \mathbf{u}_1 = \frac{\mathbf{u}_2 V_1}{Q_1 Q_3} \frac{\partial Q_1}{\partial q_3} - \frac{\mathbf{u}_3 V_1}{Q_1 Q_2} \frac{\partial Q_1}{\partial q_2} \tag{1.74}$$

and

$$\mathbf{u}_1 \times \nabla V_1 = \frac{\mathbf{u}_3}{Q_2} \frac{\partial V_1}{\partial q_2} - \frac{\mathbf{u}_2}{Q_3} \frac{\partial V_1}{\partial q_3} \tag{1.75}$$

Subtracting Eq. (1.75) from Eq. (1.74), we obtain

$$\begin{aligned} V_1 \nabla \times \mathbf{u}_1 - \mathbf{u}_1 \times \nabla V_1 &= \mathbf{u}_2 \left(\frac{V_1}{Q_1 Q_3} \frac{\partial Q_1}{\partial q_3} + \frac{1}{Q_3} \frac{\partial V_1}{\partial q_3} \right) \\ &\quad - \mathbf{u}_3 \left(\frac{V_1}{Q_1 Q_2} \frac{\partial Q_1}{\partial q_2} + \frac{1}{Q_2} \frac{\partial V_1}{\partial q_2} \right) \\ &= \mathbf{u}_2 \frac{1}{Q_1 Q_3} \frac{\partial (V_1 Q_1)}{\partial q_3} - \mathbf{u}_3 \frac{1}{Q_1 Q_2} \frac{\partial (V_1 Q_1)}{\partial q_2} \end{aligned} \tag{1.76}$$

Substituting Eq. (1.76) and the similar expressions for the other terms on the right-hand side of Eq. (1.73) into Eq. (1.73) and combining terms in determinental form, we obtain

$$\nabla \times \mathbf{V} = \frac{1}{Q_1 Q_2 Q_3} \begin{vmatrix} Q_1 \mathbf{u}_1 & Q_2 \mathbf{u}_2 & Q_3 \mathbf{u}_3 \\ \dfrac{\partial}{\partial q_1} & \dfrac{\partial}{\partial q_2} & \dfrac{\partial}{\partial q_3} \\ V_1 Q_1 & V_2 Q_2 & V_3 Q_3 \end{vmatrix} \tag{1.77}$$

We may obtain the form of the Laplacian operator in orthogonal curvilinear coordinates by simply making the substitution $\mathbf{V} = \nabla \Phi$ in Eq. (1.72), remembering that the components of $\nabla \Phi$ corresponding to the components V_1, V_2, V_3 of $\mathbf{V}$ are given by Eq. (1.60)

$$\nabla \cdot \nabla \Phi = \frac{1}{Q_1 Q_2 Q_3} \left[\frac{\partial}{\partial q_1} \left(\frac{Q_2 Q_3}{Q_1} \frac{\partial \Phi}{\partial q_1} \right) \right. \\ \left. + \frac{\partial}{\partial q_2} \left(\frac{Q_1 Q_3}{Q_2} \frac{\partial \Phi}{\partial q_2} \right) + \frac{\partial}{\partial q_3} \left(\frac{Q_1 Q_2}{Q_3} \frac{\partial \Phi}{\partial q_3} \right) \right] \tag{1.78}$$

Problem 1.3(a)   Derive the second and third equations of Eqs. (1.65) and (1.68).

1.4 Spherical Coordinates

In spherical coordinates, the position of a point P is determined by the three coordinates r, θ, φ, where r is the radius vector, θ the polar angle, and

φ the azimuth (see Fig. 1.10). The three orthogonal coordinate surfaces are concentric spheres about the origin, right circular cones with apex at the origin and axis along Z, and half-planes through the Z axis, respectively. The three unit vectors $\mathbf{r}$, $\mathbf{\theta}_1, \mathbf{\varphi}_1$ are, respectively, in the direction of increasing r, at right angles to $\mathbf{r}_1$ in the direction of increasing θ, and at right angles to both $\mathbf{r}_1$ and $\mathbf{\theta}_1$ in the direction of increasing φ in the XY plane. From Fig. 1.10 Eq. (1.48) for x, y, z in terms of r, θ, φ become

$$x = r \sin \theta \cos \varphi$$
$$y = r \sin \theta \sin \varphi \qquad (1.79)$$
$$z = r \cos \theta$$

Again from Fig. 1.10 we can also write for the distance elements of Eq. (1.52) in the directions of $\mathbf{r}_1, \mathbf{\theta}_1, \mathbf{\varphi}_1$,

$$ds_r = dr$$
$$ds_\theta = r\, d\theta \qquad (1.80)$$
$$ds_\varphi = r \sin \theta\, d\varphi$$

or

$$Q_r = 1$$
$$Q_\theta = r \qquad (1.81)$$
$$Q_\varphi = r \sin \theta$$

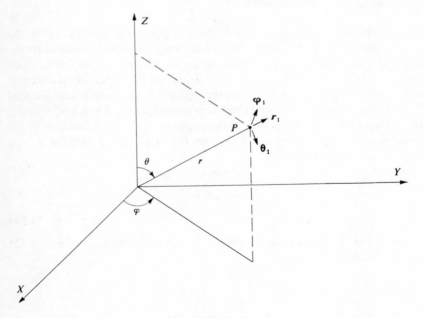

Fig. 1.10

We may then determine directly from Eq. (1.59), using Eqs. (1.81), the gradient

$$\nabla = \mathbf{r}_1 \frac{\partial}{\partial r} + \frac{\boldsymbol{\theta}_1}{r}\frac{\partial}{\partial \theta} + \frac{\boldsymbol{\varphi}_1}{r\sin\theta}\frac{\partial}{\partial \varphi} \qquad (1.82)$$

the same as we had previously in Eq. (1.22), and we may also determine the Laplacian from Eq. (1.78)

$$\nabla^2 = \frac{1}{r^2 \sin\theta}\left[\sin\theta \frac{\partial}{\partial r}\left(r^2 \frac{\partial}{\partial r}\right) + \frac{\partial}{\partial \theta}\left(\sin\theta \frac{\partial}{\partial \theta}\right) + \frac{1}{\sin\theta}\frac{\partial^2}{\partial \varphi^2}\right] \qquad (1.83)$$

Problem 1.4(a)   Show that in spherical coordinates

$$\nabla \cdot \mathbf{V} = \frac{1}{r^2 \sin\theta}\left[\sin\theta \frac{\partial}{\partial r}(r^2 V_r) + r\frac{\partial}{\partial \theta}(\sin\theta V_\theta) + r\frac{\partial V_\varphi}{\partial \varphi}\right]$$

$$(\nabla \times \mathbf{V})_r = \frac{1}{r\sin\theta}\left[\frac{\partial}{\partial \theta}(\sin\theta V_\varphi) - \frac{\partial V_\theta}{\partial \varphi}\right]$$

$$(\nabla \times \mathbf{V})_\theta = \frac{1}{r\sin\theta}\left[\frac{\partial V_r}{\partial \varphi} - \sin\theta \frac{\partial}{\partial r}(rV_\varphi)\right]$$

$$(\nabla \times \mathbf{V})_\varphi = \frac{1}{r}\left[\frac{\partial}{\partial r}(rV_\theta) - \frac{\partial V_r}{\partial \theta}\right]$$

1.5 Cylindrical Coordinates

In cylindrical coordinates, the position of a point P is determined by the three coordinates ρ, φ, z, where ρ is the distance coordinate normal to the cylindrical axis (Z axis), φ the azimuth, and z the Cartesian z coordinate (see Fig. 1.11). The three orthogonal coordinate surfaces are right circular cylinders about the Z axis, half-planes through the Z axis, and planes parallel to the XY plane, respectively. The three unit vectors $\boldsymbol{\rho}_1, \boldsymbol{\varphi}_1, \mathbf{k}$ are, respectively, in the direction of increasing ρ, at right angles to $\boldsymbol{\rho}_1$ in the direction of increasing φ, and parallel to the Z axis. For Fig. 1.11 Eqs. (1.48) for x, y, z in terms of ρ, φ, k become

$$x = \rho \cos\varphi$$
$$y = \rho \sin\varphi$$
$$z = z \qquad (1.84)$$

Again from Fig. 1.11 we can also write for the distance elements of Eq. (1.52) in the directions of $\boldsymbol{\rho}_1, \boldsymbol{\varphi}_1, \mathbf{k}$,

$$ds_\rho = d\rho$$
$$ds_\varphi = \rho \, d\varphi \qquad (1.85)$$
$$ds_z = dz$$

or

$$Q_\rho = 1$$
$$Q_\varphi = \rho \qquad (1.86)$$
$$Q_z = 1$$

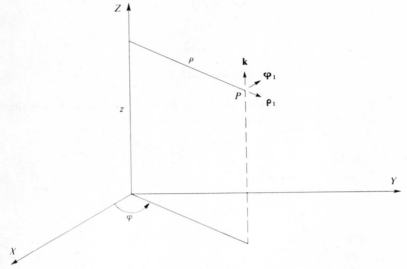

Fig. 1.11

We may then determine directly from Eq. (1.59), using Eqs. (1.86), the gradient

$$\nabla = \boldsymbol{\rho}_1 \frac{\partial}{\partial \rho} + \frac{\boldsymbol{\varphi}_1}{\rho} \frac{\partial}{\partial \varphi} + \mathbf{k} \frac{\partial}{\partial z} \qquad (1.87)$$

the same as we had previously in Eq. (1.23), and we may also determine the Laplacian from Eq. (1.78)

$$\nabla^2 = \frac{1}{\rho}\left[\frac{\partial}{\partial \rho}\left(\rho \frac{\partial}{\partial \rho}\right) + \frac{1}{\rho}\frac{\partial^2}{\partial \varphi^2} + \rho \frac{\partial^2}{\partial z^2}\right] \qquad (1.88)$$

1.6 Legendre Polynomials

We shall be interested in some of the sections in the following chapters in series expansions, or approximations, given in terms of Legendre polynomials. These will be of particular interest for problems defined in terms of spherical coordinates. We may develop the Legendre polynomials and the series related thereto by consideration of the following geometric problem.

With reference to Fig. 1.12, let us write an expression for r in terms of the coordinate distances ρ and ρ' and the angle θ. The distances ρ and ρ' are, respectively, the radial distances from the origin O to the Points P and Q

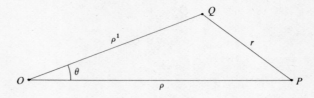

Fig: 1.12

and θ the angle between them, or the polar angle in spherical coordinates if OP is the polar axis. From the law of cosines, we have

$$r^2 = \rho^2 + \rho'^2 - 2\rho\rho' \cos \theta$$

We may rewrite

$$\frac{\rho'}{r} = \frac{1}{\sqrt{1 - 2\frac{\rho}{\rho'} \cos \theta + \frac{\rho^2}{\rho'^2}}} = \frac{1}{\sqrt{1 - 2ux + x^2}} \quad (1.89)$$

where we have set $\rho/\rho' = x$ and $\cos \theta = u$. We now wish to develop Eq. (1.89) in a power series expansion in terms of x for values of x limited to $|2ux - x^2| < 1$. From the binomial theorem, we have for such an expansion

$$(1-z)^{-1/2} = \alpha_0 + \alpha_1 z + \alpha_2 z^2 + \alpha_3 z^3 + \cdots + \alpha_n z^n + \cdots \quad (1.90)$$

where

$$\alpha_n = \frac{1 \cdot 3 \cdot \cdots \cdot (2n-1)}{2 \cdot 4 \cdot \cdots \cdot (2n)} \quad (1.91)$$

Hence

$$(1 - 2ux + x^2)^{-1/2} = \alpha_0 + \alpha_1(2ux - x^2) + \alpha_2(2ux - x^2)^2 + \alpha_3(2ux - x^2)^3$$
$$+ \cdots + \alpha_n(2ux - x^2)^n + \cdots \quad (1.92)$$

Collecting terms in x, we obtain

$$(1 - 2ux + x^2)^{-1/2} = \alpha_0 + [\alpha_1(2u)]x + [-\alpha_1 + \alpha_2(4u^2)]x^2$$
$$+ [-\alpha_2(4u) + \alpha_3(8u^3)]x^3 + \cdots$$
$$= P_0 + P_1(u)x + P_2(u)x^2 + P_3(u)x^3 + \cdots$$
$$+ P_n(u)x^n + \cdots \quad (1.93)$$

where $P_0(u)$, $P_1(u)$, $P_2(u)$, $P_3(u), \cdots, P_n(u), \cdots$ are the coefficients of x^0, x^1, x^2, $x^3, \cdots, x^n, \cdots$. The coefficients $P_n(u)$ are polynomials in u and are the desired *Legendre polynomials* for the series expansion Eq. (1.93).

We may determine the general expression for P_n by collecting the terms in x^n from Eq. (1.92). From the term $\alpha_n(2ux-x^2)^n = \alpha_n x^n(2u-x)^n$, we shall have a contribution from the first term in the binomial expansion of $(2u-x)^n$. From the term $\alpha_{n-1}(2ux-x^2)^{n-1} = \alpha_{n-1}x^{n-1}(2u-x)^{n-1}$, we shall have a contribution from the second term in the binomial expansion of $(2u-x)^{n-1}$, and so on, for the succeeding lower expression of $\alpha_{n-k}(2u-x)^{n-k}$ until we have run through the expansion of $(2u-x)^{n-k}$. Thus, we have

$$P_n(u)x^n = \alpha_n x^n[(2u)^n] + \alpha_{n-1}x^{n-1}[(n-1)(2u)^{n-2}(-x)]$$

$$+ \alpha_{n-2}x^{n-2}\left[\frac{(n-2)(n-3)}{2!}(2u)^{n-4}(-x)^2\right] + \cdots$$

$$+ \alpha_{n-k}x^{n-k}\left[\frac{(n-k)(n-k-1)\cdots(n-2k+1)}{k!}(2u)^{n-2k}(-x)^k\right]$$

$$+ \cdots$$

or

$$P_n(u) = \sum_{k=0}^{n/2} \frac{1\cdot 3\cdot\cdots\cdot(2n-2k-1)}{2\cdot 4\cdot\cdots\cdot(2n-2k)}$$

$$\left[\frac{(n-k)(n-k-1)\cdots(n-2k+1)2^{n-2k}(-1)^k}{k!}\right]u^{n-2k}$$

From the definition of the factorial, we have

$$2\cdot 4\cdot\cdots\cdot(2n-2k) = 2^{n-k}(n-k)! = 2^{n-k}(n-2k)!(n-2k+1)\cdots(n-k)$$

and substituting into the expression for $P_n(u)$, we obtain finally

$$P_n(u) = \sum_{k=0}^{n/2} \frac{1\cdot 3\cdot\cdots\cdot(2n-2k-1)}{2^k k!(n-2k)!}(-1)^k u^{n-2k} \qquad (1.94)$$

We observe that $P_n(u)$ is a polynomial of degree n and that only alternate powers of n occur in it so that the Legendre polynomials of even degree are even functions of u and those of odd degree are odd functions of u.

We now wish to derive certain recursion formulas among the P_n and their derivatives. Taking the derivative of Eq. (1.93) with respect to x, we obtain

$$\frac{u-x}{(1-2ux+x^2)^{3/2}} = P_1(u) + 2P_2(u)x + 3P_3(u)x^2 + \cdots \qquad (1.95)$$

Comparing Eq. (1.95) with Eq. (1.93), we may write, rearranging terms,

$$(u-x)[P_0 + P_1 + P_2 x^2 + \cdots + P_n x^n + \cdots]$$

$$= (1-2ux+x^2)[P_1 + 2P_2 x + 3P_3 x^2 + \cdots + (n+1)P_{n+1}x^n + \cdots]$$

24 ¶ Introduction

Equating the coefficients of x^n on both sides of this equation, which conditions must be met for the expression to be valid for all values of x, gives

$$uP_n x^n - xP_{n-1}x^{n-1} = (n+1)P_{n+1}x^n - 2uxnP_n x^{n-1} + x^2(n-1)P_{n-1}x^{n-2}$$

$$uP_n - P_{n-1} = (n+1)P_{n+1} - 2unP_n + (n-1)P_{n-1}$$

or

$$(n+1)P_{n+1} - (2n+1)uP_n + nP_{n-1} = 0 \tag{1.96}$$

Now taking the derivative of Eq. (1.93) with respect to u, we obtain

$$\frac{x}{(1-2ux+x^2)^{3/2}} = P_0'(u) + P_1'(u)x + P_2'(u)x^2 + \cdots \tag{1.97}$$

Again comparing this time Eq. (1.97) with Eq. (1.95), we may write, rearranging terms,

$$(u-x)[P_0' + P_1'x + P_2'x^2 + \cdots + P_n'x^n + \cdots]$$
$$= x[P_1 + 2P_2 x + 3P_3 x^2 + \cdots + (n+1)P_{n+1}x^n \cdots]$$

Equating the coefficients of x^n on both sides of this equation gives

$$uP_n'x^n - xP_{n-1}'x^{n-1} = nP_n x^{n-1}$$

or

$$uP_n' - P_{n-1}' = nP_n \tag{1.98}$$

From Eqs. (1.98) and (1.96) we may derive a differential equation satisfied by the Legendre polynomials. We proceed by deriving an equation for P_n and its derivatives only. Differentiating Eq. (1.96) with respect to u, we obtain

$$(n+1)P_{n+1}' - (2n+1)P_n - (2n+1)uP_n' + nP_{n-1}' = 0 \tag{1.99}$$

Substituting Eq. (1.98) into Eq. (1.99), we obtain

$$(n+1)P_{n+1}' - (2n+1)P_n - (2n+1)uP_n' + nuP_n' - n^2 P_n = 0$$
$$(n+1)P_{n+1}' - (n+1)uP_n' - (n+1)^2 P_n = 0 \tag{1.100}$$

and rewriting Eq. (1.100) for P_n rather than P_{n+1}

$$P_n' - uP_{n-1}' - nP_n = 0 \tag{1.101}$$

Now substituting Eq. (1.98) into Eq. (1.101), we obtain

$$P_n' + nuP_n - u^2 P_n' - nP_{n-1} = 0$$
$$(1-u^2)P_n' + nuP_n - nP_{n-1} = 0 \tag{1.102}$$

Differentiating Eqs. (1.102) with respect to u and substituting once more from Eq. (1.98), we obtain

$$\frac{d}{du}[(1-u^2)P_n'] + nP_n + nuP_n' + n^2 P_n - nuP_n' = 0$$

$$\frac{d}{du}[(1-u^2)P_n'] + n(n+1)P_n = 0 \tag{1.103}$$

Equation (1.103) is the desired equation. It is referred to as the *Legendre differential equation* and is a homogeneous linear differential equation of the second order.

We shall now derive an alternative expression for $P_n(u)$ to Eq. (1.94), which, in some applications, is simpler and more useful. Let us first examine the binomial expansion of the following expression:

$$(u^2-1)^n = u^{2n} + nu^{2(n-1)}(-1) + \frac{n(n-1)}{2!}u^{2(n-2)}(-1)^2$$

$$+ \frac{n(n-1)(n-2)}{3!}u^{2(n-3)}(-1)^3 + \cdots$$

$$= \sum_{k=0}^{n}(-1)^k \frac{n!}{k!(n-k)!}u^{2(n-k)}$$

Differentiating this expression n times with respect to u, we obtain

$$\frac{d^n}{du^n}(u^2-1)^n = \sum_{k=0}^{n/2}(-1)^k \frac{n!}{k!(n-k)!}\frac{(2n-2k)!}{(n-2k)!}u^{n-2k} \qquad (1.104)$$

where because of the terms dropped out due to the successive differentiations, the summation now continues only to $n/2$. Examining the factorial of $(2n-2k)$, we obtain

$$(2n-2k)! = [1 \cdot 3 \cdot 5 \cdots (2n-2k-1)][2 \cdot 4 \cdot 6 \cdots (2n-2k)]$$

$$= [1 \cdot 3 \cdot 5 \cdots (2n-2k-1)]\, 2^{n-k}(n-k)!$$

Substituting this back into Eq. (1.104) and comparing with Eq. (1.94), we obtain

$$P_n(u) = \frac{1}{2^n n!}\frac{d^n}{du^n}(u^2-1)^n \qquad (1.105)$$

This expression for $P_n(u)$ is known as *Rodrigues' formula*.

Finally, in our discussion of Legendre polynomials, we wish to demonstrate an extremely useful property of Legendre polynomials. This is *orthogonality*. A set of functions is said to be orthogonal if the integral of the product of any two of them is zero over a specified interval, and the integral of the square of any one of them is not zero over the same interval This property of Legendre polynomials is similar to that which is obtained for the Fourier series of sine and cosine. It provides us with the same useful condition for meeting boundary conditions on spherical surfaces with Legendre polynomials that are obtained through the use of Fourier series on planar surfaces.

We may proceed to demonstrate this property by multiplying the Legendre differential equation (1.103) by $P_m(u)$ and integrating from -1 to 1

$$\int_{-1}^{1} P_m(u) \frac{d}{du}[(1-u^2)P_n'(u)] \, du + n(n+1) \int_{-1}^{1} P_m(u)P_n(u) = 0$$

Integrating the first expression by parts, the integrated term vanishes at either limits, and we have

$$-\int_{-1}^{1} (1-u^2)P_m'(u)P_n'(u) \, du + n(n+1) \int_{-1}^{1} P_m(u)P_n(u) \, du = 0 \quad (1.106)$$

If we had proceeded with Eq. (1.103) in terms of $P_m(u)$ and multiplied it by $P_n(u)$, we would have obtained an expression the same as Eq. (1.106) with an interchange of m and n. Subtracting this from Eq. (1.106), we obtain

$$[n(n+1) - m(m+1)] \int_{-1}^{1} P_m(u)P_n(u) \, du = 0$$

from which we have directly our first property of orthogonality,

$$\int_{-1}^{1} P_m(u)P_n(u) \, du = 0 \quad (m \neq n) \quad (1.107)$$

Through the use of Rodrigues' formula, we can obtain our second orthogonality property that

$$\int_{-1}^{1} P_n^2(u) \, du = \frac{2}{2n+1} \quad (1.108)$$

We now are in a position to determine the coefficients in the development of a given function in a series of Legendre polynomials. If we have such a series,

$$V(u) = c_0 P_0(u) + c_1 P_1(u) + c_2 P_2(u) + \cdots$$

$$= \sum_{n=0}^{\infty} c_n P_n(u) \quad (1.109)$$

multiplication by $P_m(u)$ and integrating from -1 to 1 will give, with the use of Eqs. (1.107) and (1.108),

$$\int_{-1}^{1} V(u) P_m(u) \, du = c_m \int_{-1}^{1} P_m^2(u) \, du = c_m \frac{2}{2m+1}$$

or

$$c_m = \frac{2m+1}{2} \int_{-1}^{1} V(u) P_m(u) \, du \quad (1.110)$$

For some problems involving boundary conditions on a sphere, it is more convenient to express Eq. (1.110) in terms of the incremental solid angle $d\omega$. Remembering that we have from Fig. 1.13

$$d\omega = 2\pi \sin \theta \, d\theta = -2\pi \, du$$

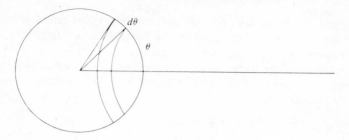

Fig. 1.13

so that Eq. (1.110) becomes, reversing the limits of integration,

$$c_m = \frac{2m+1}{4\pi} \int_0^\pi V(u) P_m(u) \, d\omega \tag{1.111}$$

Problem 1.6(a)   Determine by expansion of Eq. (1.93) that the first five Legendre polynomials are

$$P_0 = 1$$
$$P_1 = \cos\theta$$
$$P_2 = \tfrac{3}{2}(\cos^2\theta - \tfrac{1}{3})$$
$$P_3 = \tfrac{5}{2}(\cos^3\theta - \tfrac{3}{5}\cos\theta)$$
$$P_4 = \tfrac{3\cdot 5}{8}(\cos^4\theta - \tfrac{6}{7}\cos^2\theta + \tfrac{3}{35})$$

Problem 1.6(b)   Using Rodrigues' formula and integrating by parts, obtain Eq. (1.108).

1.7 Laplace's Equation

Laplace's equation is obtained by setting the Laplacian operator, Eq. (1.34), of a scalar equal to zero, or

$$\nabla \cdot \nabla \Phi = 0 \tag{1.112}$$

Laplace's equation is of great importance in theoretical physics. It appears in problems occurring in gravity, physical oceanography, thermodynamics, seismology, and geomagnetism. It is indeed the most often occurring partial differential equation in theoretical geophysics. It is of advantage for us to look at some of its solutions here.

We recall that in dealing with solutions of partial differential equations, we can be confronted with a variety of functions that satisfy the determining equation. We shall restrict ourselves to consideration of solutions that will be of use to us in later problems, that is, to solutions that may be made unique by satisfying the necessary and sufficient boundary and initial conditions of these problems. To obtain these solutions, we shall use the familiar method of solution known as the separation of variables.

To start, let us consider two rather obvious characteristics of solutions of Laplace's equation. First, if $\Phi_1, \Phi_2, \cdots, \Phi_n$ are individual solutions of Laplace's equation, then

$$\Phi = A_1\Phi_1 + A_2\Phi_2 + \cdots + A_n\Phi_n$$

is a solution where $A_1, A_2, \cdots, A_n$ are any constants. For

$$\nabla \cdot \nabla \Phi = A_1 \nabla \cdot \nabla \Phi_1 + A_2 \nabla \cdot \nabla \Phi_2 + \cdots + A_n \nabla \cdot \nabla \Phi_n = 0$$

as each term contains a vanishing factor. Second, if Φ is a solution of Laplace's equation, any partial derivative of Φ of any order with respect to the coordinates x, y, z is a solution. Writing out $\nabla \cdot \nabla$ in terms of partial derivatives with respect to x, y, z, Laplace's equation takes the form

$$\frac{\partial^2 \Phi}{\partial x^2} + \frac{\partial^2 \Phi}{\partial y^2} + \frac{\partial^2 \Phi}{\partial z^2} = 0$$

and differentiating with respect to x, we obtain

$$\frac{\partial^2}{\partial x^2}\left(\frac{\partial \Phi}{\partial x}\right) + \frac{\partial^2}{\partial y^2}\left(\frac{\partial \Phi}{\partial x}\right) + \frac{\partial^2}{\partial z^2}\left(\frac{\partial \Phi}{\partial x}\right) = 0$$

Therefore, if Φ is a solution of Laplace's equation, $\partial \Phi / \partial x$ is also. In similar fashion, it may be shown that

$$\frac{\partial \Phi}{\partial y}, \frac{\partial \Phi}{\partial z}, \frac{\partial^2 \Phi}{\partial x^2}, \frac{\partial^2 \Phi}{\partial x \partial y}, \cdots$$

and also the vector $\nabla \Phi$ are solutions.

Let us examine the two-dimensional form of Laplace's equation in rectangular coordinates,

$$\frac{\partial^2 \Phi}{\partial x^2} + \frac{\partial^2 \Phi}{\partial y^2} = 0 \tag{1.113}$$

To use the method of separation of variables, let us make the assumption, justifiable by its success, that a particular solution of Φ may be written in the form

$$\Phi = X(x)Y(y) \tag{1.114}$$

where X and Y are functions of only one independent variable, x and y, respectively. Substituting into Eq. (1.113), we obtain

$$X''Y + XY'' = 0$$

or

$$\frac{X''}{X} + \frac{Y''}{Y} = 0 \tag{1.115}$$

If Eq. (1.115) is to have any solution at all, then each of the terms on the left-hand side must separately be equal to a constant; for a change in x would not alter the value of the second term, and a change in y would not alter the value of the first term. One may, therefore, conclude that

$$\frac{X''}{X} = k^2 \qquad \frac{Y''}{Y} = -k^2 \qquad (1.116)$$

where the constant parameter k^2 may have any value real, imaginary, or complex. Equations (1.116) are two ordinary differential equations whose solutions are simply

$$X = a_1 e^{\pm kx} \qquad Y = a_2 e^{\pm iky} \qquad (1.117)$$

which may be seen by substitution back into Eqs. (1.116). Our solution of the form of Eq. (1.114) in terms of the parameter k is then

$$\Phi_k = c_k e^{\pm k(x+iy)} \qquad (1.118)$$

or since k is undetermined and may have any constant value, a more general solution is

$$\Phi = \sum_k c_k e^{\pm k(x+iy)} \qquad (1.119)$$

Let us now examine Laplace's equation in spherical coordinates. From Eq. (1.83) we have

$$\frac{1}{r^2}\frac{\partial}{\partial r}\left(r^2 \frac{\partial \Phi}{\partial r}\right) + \frac{1}{r^2 \sin \theta}\frac{\partial}{\partial \theta}\left(\sin \theta \frac{\partial \Phi}{\partial \theta}\right) + \frac{1}{r^2 \sin^2 \theta}\frac{\partial^2 \Phi}{\partial \varphi^2} = 0 \quad (1.120)$$

If we put u for $\cos \theta$, we shall have

$$\frac{\partial}{\partial \theta} = \frac{\partial}{\partial u}\frac{du}{d\theta} = -\sin \theta \frac{\partial}{\partial u}$$

so that Eq. (1.120) becomes

$$\frac{\partial}{\partial r}\left(r^2 \frac{\partial \Phi}{\partial r}\right) + \frac{\partial}{\partial u}\left[(1-u^2)\frac{\partial \Phi}{\partial u}\right] + \frac{1}{1-u^2}\frac{\partial^2 \Phi}{\partial \varphi^2} = 0 \qquad (1.121)$$

We shall be interested here in solutions that are functions of r and θ only, and not of the azimuth φ. Then the last term in Eq. (1.121) vanishes, and we have

$$\frac{\partial}{\partial r}\left(r^2 \frac{\partial \Phi}{\partial r}\right) + \frac{\partial}{\partial u}\left[(1-u^2)\frac{\partial \Phi}{\partial u}\right] = 0 \qquad (1.122)$$

We shall again use the method of separation of variables and assume a solution of the form

$$\Phi = R(r)U(u) \qquad (1.123)$$

Substituting into Eq. (1.122), we obtain

$$\frac{\partial}{\partial r}(r^2 R' U) + \frac{\partial}{\partial u}[(1-u^2)RU'] = 0$$

or

$$\frac{1}{R}\frac{\partial}{\partial r}(r^2 R') + \frac{1}{U}\frac{\partial}{\partial u}[(1-u^2)U'] = 0 \qquad (1.124)$$

We see that this is of the same form as Eq. (1.115) with a separation of the variables r and u. We shall choose, for convenience, in this case our constant parameter to be $n(n+1)$. We then obtain the two ordinary differential equations, corresponding to Eqs. (1.116),

$$\frac{d}{dr}(r^2 R') - n(n+1)R = 0 \qquad (1.125)$$

and

$$\frac{d}{du}[(1-u^2)U'] + n(n+1)U = 0 \qquad (1.126)$$

Examining Eq. (1.125), we have

$$r^2 R'' + 2rR' - n(n+1)R = 0$$

This is a familiar ordinary differential equation, whose solution is of the form $R = r^\alpha$. Substituting, we obtain the conditional equation for α,

$$\alpha(\alpha-1) + 2\alpha - n(n+1) = 0$$

or

$$(\alpha-n)(\alpha+n+1) = 0$$

whose solutions are

$$\alpha = n, \; -(n+1)$$

Hence

$$R = a_1 r^n + a_2 r^{-(n+1)} \qquad (1.127)$$

Returning to Eq. (1.126), we notice that it is the Legendre differential equation (1.103) whose solutions are the Legendre polynomials P_n. We may then write, for our final solution in the form of Eq. (1.119),

$$\Phi = \sum_n [a_n r^n + b_n r^{-(n+1)}] P_n(\cos\theta) \qquad (1.128)$$

Let us now examine Laplace's equation in cylindrical coordinates. From Eq. (1.88) we have

$$\frac{1}{\rho}\frac{\partial}{\partial \rho}\left(\rho \frac{\partial \Phi}{\partial \rho}\right) + \frac{1}{\rho^2}\frac{\partial^2 \Phi}{\partial \varphi^2} + \frac{\partial^2 \Phi}{\partial z^2} = 0 \qquad (1.129)$$

We shall be interested here in solutions that are functions of ρ and z only, and not of the azimuth φ. The second term in Eq. (1.129) vanishes, and we have

$$\frac{\partial^2 \Phi}{\partial \rho^2} + \frac{1}{\rho}\frac{\partial \Phi}{\partial \rho} + \frac{\partial^2 \Phi}{\partial z^2} = 0 \qquad (1.130)$$

Again, using the method of separation of variables, we assume a solution of the form

$$\Phi = M(\rho)Z(z) \qquad (1.131)$$

Substituting into Eq. (1.130), we obtain

$$M''Z + \frac{1}{\rho}M'Z + MZ'' = 0$$

or

$$\frac{M''}{M} + \frac{1}{\rho}\frac{M'}{M} + \frac{Z}{Z} = 0 \qquad (1.132)$$

We see that Eq. (1.132) has the proper separation of variables, and we shall choose our constant parameter to be k^2. We obtain as before

$$M'' + \frac{1}{\rho}M' + k^2 M = 0$$

or

$$\rho^2 M'' + \rho M' + \rho^2 k^2 M = 0 \qquad (1.133)$$

and

$$Z'' - k^2 Z = 0 \qquad (1.134)$$

Examining Eq. (1.134), we see that it is the same as the first equation of Eqs. (1.116) and has the solution

$$Z = a_1 e^{\pm kz} \qquad (1.135)$$

In Eq. (1.133) if we make a change of variable to x such that $x = k\rho$, we shall have

$$\frac{\partial}{\partial \rho} = \frac{\partial}{\partial x}\frac{dx}{d\rho} = k\frac{\partial}{\partial x}$$

so that Eq. (1.133) becomes,

$$x^2 \frac{d^2 M}{dx^2} + x\frac{dM}{dx} + x^2 M = 0 \qquad (1.136)$$

which is a reduced form of the *Bessel differential equation*. We shall look for a solution of Eq. (1.136) in the form of a convergent series in x. We shall assume

$$M = \sum_{\lambda=0}^{\infty} a_\lambda x^\lambda = a_0 + a_1 x + a_2 x^2 + a_3 x^3 + \cdots + a_\lambda x^\lambda + \cdots \qquad (1.137)$$

Substituting into Eq. (1.136), we obtain

$$\sum_\lambda a_\lambda \lambda(\lambda-1)x^\lambda + \sum_\lambda a_\lambda \lambda x^\lambda + \sum_\lambda a_\lambda x^{\lambda+2} = 0$$

This equation must hold for every value of x, and this can only be true if the coefficient of every power of x is identically zero. Equating the coefficients of x^λ to zero, we obtain

$$a_\lambda \lambda(\lambda-1) + a_\lambda \lambda + a_{\lambda-2} = 0$$

or, increasing λ to $\lambda+2$,

$$a_{\lambda+2}(\lambda+2)(\lambda+1) + a_{\lambda+2}(\lambda+2) + a_\lambda = 0$$

$$a_{\lambda+2}(\lambda+2)^2 = -a_\lambda$$

$$a_{\lambda+2} = \frac{-1}{(\lambda+2)^2} a_\lambda \qquad (1.138)$$

This gives us a recursion relation for $a_{\lambda+2}$ in terms of a_λ. Thus, if the coefficients a_0 and a_1 are specified, all other a_λ can be determined from Eq. (1.138). We are particularly interested in the solution that occurs if we take a_0 equal to a constant, chosen for convenience as unity, and a_1 equal to zero. This gives us a convergent series for all values of x and is one of the two particular solutions of Eq. (1.136). It is the one that will be of interest to us in later problems. Thus, Eq. (1.137) becomes

$$M = 1 - \frac{1}{2^2}x^2 + \frac{1}{2^2 \cdot 4^2}x^4 - \frac{1}{2^2 \cdot 4^2 \cdot 6^2}x^6 + \cdots$$

$$+ \frac{(-1)^m}{2^2 \cdot 4^2 \cdot 6^2 \cdots \cdot m^2}x^{2m} + \cdots$$

This is the *Bessel function of zero order* $J_0(x)$, which, remembering the definition of the factorial, becomes in summary form

$$M = J_0(x) = \sum_{\lambda=0}^{\infty} \frac{(-1)^\lambda}{\lambda! \lambda!} \left(\frac{x}{2}\right)^{2\lambda} \qquad (1.139)$$

The function $J_0(x)$ oscillates back and forth across the X axis, similar to the sine and cosine functions, but with decreasing amplitude. For our final solution in the form of Eq. (1.119), we may then write

$$\Phi = \sum_k c_k J_0(k\rho) e^{\pm kz} \qquad (1.140)$$

We should now like to look at the conditions required for any of the particular solutions developed above to represent the complete solution to a problem for a region in which there are no singularities. We shall show that if Φ is given at all points on a surface surrounding a region in which Laplace's equation holds, then a solution of this equation that satisfies the assigned boundary conditions is unique. Let Φ_1 be a solution of Laplace's equation satisfying the boundary conditions. Assume that another distinct solution Φ_2, which satisfies the same boundary conditions, exists. We shall prove that Φ_2 must be identical with Φ_1. Let $\Psi = \Phi_1 - \Phi_2$; then Ψ satisfies Laplace's equation in the region under consideration, since Φ_1 and Φ_2 satisfy it individually. Expanding the expression $\nabla \cdot (\Psi \nabla \Psi)$, we obtain

$$\nabla \cdot (\Psi \nabla \Psi) = \nabla \Psi \cdot \nabla \Psi + \Psi \nabla \cdot \nabla \Psi$$

or

$$\nabla \Psi \cdot \nabla \Psi = \nabla \cdot (\Psi \nabla \Psi) - \Psi \nabla \cdot \nabla \Psi$$

Integrating this expression over the region τ in which Laplace's equation holds, we have

$$\int_\tau (\nabla \Psi)^2 \, d\tau = \int_\tau \nabla \cdot (\Psi \nabla \Psi) \, d\tau - \int_\tau \Psi \nabla \cdot \nabla \Psi \, d\tau$$

As Ψ satisfies Laplace's equation, the integrand of the last term on the right vanishes throughout the region τ. The first term on the right may be transformed into a surface integral by Gauss' theorem. Hence,

$$\int_\tau (\nabla \Psi)^2 \, d\tau = \int_\sigma (\Psi \nabla \Psi) \cdot d\boldsymbol{\sigma}$$

But as $\Phi_1 = \Phi_2$ everywhere on the surface σ bounding the volume τ, the right-hand side of this equation vanishes. Therefore, as the integrand of the volume integral is a sum of squares, the integrand itself must vanish; that is,

$$(\nabla \Psi)^2 = \left(\frac{\partial \Psi}{\partial x}\right)^2 + \left(\frac{\partial \Psi}{\partial y}\right)^2 + \left(\frac{\partial \Psi}{\partial z}\right)^2$$

and again for this equation to be zero for all x, y, z, each of its terms must be zero, or

$$\frac{\partial \Psi}{\partial x} = 0 \qquad \frac{\partial \Psi}{\partial y} = 0 \qquad \frac{\partial \Psi}{\partial z} = 0$$

Therefore, Ψ is a constant. But, since Ψ vanishes on the boundary, this constant is zero. Therefore,

$$\Phi_1 = \Phi_2$$

and the two solutions of Laplace's equation are the same.

Suppose, now, that in solving a physical problem, we find that a certain function Φ of the coordinates must satisfy Laplace's equation inside a region τ and have certain assigned values over a surface bounding this region. Then if we find a solution of Laplace's equation that satisfies the given boundary conditions, we know that we have the correct solution to the problem. We need not ponder the possibility that there may exist some other solution of Laplace's equation that satisfies the assigned boundary conditions and also fulfills further necessary conditions that we have overlooked and that are not fulfilled by the first solution. For there is only one solution of Laplace's equation that can satisfy the given boundary conditions. In mathematical language, Laplace's equation and the assigned boundary conditions are sufficient to determine the function Φ.

1.8 Fourier Series

It is sometimes useful to express a given function $f(x)$ in terms of a series of trigonometric functions, as contrasted, say, with a power series expansion on an expansion in terms of Legendre polynomials. Such a series, known as *Fourier series*, may be written in general form:

$$f(x) = \frac{a_0}{2} + a_1 \cos x + a_2 \cos 2x + \cdots + a_n \cos nx + \cdots$$
$$+ b_1 \sin x + b_2 \sin 2x + \cdots + b_n \sin nx + \cdots \quad (1.141)$$

The constant term has been written $a_0/2$ rather than a_0 for subsequent convenience in evaluation of the a_n. We note immediately that since $\cos x$ and $\sin x$, and consequently $\cos 2x$, $\sin 2x, \cdots$, $\cos nx$, $\sin nx, \cdots$ are periodic with a period 2π, the function $f(x)$ is necessarily restricted to being a periodic function with a period of 2π; in a physical sense, this means that we can define $f(x)$ only over a 2π interval. We shall see later that this seemingly severe restriction can be relieved sufficiently so that Fourier series are quite useful in problems of theoretical geophysics. Solutions in terms of Fourier series, and logical extensions thereof, are often useful in problems having to do with vibration or wave propagation.

The trigonometric functions of a Fourier series exhibit the same useful property of orthogonality over the interval 2π as the Legendre polynomials. For convenience, we shall take our 2π interval from $-\pi$ to π. We see immediately that

$$\int_{-\pi}^{\pi} \sin mx \cos nx \, dx = \int_{-\pi}^{\pi} \sin mx \sin nx \, dx$$
$$= \int_{-\pi}^{\pi} \cos mx \cos nx \, dx = 0 \quad (1.142)$$

for $m \neq n$ for the last two integrals, from the trigonometric identities

$$\sin mx \cos nx = \tfrac{1}{2}[\sin (m-n)x + \sin (m+n)x]$$
$$\sin mx \sin nx = \tfrac{1}{2}[\cos (m-n)x - \cos (m+n)x]$$
$$\cos mx \cos nx = \tfrac{1}{2}[\cos (m-n)x + \cos (m+n)x]$$

and also that

$$\int_{-\pi}^{\pi} \sin^2 nx \, dx = \tfrac{1}{2} \int_{-\pi}^{\pi} (1-\cos 2nx) \, dx = \pi$$
$$\int_{-\pi}^{\pi} \cos^2 nx \, dx = \tfrac{1}{2} \int_{-\pi}^{\pi} (1+\cos 2nx) \, dx = \pi \tag{1.143}$$

From these orthogonal relations, we can determine the coefficients a_n and b_n in Eq. (1.141). We proceed as follows in a manner similar to that for the determination of the coefficients in a Legendre polynomial series. Multiply Eq. (1.141) by $\cos mx$ and integrate from $-\pi$ to π. From Eqs. (1.142) and (1.143) we shall obtain

$$\int_{-\pi}^{\pi} f(x) \cos mx \, dx = a_m \int_{-\pi}^{\pi} \cos^2 mx \, dx = a_m \pi$$

or

$$a_m = \frac{1}{\pi} \int_{-\pi}^{\pi} f(x) \cos mx \, dx \tag{1.144}$$

including $m = 0$. Similarly, we obtain

$$b_m = \frac{1}{\pi} \int_{-\pi}^{\pi} f(x) \sin mx \, dx \tag{1.145}$$

Let us examine for the moment the properties of a function referred to as even and odd. By definition, a function $f(x)$ is called an *even function* if $f(-x) = f(x)$. We see then that $\cos nx$ is an even function since $\cos(-nx) = \cos(nx)$. Also, by definition, a function $f(x)$ is called an *odd* function if $f(-x) = -f(x)$. We see then that $\sin nx$ is an odd function since $\sin(-nx) = -\sin(nx)$. Now, if perchance $f(x)$ in Eq. (1.141) is an even function, we see that the coefficients of the trigonometric expansion are given by

$$a_m = \frac{2}{\pi} \int_0^{\pi} f(x) \cos mx \, dx \tag{1.146}$$

$$b_m = 0$$

since the contribution from $-\pi$ to 0 in Eq. (1.145) is canceled by the contribution from 0 to π. For an even function $f(x)$, the Fourier series expansion is given in terms of cosine functions only. Similarly, if $f(x)$ in Eq. (1.141) is odd, we have, for the coefficients of the trigonometric expansion,

$$a_m = 0$$

$$b_m = \frac{2}{\pi} \int_0^{\pi} f(x) \sin mx \, dx \tag{1.147}$$

For an odd function, $f(x)$, the Fourier series expansion is given in terms of sine functions only.

Often in geophysical problems, we are interested in defining or in obtaining values of $f(x)$ over positive values of x. We may then wish to have $f(x)$ defined only over the interval 0 to π, that is, half the interval previously considered. We may then arbitrarily define $f(x)$ over the interval $-\pi$ to 0 to be an even function, in which case $f(x)$ will be given by a cosine series only, whose coefficients are determined by Eqs. (1.146). Alternatively, we may arbitrarily define $f(x)$ over the interval $-\pi$ to 0 to be an odd function, in which case $f(x)$ will be given by a sine series only, whose coefficients are determined by Eqs. (1.147). Such series are known as half-range series.

In general, we do not wish to restrict ourselves to an interval $-\pi$ to π or 0 to π. It is desirable to develop methods and formulas that will enable us to expand a function over an arbitrary interval $-l$ to l or 0 to l. Consider a Fourier series expansion of $F(z)$ in the form of Eq. (1.141)

$$F(z) = \frac{a_0}{2} + a_1 \cos z + a_2 \cos 2z + \cdots + a_n \cos nz + \cdots$$
$$+ b_1 \sin z + b_2 \sin 2z + \cdots + b_n \sin nz + \cdots \quad (1.148)$$

Let us make the substitution in Eq. (1.148) of

$$z = \frac{\pi x}{l} \quad (1.149)$$

from which we shall define $f(x)$ by

$$F(z) = F\left(\frac{\pi x}{l}\right) = f(x) \quad (1.150)$$

Substituting Eqs. (1.149) and (1.150) into Eq. (1.148), we obtain

$$f(x) = \frac{a_0}{2} + a_1 \cos \frac{\pi x}{l} + a_2 \cos \frac{2\pi x}{l} + \cdots + a_n \cos \frac{n\pi x}{l} + \cdots$$
$$+ b_1 \sin \frac{\pi x}{l} + b_2 \sin \frac{2\pi x}{l} + \cdots + b_n \sin \frac{n\pi x}{l} + \cdots \quad (1.151)$$

Evaluating the coefficients of Eqs. (1.148) and (1.151) by Eqs. (1.144) and (1.145), we obtain

$$a_m = \frac{1}{\pi} \int_{-\pi}^{\pi} F(z) \cos mz \, dz$$
$$= \frac{1}{l} \int_{-l}^{l} f(x) \cos \frac{m\pi x}{l} \, dx \quad (1.152)$$

and

$$b_m = \frac{1}{l} \int_{-l}^{l} f(x) \sin \frac{m\pi x}{l} \, dx \quad (1.153)$$

For the coefficients of a half-range cosine series, we obtain

$$a_m = \frac{2}{l}\int_0^l f(x)\cos\frac{m\pi x}{l}\,dx \qquad (1.154)$$

and for the coefficients of a half-range sine series, we obtain

$$b_m = \frac{2}{l}\int_0^l f(x)\sin\frac{m\pi x}{l}\,dx \qquad (1.155)$$

In some instances, it is more convenient to have our Fourier series expansion expressed in complex form. If the cosine and sine functions are written in their exponential form, we can obtain for Eq. (1.151)

$$f(x) = \sum_{-\infty}^{\infty} c_n e^{i(n\pi x/l)} \qquad (1.156)$$

where the coefficients c_n are complex and are given by

$$c_m = \frac{1}{2l}\int_{-l}^{l} f(\xi)\, e^{-i(m\pi\xi/l)}\,d\xi \qquad (1.157)$$

and where, for convenience, we have used the variable ξ for the variable of integration in the determination of our coefficients c_m to distinguish from the variable x in our final expression.

Problem 1.8(a)   Determine the half-range cosine and sine coefficients, respectively, and the Fourier series expansions for $f(x) = 1$ defined over the interval 0 to l.

Problem 1.8(b)   Determine the half-range cosine and sine coefficients, respectively, and the Fourier series expansions for $f(x) = x$ defined over the interval 0 to l.

Problem 1.8(c)   Determine the half-range cosine and sine coefficients, respectively, and the Fourier series expansions for $f(x) = x^2$ defined over the interval 0 to l.

Problem 1.8(d)   Prove that $e^{i(n\pi x/l)}$ and $e^{-i(m\pi x/l)}$ are orthogonal over the interval $-l$ to l.

Problem 1.8(e)   Derive expressions (1.156) and (1.157).

1.9 Fourier Integrals

In the previous section, we have seen how we can express a function

38 ¶ Introduction

$f(x)$ defined over an interval from $-\pi$ to π in terms of a Fourier series, and we have extended this to an arbitrary interval from $-l$ to l. It is of interest in this section to extend this analysis to a function defined over the entire real axis from $-\infty$ to ∞.

From Eqs. (1.151), (1.152), and (1.153) we may write, for our function $f(x)$ defined over the interval $-l$ to l,

$$f(x) = \frac{1}{2l}\int_{-l}^{l} f(\xi)\,d\xi + \frac{1}{l}\sum_{n=1}^{\infty}\left[\cos\frac{n\pi x}{l}\int_{-l}^{l} f(\xi)\cos\frac{n\pi \xi}{l}\,d\xi\right.$$

$$\left. + \sin\frac{n\pi x}{l}\int_{-l}^{l} f(\xi)\sin\frac{n\pi \xi}{l}\,d\xi\right] \quad (1.158)$$

where we have used ξ as the variable of integration in the determination of the coefficients a_n and b_n for convenience to distinguish it from the variable x in our final expression. From the trigonometric identity

$$\cos\frac{n\pi \xi}{l}\cos\frac{n\pi x}{l} + \sin\frac{n\pi \xi}{l}\sin\frac{n\pi x}{l} = \cos\frac{n\pi}{l}(\xi-x)$$

we may write for Eq. (1.158)

$$f(x) = \frac{1}{2l}\int_{-l}^{l} f(\xi)\,d\xi + \frac{1}{l}\sum_{n=1}^{\infty}\int_{-l}^{l} f(\xi)\cos\frac{n\pi}{l}(\xi-x)\,d\xi \quad (1.159)$$

Now if l is taken large enough, we may neglect the first term in Eq. (1.159) provided that the integral in this term converges as $l \to \infty$. We are then left with the second term, which we can express as

$$\frac{1}{l}\sum_{n=1}^{\infty}\int_{-l}^{l} f(\xi)\cos\frac{n\pi}{l}(\xi-x)\,d\xi$$

$$= \frac{1}{\pi}\left[\delta k \int_{-l}^{l} f(\xi)\cos\delta k(\xi-x)\,d\xi + \delta k \int_{-l}^{l} f(\xi)\cos 2\delta k(\xi-x)\,d\xi + \cdots\right]$$

$$= \frac{1}{\pi}\int_{-l}^{l} f(\xi)\,d\xi\,\{[\cos\delta k(\xi-x)+\cos 2\delta k(\xi-x)+\cdots]\,\delta k\}\,d\xi$$
(1.160)

where we have substituted for l in the integrand

$$\delta k = \frac{\pi}{l}$$

The expression in the braces of Eq. (1.160) is simply a series approximation to the definite integral

$$\int_0^{\infty} \cos k(\xi-x)\,dk = \lim_{k \to 0}\{[\cos\delta k(\xi-x)+\cos 2\delta k(\xi-x)+\cdots]\,\delta k\}$$

Assuming that this sum has a limit as $l \to \infty$, we may then write for $f(x)$, now defined over the interval $-\infty$ to ∞,

$$f(x) = \frac{1}{\pi} \int_{-\infty}^{\infty} f(\xi) \, d\xi \int_{0}^{\infty} \cos k \, (\xi - x) \, dk$$

$$= \frac{1}{\pi} \int_{0}^{\infty} dk \int_{-\infty}^{\infty} f(\xi) \cos k \, (\xi - x) \, d\xi \qquad (1.161)$$

Equation (1.161) is known as the *Fourier integral theorem*. The condition that applies to $f(x)$ for this integral to exist, which we shall state but not prove, is that

$$\int_{-\infty}^{\infty} |f(x)| \, dx$$

exists.

By the same method as above or directly from Eq. (1.161), we may obtain the half-range cosine series, where $f(x)$ is defined from 0 to ∞ and $f(-x) = f(x)$,

$$f(x) = \frac{2}{\pi} \int_{0}^{\infty} dk \int_{0}^{\infty} f(\xi) \cos k\xi \cos kx \, d\xi \qquad (1.162)$$

Similarly, for the half-range sine series, where $f(x)$ is defined from 0 to ∞ and $f(-x) = -f(x)$,

$$f(x) = \frac{2}{\pi} \int_{0}^{\infty} dk \int_{0}^{\infty} f(\xi) \sin k\xi \sin kx \, d\xi \qquad (1.163)$$

Equations (1.162) and (1.163) may be written in complementary form as

$$f(x) = \sqrt{\frac{2}{\pi}} \int_{0}^{\infty} g(k) \cos kx \, dk$$

$$g(k) = \sqrt{\frac{2}{\pi}} \int_{0}^{\infty} f(\xi) \cos k\xi \, d\xi \qquad (1.164)$$

for the half-range cosine series, and as

$$f(x) = \sqrt{\frac{2}{\pi}} \int_{0}^{\infty} h(k) \sin kx \, dk$$

$$h(k) = \sqrt{\frac{2}{\pi}} \int_{0}^{\infty} f(\xi) \sin k\xi \, d\xi \qquad (1.165)$$

for the half-range sine series. Expressions (1.164) and (1.165) are known as the *Fourier cosine and sine transforms*.

If we had followed through the analysis of this section for our Fourier series expansion expressed in complex form, we would have obtained for the Fourier integral theorem

40 ¶ Introduction

$$f(x) = \frac{1}{2\pi} \int_{-\infty}^{\infty} dk \int_{-\infty}^{\infty} f(\xi) \, e^{ik(x-\xi)} \, d\xi \qquad (1.166)$$

and for the *Fourier transforms*

$$f(x) = \frac{1}{\sqrt{2\pi}} \int_{-\infty}^{\infty} g(k) \, e^{ikx} \, dk$$

$$g(k) = \frac{1}{\sqrt{2\pi}} \int_{-\infty}^{\infty} f(\xi) \, e^{-ik\xi} \, d\xi \qquad (1.167)$$

The Fourier transforms, either Eqs. (1.164) or (1.165) or more generally Eqs. (1.167), have a useful application in the description of the source function for problems in vibration or wave propagation. Often, we wish to consider a source function $f(t)$, which is usually given as some function of the pressure, velocity, or displacement with respect to time in terms of its *simple harmonic components*. By a simple harmonic component, we mean an oscillation with respect to time of a cosine or sine expression. Let us consider such a component in its most general form

$$y = a \cos(\omega t - \delta) \qquad (1.168)$$

Its graph will be as shown in Fig. 1.14. It oscillates back and forth across the

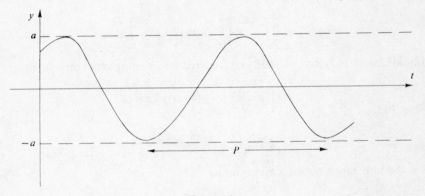

Fig. 1.14

t axis between maximum values of a and $-a$. The maximum displacement a is referred to as the *amplitude* of the motion. The *period* P of the motion is the time required to execute a complete oscillation. From Eq. (1.168) this will be simply

$$[\omega(t+P) - \delta] - [\omega t - \delta] = 2\pi$$

or

$$P = \frac{2\pi}{\omega} \qquad (1.169)$$

The *frequency* ν is defined as the number of oscillations per unit time. It is, therefore, simply the reciprocal of the period

$$\nu = \frac{1}{P} = \frac{\omega}{2\pi}$$

or (1.170)

$$\omega = 2\pi\nu$$

The quantity ω is referred to as the *circular* or *angular frequency*, and the quantity δ is referred to as the *initial phase*. The usefulness of the concept of simple harmonic motion is, in part, that such functions are often solutions to the partial differential equations of physics.

We should like now to generalize our simple harmonic motion components to an arbitrary source function. We can do this through the Fourier transforms. Let us consider, for example, a source function $f(t)$, which is an even function of t. In Eqs. (1.164) let the time t replace the variable x, τ replace the variable of integration ξ, the angular frequency ω replace the variable k, and the function $s(\omega)$ replace the function $(2/\pi)^{1/2} g(\omega)$. Then we shall have

$$f(t) = \int_0^\infty s(\omega) \cos \omega t \, d\omega$$
$$s(\omega) = \frac{2}{\pi} \int_0^\infty f(\tau) \cos \omega\tau \, d\tau$$
(1.171)

our desired result. The source function $f(t)$ is given by a summation over all frequencies ω of the simple harmonic components, each with an amplitude $s(\omega)$. The amplitude of the components is determined from the second of Eq. (1.171). The quantity $s(\omega)$ is referred to as the *spectrum* of frequencies which, when added together, give the desired impulse $f(t)$.

In complex form, these same equations become from Eqs. (1.167)

$$f(t) = \int_{-\infty}^\infty s(\omega) e^{i\omega t} \, d\omega$$
$$s(\omega) = \frac{1}{2\pi} \int_{-\infty}^\infty f(\tau) e^{-i\omega\tau} \, d\tau$$
(1.172)

Usually, the integrand of the first integral of Eqs. (1.172) is an even function, which conveniently obviates the necessity of considering negative frequencies. The spectrum function is then $2s(\omega)$. In this case, $s(\omega)$ is, in general, complex expressing both the amplitude and phase of the simple harmonic components.

Problem 1.9(a) Derive expressions (1.166) and (1.167).

Problem 1.9(b) Show that the Fourier transform of $f(x) = e^{-x^2/2}$ is $g(k) = e^{-k^2/2}$.

42 ¶ *Introduction*

Problem 1.9(c) Show that the Fourier transform of the step function

$$f(x) = \frac{\sqrt{2\pi}}{2l} \quad \text{for } |x| < l$$
$$= 0 \quad \text{for } |x| > l$$

is $g(k) = \sin kl/kl$.

Problem 1.9(d) Extend the results of Problem 1.9(c) to show that the Fourier transform of the unit impulse, $f(x) = \infty$ at $x = 0$ and $f(x) = 0$ for all other values of x, is $g(k) = 1$.

Problem 1.9(e) Derive the spectrum function $s(\omega)$ for

$$f(t) = \cos \gamma t \quad \text{for } |t| < \tfrac{1}{2}T$$
$$= 0 \quad \text{for } |t| > \tfrac{1}{2}T$$

and discuss the results.

Problem 1.9(f) Derive the spectrum function for

$$f(t) = e^{-\lambda t} \quad \text{for } t > 0$$
$$= 0 \quad \text{for } t < 0$$

and discuss the results.

1.10 Wave Equation

Another important partial differential equation in geophysics is the *wave equation*. It is obtained by setting the second partial derivative of a function, Ψ, with respect to time equal to a factor times the Laplacian of Ψ, or

$$c^2 \nabla^2 \Psi = \frac{\partial^2 \Psi}{\partial t^2} \tag{1.173}$$

It is convenient to denote the factor of proportionality by c^2, for as we shall soon find out, c is the *velocity* of propagation of Ψ. In general, the factor c will be a function of one or more of the space coordinates. For our discussion in this section, we shall restrict ourselves to considering c a constant.

Let us consider briefly here the properties of one type of general solution of this equation. For one-dimensional motion, $\Psi = \Psi(x, t)$, and Eq. (1.173) becomes

$$c^2 \frac{\partial^2 \Psi}{\partial x^2} = \frac{\partial^2 \Psi}{\partial t^2} \tag{1.174}$$

A general solution of this equation is

$$\Psi = f(x - ct) + F(x + ct) \tag{1.175}$$

as may be seen by substitution back into Eq. (1.174) and where f and F are arbitrary functions. Let us examine what significance these functions have. If, in the first function, the time t is increased by an amount δt, the value of f will remain unchanged if x is increased by an amount $c\delta t$. We see then that a particular value of the function f, which was observed at position x at time t, will be found at position $x + c\delta t$ at time $t + \delta t$ and that this is valid in a corresponding manner for all other values of x. The function f represents a disturbance propagated in the positive x direction with a velocity c. Similarly, the function F remains unchanged at a time $t + \delta t$ by a decrease in x of $c\delta t$; it represents a disturbance propagated in the negative x direction with a velocity c. These relations are illustrated in Fig. 1.15.

Fig. 1.15

Next, let us consider a disturbance that is propagated in three dimensions and is symmetrical about a center O. Then $\Psi = \Psi(r, t)$, and the Laplacian of Eq. (1.83) in spherical coordinates and Eq. (1.173) will reduce to

$$\frac{c^2}{r^2} \frac{\partial}{\partial r}\left(r^2 \frac{\partial \Psi}{\partial r}\right) = \frac{\partial^2 \Psi}{\partial t^2}$$

$$c^2 \left(\frac{\partial^2 \Psi}{\partial r^2} + \frac{2}{r} \frac{\partial \Psi}{\partial r}\right) = \frac{\partial^2 \Psi}{\partial t^2}$$

$$c^2 \frac{\partial^2 (r\Psi)}{\partial r^2} = \frac{\partial^2 (r\Psi)}{\partial t^2} \tag{1.176}$$

which is of the same form as Eq. (1.174), giving for Ψ

$$\Psi = \frac{1}{r} f(r - ct) + \frac{1}{r} F(r + ct) \tag{1.177}$$

The function $f(r-ct)$ represents a spherical wave diverging from O, and the function $F(r+ct)$ represents a spherical wave converging on O. The amplitude of $f(r-ct)$ decreases as r^{-1}. The amplitude factor would be expected from physical considerations; for as a spherical wave is propagated outward, the area of the wave front (sphere) increases as r^2. Consequently, the energy flow per unit area will decrease as r^{-2}. As will be shown later, the energy flow per unit area is proportional to Ψ^2.

It is of interest to examine, at this point, the type of solution of the wave equation that results from the use of the method of separation of variables. For simplicity, we shall examine the type of solution obtained from the one-dimensional wave equation (1.174). As with our examples in Section 1.7, we shall assume a solution of the form

$$\Psi = X(x)T(t) \tag{1.178}$$

Substituting Eq. (1.178) into Eq. (1.174), we obtain

$$c^2 X''T = XT''$$

$$c^2 \frac{X''}{X} - \frac{T''}{T} = 0 \tag{1.179}$$

The separation is as before. We shall, for convenience, designate this constant as $-\omega^2$ for the first term, and necessarily ω^2 for the second term. Equation (1.179) then gives us the following two equations:

$$T'' + \omega^2 T = 0 \tag{1.180}$$

and

$$X'' + \frac{\omega^2}{c^2} X = 0$$

$$X'' + k^2 X = 0 \tag{1.181}$$

where we have made the substitution

$$k = \frac{\omega}{c} \tag{1.182}$$

in the second equation. Solutions of these two ordinary differential equations are simply

$$T = a_1 e^{i\omega t} + a_2 e^{-i\omega t} \tag{1.183}$$

and

$$X = b_1 e^{ikx} + b_2 e^{-ikx} \tag{1.184}$$

in complex form.

Let us look at one of the four real-value components of this solution. We shall take, for convenience, from Eqs. (1.183) and (1.184)

$$\Psi = XT = a \cos(kx - \omega t) \tag{1.185}$$

This is an expression of *simple harmonic motion*. If we were to look at the motion at a particular instant of time, it would be represented by an oscillation back and forth across the X axis. The distance between two successively repeated points, such as the distance between two troughs as indicated in Fig. 1.16, is defined as the *wavelength*. From Eq. (1.185) the wavelength λ will then be simply

$$kx = k\lambda = 2\pi$$

$$\lambda = \frac{2\pi}{k} \tag{1.186}$$

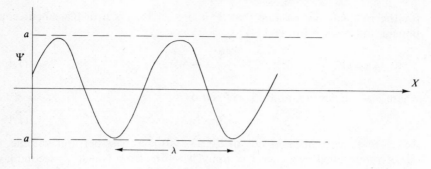

Fig. 1.16

The quantity k is referred to as the *wave number*. Similarly, if we were to stand at a particular point in space and observe the motion as a function of time, it would be a series of oscillations as shown in Fig. 1.14. From Eq. (1.185) we would have the same relation for period P and frequency ν, as before, or

$$P = \frac{2\pi}{\omega} \qquad (1.187)$$

and

$$\nu = \frac{\omega}{2\pi} \qquad (1.188)$$

We see that again our constant ω is the angular frequency. From Eqs. (1.182), (1.186), and (1.188) we obtain the familiar relation

$$c = \frac{\omega}{k} = \nu\lambda \qquad (1.189)$$

which states that the velocity of propagation of simple harmonic motion is equal to the product of frequency and wavelength. This is physically understandable in that if we are at a fixed point in space and observe the number of cycles that pass in a unit time, the velocity of propagation of the wave motion will be the product of this number by the length of one cycle, or Eq. (1.189). It we were to examine the other three terms of Eqs. (1.183) and (1.184), we would obtain the same results for motion propagating in either the positive or negative X direction.

Let us look, for the moment, at the energy partition occurring in simple harmonic motion. As we shall see in later sections, one associates two types of energy with wave or particle motion. One is the energy associated with the motion, or kinetic energy, and the other is the energy associated with the state of stress, or potential energy. The potential energy is the negative of the work that would have to be done to reach the particular state of stress. If, for the moment, we assume that Ψ in Eq. (1.185) is a displacement component and rewrite Eq. (1.174) as

$$\mu \frac{\partial^2 \Psi}{\partial x^2} = \rho \frac{\partial^2 \Psi}{\partial t^2} \tag{1.190}$$

where ρ is the density, and μ is defined by

$$\mu = c^2 \rho \tag{1.191}$$

we see that the right-hand side of Eq. (1.190) is a mass per unit volume, or density, multiplied by an acceleration. Therefore, from Newton's second law, the left-hand side must be a force per unit volume. We may then write, for the kinetic energy per unit volume, T,

$$T = \tfrac{1}{2}\rho \dot{\Psi}^2 = \tfrac{1}{2}\rho \omega^2 a^2 \sin^2 (kx - wt) \tag{1.192}$$

and for the potential energy per unit volume, V,

$$V = -\mu \int_0^\Psi \frac{\partial^2 \Psi}{\partial x^2} d\Psi = \mu \int_0^\Psi k^2 a \cos (kx - \omega t)\, d\Psi = \mu k^2 \int_0^\Psi \Psi\, d\Psi$$

$$= \tfrac{1}{2} \mu k^2 \Psi^2 = \tfrac{1}{2} \mu k^2 a^2 \cos^2 (kx - \omega t) \tag{1.193}$$

Since simple harmonic motion is repetitive over one cycle, we may find the average kinetic and potential energies per unit volume $\overline{T}$, and $\overline{V}$, by integrating over one period, or

$$\overline{T} = \tfrac{1}{2}\rho \omega^2 a^2 \left[\frac{1}{P} \int_0^P \sin^2 (kx - \omega t)\, dt \right]$$

$$= \tfrac{1}{2}\rho \omega^2 a^2 \left\{ \frac{-1}{2\pi} [\tfrac{1}{2}(kx - \omega t) - \tfrac{1}{4} \sin 2(kx - \omega t)] \right\}_0^{2\pi/\omega}$$

$$= \tfrac{1}{4}\rho \omega^2 a^2 \tag{1.194}$$

and

$$\overline{V} = \tfrac{1}{2}\mu k^2 a^2 \left[\frac{1}{P} \int_0^P \cos^2 (kx - wt)\, dt \right]$$

$$= \tfrac{1}{2}\mu k^2 a^2 \left\{ \frac{-1}{2\pi} [\tfrac{1}{2}(kx - \omega t) + \tfrac{1}{4} \sin 2(kx - wt)] \right\}_0^{2\pi/\omega}$$

$$= \tfrac{1}{4}\mu k^2 a^2 = \tfrac{1}{4}\rho \omega^2 a^2 \tag{1.195}$$

using Eqs. (1.187), (1.189), and (1.191). The average kinetic and potential energies are equal for simple harmonic motion. The total energy is the sum of the kinetic and potential energies. We then have, for the average total energy, $\overline{U}$, per unit volume,

$$\overline{U} = \overline{T} + \overline{V} = \tfrac{1}{2}\rho \omega^2 a^2 \tag{1.196}$$

We see that this relation between average kinetic and potential energies holds whether Ψ is the displacement or the displacement is given by some space or time derivative of Ψ.

It is easy to extend the separation of variables solution to two and three dimensions in rectangular coordinates. For the two-dimensional wave equation, we obtain

$$\Psi = a e^{\pm i k_x x} e^{\pm i k_y y} e^{\pm i \omega t} \tag{1.197}$$

where

$$k^2 = k_x^2 + k_y^2 \tag{1.198}$$

Following the same reasoning as before, when k_x and k_y are real, Eq. (1.197) represents simple harmonic motion propagating in a direction $\mathbf{r} = \mathbf{i} x + \mathbf{j} y$, whose wave number components in the X and Y directions are k_x and k_y. From Fig. 1.17 we can see the physical significance of Eq. (1.198). Let AB

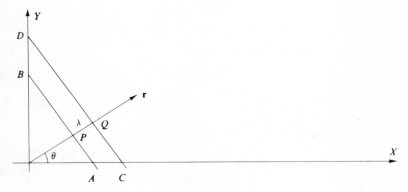

Fig. 1.17

and CD be the traces of the two successive crests of a plane wave propagating in the direction $\mathbf{r}$. Then the wavelength λ will be the distance PQ as shown, and the wavelengths λ_x and λ_y, measured along the X and Y directions, will be the distances AC and BD, respectively. From Fig. 1.17 and Eq. (1.186) we have

$$k_x = \frac{2\pi}{\lambda_x} = \frac{2\pi \cos \theta}{\lambda}$$

and (1.199)

$$k_y = \frac{2\pi}{\lambda_y} = \frac{2\pi \sin \theta}{\lambda}$$

or

$$k^2 = k_x^2 + k_y^2$$

48 ¶ *Introduction*

For the three-dimensional wave equation in rectangular coordinates, our solution will be, following the same reasoning,

$$\Psi = ae^{\pm ik_x x} e^{\pm ik_y y} e^{\pm ik_z z} e^{\pm i\omega t} \tag{1.200}$$

where

$$k^2 = k_x^2 + k_y^2 + k_z^2 \tag{1.201}$$

We may think of Eq. (1.201) as defining a vector **k**, whose components are k_x, k_y, and k_z. Equation (1.200) can then be written

$$\Psi = be^{\pm i(\mathbf{k}\cdot\mathbf{r} \pm \omega t)} \tag{1.202}$$

We are interested in looking at one other special type of solution of the wave equation by the method of separation of variables. This is for the Laplacian expressed in cylindrical coordinates and for which Ψ is a function of ρ and z only, and not of the azimuth φ. From Eqs. (1.88) and (1.173) we obtain

$$c^2 \left[\frac{\partial^2 \Psi}{\partial \rho^2} + \frac{1}{\rho} \frac{\partial \Psi}{\partial \rho} + \frac{\partial^2 \Psi}{\partial z^2} \right] = \frac{\partial^2 \Psi}{\partial t^2} \tag{1.203}$$

Choosing a solution of the form

$$\Psi = M(\rho)Z(z)T(t) \tag{1.204}$$

we obtain, upon substituting into Eq. (1.203),

$$c^2 \left[M''ZT + \frac{1}{\rho} M'ZT + MZ''T \right] - MZT'' = 0$$

$$c^2 \left[\frac{M''}{M} + \frac{1}{\rho} \frac{M'}{M} + \frac{Z''}{Z} \right] - \frac{T''}{T} = 0 \tag{1.205}$$

Again, we shall designate $-\omega^2$ as the constant for the first term and ω^2 for the second term. This operation yields Eq. (1.180) for the second term, whose solution is Eq. (1.183), and

$$\left[\frac{M''}{M} + \frac{1}{\rho} \frac{M'}{M} \right] + \left[\frac{Z''}{Z} + k^2 \right] = 0 \tag{1.206}$$

for the first term. We again have a separation of variables, and we shall choose $-\kappa^2$ as the constant for the first term and κ^2 for the second term, which lead to

$$\rho^2 M'' + \rho M' + \rho^2 \kappa^2 M = 0 \tag{1.207}$$

and

$$Z'' + (k^2 - \kappa^2)Z = 0 \tag{1.208}$$

Equation (1.207) is the same as Eq. (1.133) whose solution is (1.139), and

Eq. (1.208) is of the same form as Eq. (1.181). We may then write, for our solution in general form,

$$\Psi = \sum_\omega \sum_\kappa a_{\omega\kappa} J_0(\kappa\rho) \, e^{\pm i\sqrt{k^2-\kappa^2}\,z} e^{\pm i\omega t} \quad (1.209)$$

where the summation is taken over both ω (or k) and κ. When κ is real and less than k, the following physical significance can be attached to it. The z component of Eq. (1.209) is of a form that shows that the wave number in the Z direction is the radical in the exponent of that term. The wave number in the ρ direction will then be, from Eq. (1.198),

$$k_\rho^2 = k^2 - k_z^2 = k^2 - (k^2 - \kappa^2) = \kappa^2 \quad (1.210)$$

The wave number in the ρ direction is the constant κ.

Problem 1.10(a) Carry through the reductions to obtain Eqs. (1.197) and (1.198).

1.11 Associated Legendre Polynomials

Although we shall be concerned almost entirely, at the level of this book, with azimuthal symmetry for those problems utilizing spherical coordinates, it is of interest to develop the functions that occur in the three-dimensional solution of Laplace's equation in spherical coordinates. Such solutions involving these functions necessarily occur in many geophysical problems involving solutions of Laplace's equation for gravitational, hydrodynamic, and magnetic potentials. These relations can be developed readily from the results already obtained in Sections 1.6 and 1.7.

To proceed, we know from Section 1.6 that the Legendre polynomials P_n are solutions of the Legendre differential equation (1.103), written as

$$(1-u^2)P_n'' - 2uP_n' + n(n+1)P_n = 0 \quad (1.211)$$

Differentiating this equation m times with respect to u, we obtain sequentially

$$(1-u^2)\frac{d}{du}P_n'' - 2uP_n'' - 2u\frac{d}{du}P_n' - 2P_n' + n(n+1)\frac{d}{du}P_n = 0$$

$$(1-u^2)\frac{d}{du}P_n'' - 2(2)u\frac{d}{du}P_n' + [n(n+1)-2]\frac{d}{du}P_n = 0$$

for the first differentiation,

$$(1-u^2)\frac{d^2}{du^2}P_n'' - 2(3)u\frac{d^2}{du^2}P_n' + [n(n+1)-2(3)]\frac{d^2}{du^2}P_n = 0$$

for the second differentiation,

$$(1-u^2)\frac{d^3}{du^3}P_n''-2(4)u\frac{d^3}{du^3}P_n'+[n(n+1)-3(4)]\frac{d^3}{du^3}P_n = 0$$

for the third differentiation, and

$$(1-u^2)\frac{d^m}{du^m}P_n''-2(m+1)u\frac{d^m}{du^m}P_n'+[n(n+1)-m(m+1)]\frac{d^m}{dn^m}P_n = 0 \quad (1.212)$$

for the mth differentiation. Now let the function $\Theta(u)$ be defined by the relation

$$\frac{d^m}{du^m}P_n(u) = (1-u^2)^{-m/2}\Theta(u) \quad (1.213)$$

and let us see what differential equation Θ satisfies. We have then

$$\frac{d^m}{du^m}P_n'(u) = (1-u^2)^{-m/2}\Theta' + mu(1-u^2)^{-(m/2)-1}\Theta$$

$$= (1-u^2)^{-(m/2)-1}[(1-u^2)\Theta' + mu\Theta]$$

$$\frac{d^m}{du^m}P_n''(u) = (1-u^2)^{-m/2}\Theta'' + 2mu(1-u^2)^{-(m/2)-1}\Theta' + m(1-u^2)^{-(m/2)-1}\Theta$$

$$+ m(m+2)u^2(1-u^2)^{-(m/2)-2}\Theta$$

$$= (1-u^2)^{-(m/2)-1}\left[(1-u^2)\Theta'' + 2mu\Theta' + m\Theta + \frac{m(m+2)u^2}{1-u^2}\Theta\right]$$

Substituting these relations into Eq. (1.212), we obtain

$$(1-u^2)^2\Theta'' + 2mu(1-u^2)\Theta' + m(1-u^2)\Theta + m(m+2)u^2\Theta$$
$$-2(m+1)u(1-u^2)\Theta' - 2m(m+1)u^2\Theta + [n(n+1)-m(m+1)](1-u^2)\Theta = 0$$

or after algebraic reduction

$$(1-u^2)\Theta'' - 2u\Theta' + \left[n(n+1) - \frac{m^2}{1-u^2}\right]\Theta = 0 \quad (1.214)$$

which is our desired result. The functions $\Theta(u)$, usually designated by $P_n^m(u)$, given from Eq. (1.213) as

$$\Theta(u) = P_n^m(u) = (1-u^2)^{m/2}\frac{d^m}{du^m}P_n(u) \quad (1.215)$$

are known as the *associated Legendre polynomials* or *associated spherical harmonics*.

Laplace's equation in spherical coordinates is given by Eq. (1.121). Again, using the method of separation of variables of Section 1.7, we assume a solution of the form

$$\Phi = R(r)\Theta(u)H(\varphi) \quad (1.216)$$

Substituting into Eq. (1.212), we obtain

$$\frac{\partial}{\partial r}(r^2 R'\Theta H) + \frac{\partial}{\partial u}[(1-u^2)R\Theta'H] + \frac{1}{1-u^2}R\Theta H'' = 0$$

or

$$\frac{1-u^2}{R}\frac{\partial}{\partial r}(r^2 R') + \frac{1-u^2}{\Theta}\frac{\partial}{\partial u}[(1-u^2)\Theta'] + \frac{H''}{H} = 0 \quad ((1.217)$$

Setting the third term equal to $-m^2$, we have the same equation as the second of Eqs. (1.116), whose solution [Eqs. (1.117)] is

$$H = ce^{\pm im\varphi} \tag{1.218}$$

Equation (1.217) then becomes

$$\frac{1}{R}\frac{\partial}{\partial r}(r^2 R') + \frac{1}{\Theta}\frac{\partial}{\partial u}[(1-u^2)\Theta'] - \frac{m^2}{1-u^2} = 0 \tag{1.219}$$

Equation (1.219) is similar to Eq. (1.124), and using again the quantity $n(n+1)$ as the constant parameter, Eq. (1.219) reduces to Eq. (1.125), whose solution is Eq. (1.127), and to

$$\frac{d}{du}[(1-u^2)\Theta'] - \frac{m^2}{1-u^2}\Theta + n(n+1)\Theta = 0$$

or

$$(1-u^2)\Theta'' - 2u\Theta' + \left[n(n+1) - \frac{m^2}{1-u^2}\right]\Theta = 0$$

the equation satisfied by the associated Legendre polynomials. Our final answer then, of the form of Eq. (1.128), is

$$\Phi = \sum_{mn}[a_{mn}r^n + b_{mn}r^{-(n+1)}]P_n^m(\cos\theta)e^{\pm im\varphi} \tag{1.220}$$

In geophysical potential theory problems, the Legendre polynomials P_n are often referred to as *zonal harmonics*, and the associated Legendre polynomials P_n^m referred to as *tesseral harmonics* when $m \neq n$ and *sectorial harmonics* when $m = n$.

Problem 1.11(a) Determine the tesseral and sectorial harmonics for P_1, P_2, P_3.

Ans. $P_1^1 = \sin\theta$
$P_2^1 = 3\sin\theta\cos\theta$
$P_2^2 = 3\sin^2\theta$
$P_3^1 = \frac{15}{2}\sin\theta(\cos^2\theta - \frac{1}{5})$
$P_3^2 = 15\sin^2\theta\cos\theta$
$P_3^3 = 15\sin^3\theta$

1.12 Bessel Functions

One of the more important functions that appears in solutions of partial differential equations in geophysics is the Bessel function. In particular, it occurs in solutions of problems involving cylindrical symmetry and in problems involving solutions of the heat conduction, diffusion, and electromagnetic induction equations, all of which are given in the form of the Laplacian of a quantity being equal to the partial time derivative of the same quantity. Although little use will be made of such solutions at the level of this text, it is necessary to derive some of the basic relations involving Bessel functions.

The *Bessel function of order n*, $J_n(x)$, is defined as

$$y = J_n(x) = \frac{x^n}{2^n n!}\left(1 - \frac{x^2}{2(2n+2)} + \frac{x^4}{2 \cdot 4(2n+2)(2n+4)} + \cdots \right.$$
$$\left. + \frac{(-1)^m x^{2m}}{2 \cdot 4 \cdots 2n(2n+2)(2n+4)\cdots(2n+2m)} + \cdots \right)$$
$$= \sum_{\lambda=0}^{\infty} \frac{(-1)^\lambda}{(\lambda)!(\lambda+n)!}\left(\frac{x}{2}\right)^{n+2\lambda} \qquad (1.221)$$

Taking the first and second derivative of $J_n(x)$ with respect to x, we shall have

$$J_n'(x) = \sum_{\lambda=0}^{\infty} \frac{(-1)^\lambda}{\lambda!(\lambda+n)!} \frac{n+2\lambda}{2}\left(\frac{x}{2}\right)^{n-1+2\lambda}$$

and

$$J_n''(x) = \sum_{\lambda=0}^{\infty} \frac{(-1)^\lambda}{\lambda!(\lambda+n)!} \frac{(n+2\lambda)(n-1+2\lambda)}{2^2}\left(\frac{x}{2}\right)^{n-2+2\lambda}$$

Multiplying the second derivative by x^2 and the first derivative by x and adding and subtracting n^2 times $J_n(x)$, we shall have, for the coefficient of the $(x/2)^{n+2\lambda}$ term,

$$\frac{(-1)^\lambda}{\lambda!(\lambda+n)!}[(n+2\lambda)(n-1+2\lambda)+(n+2\lambda)-n^2] = \frac{(-1)^\lambda}{\lambda!(\lambda+n)!}[(n+2\lambda)^2-n^2]$$
$$= \frac{(-1)^\lambda}{\lambda!(\lambda+n)!}[4\lambda(\lambda+n)]$$

If we were to multiply $J_n(x)$ by x^2, we would have, for the coefficient of the $(x/2)^{n+2\lambda}$ term of that expression

$$\frac{(-1)^{\lambda-1}}{(\lambda-1)!(\lambda+n-1)!}[4] = -\frac{(-1)^\lambda}{\lambda!(\lambda+n)!}[4\lambda(\lambda+n)]$$

We see then that $J_n(x)$ satisfies the differential equation

$$x^2 J_n''(x) + x J_n'(x) - n^2 J_n(x) = -x^2 J_n(x)$$

or

$$x^2 \frac{d^2 y}{dx^2} + x \frac{dy}{dx} + (x^2 - n^2) y = 0 \qquad (1.222)$$

Equation (1.222) is known as *Bessel's differential equation of order n*. We see that $J_n(x)$ is one of the two particular solutions to this equation. Referring back to Eqs. (1.136) and (1.139), we see from Eqs. (1.222) and (1.221) that they are the Bessel equation of zero order and the zero order Bessel function.

We now wish to obtain the recurrence formulas among Bessel functions of different orders. For $J_{n-1}(x)$ and $J_{n+1}(x)$, we shall have

$$J_{n-1}(x) = \sum_{\lambda=0}^{\infty} \frac{(-1)^\lambda}{\lambda!(\lambda+n-1)!} \left(\frac{x}{2}\right)^{n-1+2\lambda}$$

and

$$J_{n+1}(x) = \sum_{\lambda=0}^{\infty} \frac{(-1)^\lambda}{\lambda!(\lambda+n+1)!} \left(\frac{x}{2}\right)^{n+1+2\lambda}$$

Adding these two expressions, we shall have, for the term of order $(x/2)^{n-1+2\lambda}$,

$$\left[\frac{(-1)^\lambda}{\lambda!(\lambda+n-1)!} + \frac{(-1)^{\lambda-1}}{(\lambda-1)!(\lambda+n)!}\right] \left(\frac{x}{2}\right)^{n-1+2\lambda}$$

$$= \frac{(-1)^\lambda(\lambda+n) + (-1)^{\lambda-1}(\lambda)}{\lambda!(\lambda+n)!} \left(\frac{x}{2}\right)^{n-1+2\lambda}$$

$$= \frac{(-1)^\lambda}{\lambda!(\lambda+n)!} \frac{2n}{x} \left(\frac{x}{2}\right)^{n+2\lambda}$$

Comparing this with the expression (1.221) for the definition of $J_n(x)$, we see that

$$\frac{2n}{x} J_n(x) = J_{n-1}(x) + J_{n+1}(x) \qquad (1.223)$$

Subtracting $J_{n+1}(x)$ from $J_{n-1}(x)$, we shall have, for the term of order $(x/2)^{n-1+2\lambda}$,

$$\frac{(-1)^\lambda}{\lambda!(\lambda+n)!} (n+2\lambda) \left(\frac{x}{2}\right)^{n-1+2\lambda}$$

Comparing this with the expression for $J_n'(x)$, we see that

$$2 J_n'(x) = J_{n-1}(x) - J_{n+1}(x) \qquad (1.224)$$

Eliminating $J_{n-1}(x)$ from Eqs. (1.223) and (1.224), we obtain

$$\frac{n}{x}J_n(x)-J'_n(x) = J_{n+1}(x) \tag{1.225}$$

or

$$\frac{d}{dx}[x^{-n}J_n(x)] = -x^{-n}J_{n+1}(x) \tag{1.226}$$

Eliminating $J_{n+1}(x)$ from Eqs. (1.223) and (1.224), we obtain

$$\frac{n}{x}J_n(x)+J'_n(x) = J_{n-1}(x) \tag{1.227}$$

or

$$\frac{d}{dx}[x^n J_n(x)] = x^n J_{n-1}(x) \tag{1.228}$$

We shall have some interest in solutions of Bessel's equation when n is half an odd integer, that is, when n is replaced by $p+\frac{1}{2}$ in Eq. (1.222), p being an integer. In this case, the solutions have a particularly simple form related to the trigonometric functions. To proceed, let us solve Eq. (1.222) for the case when $p = 0$, or

$$x^2 \frac{d^2y}{dx^2} + x\frac{dy}{dx} + (x^2-\tfrac{1}{4})y = 0 \tag{1.229}$$

Making a change of dependent variable from y to z, defined by

$$y = x^{-1/2}z$$

we have

$$y' = x^{-1/2}(z' - \tfrac{1}{2}x^{-1}z)$$

and

$$y'' = x^{-1/2}(z'' - x^{-1}z' + \tfrac{3}{4}x^{-2}z)$$

Substituting these expressions into Eq. (1.229), we obtain

$$z'' + z = 0$$

whose solutions in terms of y are simply

$$y = ax^{-1/2}\sin x + bx^{-1/2}\cos x \tag{1.230}$$

By choosing the value of the constants of integration to be $a = b = (2/\pi)^{1/2}$, it can be shown that the solutions [Eq. (1.230)] are equal to the reduced form of the Bessel function [Eq. (1.221)] for $n = \tfrac{1}{2}$ and $n = -\tfrac{1}{2}$, respectively, or

$$J_{1/2}(x) = \left(\frac{2}{\pi x}\right)^{1/2} \sin x$$

$$J_{-1/2}(x) = \left(\frac{2}{\pi x}\right)^{1/2} \cos x \tag{1.231}$$

The other half-order Bessel functions can then be found simply by application of the recurrence formulas (1.226) and (1.228).

We should also like to examine briefly the relations among Bessel functions corresponding to the relations between trigonometric and hyperbolic functions. In Eq. (1.222), let us replace x by kx, where k is a constant, for which we shall have then

$$x^2 \frac{d^2y}{dx^2} + x\frac{dy}{dx} + (k^2x^2 - n^2)y = 0 \qquad (1.232)$$

whose solution is simply $J_n(kx)$. Let us now set k equal to the imaginary, $k = i = \sqrt{-1}$, for which Eq. (1.232) will become

$$x^2 \frac{d^2y}{dx^2} + x\frac{dy}{dx} - (x^2 + n^2)y = 0 \qquad (1.233)$$

Equation (1.233) is called the *modified Bessel equation of order n*. We shall take its solution in the form

$$y = I_n(x) = \frac{x^n}{2^n n!}\left(1 + \frac{x^2}{2(2n+2)} + \frac{x^4}{2\cdot 4(2n+2)(2n+4)} + \cdots \right.$$

$$\left. + \frac{x^{2m}}{2\cdot 4 \cdot \cdots \cdot 2n(2n+2)(2n+4)\cdot \cdots \cdot(2n+2m)} + \cdots \right) \qquad (1.234)$$

corresponding to Eq. (1.221). From Eq. (1.221) we note that

$$I_n(x) = i^{-n} J_n(ix) \qquad (1.235)$$

The function $I_n(x)$ is known as the *modified Bessel function*. In particular, from Eqs. (1.235) and (1.231) or from the derivation of the last paragraph, we find that

$$I_{1/2}(x) = \left(\frac{2}{\pi x}\right)^{1/2} \sinh x$$
$$I_{-1/2}(x) = \left(\frac{2}{\pi x}\right)^{-1/2} \cosh x \qquad (1.236)$$

Problem 1.12(a) Reduce the expression (1.221) to Eqs. (1.231) for $n = \pm\frac{1}{2}$.

Hint: Replace the factorial by its generalization in terms of the gamma function.

Problem 1.12(b) Derive the recurrence formulas for $I_n(x)$.

Ans. $\dfrac{d}{dx}[x^{-n}I_n(x)] = x^{-n}I_{n+1}(x)$

$\dfrac{d}{dx}[x^n I_n(x)] = x^n I_{n-1}(x)$

Problem 1.12(c) Derive the expressions for $J_{3/2}(x)$, $J_{-3/2}(x)$, and $J_{5/2}(x)$.

Ans. $$J_{3/2}(x) = \left(\frac{2}{\pi x}\right)^{1/2}\left(-\cos x + \frac{\sin x}{x}\right)$$

$$J_{-3/2}(x) = \left(\frac{2}{\pi x}\right)^{1/2}\left(-\sin x - \frac{\cos x}{x}\right)$$

$$J_{5/2}(x) = \left(\frac{2}{\pi x}\right)^{1/2}\left[\left(\frac{3}{x^2} - 1\right)\sin x - \frac{3}{x}\cos x\right]$$

Problem 1.12(d) Derive the expressions for $I_{3/2}(x)$ and $I_{-3/2}(x)$.

Ans. $$I_{3/2}(x) = \left(\frac{2}{\pi x}\right)^{1/2}\left(\cosh x - \frac{\sinh x}{x}\right)$$

$$I_{-3/2}(x) = \left(\frac{2}{\pi x}\right)^{1/2}\left(\sinh x - \frac{\cosh x}{x}\right)$$

1.13 Velocity and Acceleration Referred to Moving Axes

We should now like to consider one final background item in this introductory chapter. That has to do with vector relations in a moving coordinate system and, in particular, with those having to do with velocity and acceleration. Our equations of motion refer back to an inertial, reference coordinate system. However, for earth problems we shall often use a coordinate system fixed to the earth which, due to the earth's rotation and revolution, will not be an inertial system. We should like to find the relations between velocity and acceleration as referred to the moving and inertial coordinate systems.

Consider first a vector **A** defined by its coordinates A_x, A_y, A_z in a rotating coordinate system. There, the unit vectors **i**, **j**, **k** will no longer be fixed in direction, and we shall have, for the time derivative of **A**,

$$\frac{d\mathbf{A}}{dt} = \mathbf{i}\frac{dA_x}{dt} + \mathbf{j}\frac{dA_y}{dt} + \mathbf{k}\frac{dA_z}{dt} + A_x\frac{d\mathbf{i}}{dt} + A_y\frac{d\mathbf{j}}{dt} + A_z\frac{d\mathbf{k}}{dt} \quad (1.237)$$

Referring to Fig. 1.18, we see that the magnitude of the increment $d\mathbf{i}$ is given by

$$|d\mathbf{i}| = |\mathbf{i}|\sin\alpha\, d\theta = |\mathbf{i}|\sin\alpha\, \omega\, dt \quad (1.238)$$

where **ω** is the angular velocity of rotation of the coordinate system and that its direction is at right angles to **ω** and **i** so that in vector form we shall have

$$\frac{d\mathbf{i}}{dt} = \boldsymbol{\omega} \times \mathbf{i} \quad (1.239)$$

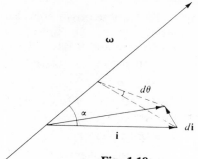

Fig. 1.18

and similarly for $d\mathbf{j}/dt$ and $d\mathbf{k}/dt$. We also see that the first three terms in Eq. (1.237) are simply the time derivative of $\mathbf{A}$ referred to the rotating coordinates axes so that we may write

$$\frac{d\mathbf{A}}{dt} = \frac{d\mathbf{A}}{d\tau} + \boldsymbol{\omega} \times \mathbf{A} \qquad (1.240)$$

or the operator

$$\frac{d}{dt} = \frac{d}{d\tau} + \boldsymbol{\omega} \times \qquad (1.241)$$

where d/dt is used to indicate a time derivation with respect to an inertial system and $d/d\tau$ with respect to the rotating system.

Now for the velocity of a point P, as shown in Fig. 1.19, referred to a

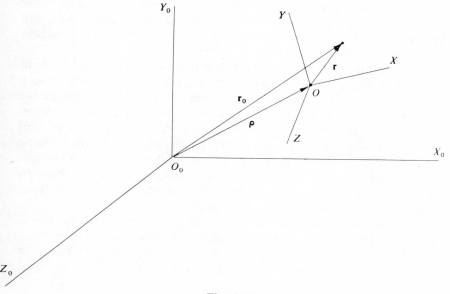

Fig. 1.19

rotating coordinate system, X, Y, Z whose origin O is also translating with respect to an inertial coordinate system X_0, Y_0, Z_0, we shall have first that the position vectors $\mathbf{r}_0$ and $\mathbf{r}$ of the point P referred to the inertial and rotating coordinate systems are related by

$$\mathbf{r}_0 = \boldsymbol{\rho} + \mathbf{r} \tag{1.242}$$

where $\boldsymbol{\rho}$ is the position vector of O with respect to O_0. Applying Eq. (1.241), we shall then have for the velocities

$$\mathbf{v}_0 = \dot{\boldsymbol{\rho}} + \mathbf{v} + \boldsymbol{\omega} \times \mathbf{r} \tag{1.243}$$

since the unit vectors $\mathbf{i}_0, \mathbf{j}_0, \mathbf{k}_0$ of the components of $\boldsymbol{\rho}$ are fixed and where $\mathbf{v}_0$ is the velocity of P referred to X_0, Y_0, Z_0, $\mathbf{v}$ the velocity of P referred to X, Y, Z, and $\dot{\boldsymbol{\rho}}$ the velocity of the moving origin. This, of course, is exactly the result that we would expect; for referring to Fig. 1.9, we see that the velocity $\mathbf{v}_0$ will be composed of three parts: the velocity $\mathbf{v}$ of P with respect to the rotating coordinates X, Y, Z, the linear velocity $\boldsymbol{\omega} \times \mathbf{r}$ of P as part of the rotating coordinate system, and the translation velocity $\dot{\boldsymbol{\rho}}$ of O with respect to O_0.

To find the acceleration, we simply apply Eq. (1.241) to Eq. (1.243), obtaining

$$\begin{aligned}\mathbf{f}_0 &= \ddot{\boldsymbol{\rho}} + \mathbf{f} + \boldsymbol{\omega} \times \mathbf{v} + \dot{\boldsymbol{\omega}} \times \mathbf{r} + \boldsymbol{\omega} \times \mathbf{v} + \boldsymbol{\omega} \times (\boldsymbol{\omega} \times \mathbf{r}) \\ &= \ddot{\boldsymbol{\rho}} + \boldsymbol{\omega} \times (\boldsymbol{\omega} \times \mathbf{r}) + \dot{\boldsymbol{\omega}} \times \mathbf{r} + 2\boldsymbol{\omega} \times \mathbf{v} + \mathbf{f}\end{aligned} \tag{1.244}$$

So the acceleration of the point P is made up of five terms. The first represents the acceleration of the origin of the moving axes, the second the centripetal acceleration due to the rotation of the moving axes, the third the linear acceleration due to the angular acceleration $\dot{\boldsymbol{\omega}}$ of the moving axes, the fourth the *Coriolis* acceleration, and the fifth the apparent acceleration of P relative to the moving axes.

Let us consider, in some more detail, certain of the individual terms of which the acceleration $\mathbf{f}_0$ is composed. If P is rigidly attached to the moving axes, $\mathbf{v}$ and $\mathbf{f}$ vanish and $\mathbf{f}_0$ reduces to the sum of the first three terms on the right of Eq. (1.244). Therefore, the sum of these three terms is said to represent the *acceleration of transport* of the moving space XYZ. The term $2\boldsymbol{\omega} \times \mathbf{v}$ exists only when the point is moving relative to the rotating axes. It is directed perpendicular to the plane of $\boldsymbol{\omega}$ and $\mathbf{v}$ in the sense of advance of a right-handed screw rotated from $\boldsymbol{\omega}$ to $\mathbf{v}$ and is equal in magnitude to twice the product of $\boldsymbol{\omega}$ by the component of $\mathbf{v}$ perpendicular to $\boldsymbol{\omega}$. Therefore, if $\mathbf{v}$ is parallel to $\boldsymbol{\omega}$, the Coriolis acceleration vanishes.

When referred to the earth, the angular velocity of the earth's rotation is constant so that the term in $\dot{\boldsymbol{\omega}}$ disappear, and the linear acceleration of the earth's center is so small that the term in $\ddot{\boldsymbol{\rho}}$ can be neglected. We may then write for our general equation of motion, $m\mathbf{f}_0 = \mathbf{F}$, where $\mathbf{F}$ is the resultant force acting on the particle P,

$$m\mathbf{f} = \mathbf{F} - m\boldsymbol{\omega} \times (\boldsymbol{\omega} \times \mathbf{r}) - 2m\boldsymbol{\omega} \times \mathbf{v} \tag{1.245}$$

referred to an origin at the center of the earth. Now if the only impressed force acting on the particle is the force $\mathbf{F}_g$ due to the gravitational attraction of the earth, Eq. (1.245) will be simply

$$m\mathbf{f} = \mathbf{F}_g - m\boldsymbol{\omega} \times (\boldsymbol{\omega} \times \mathbf{r}) - 2m\boldsymbol{\omega} \times \mathbf{v} \qquad (1.246)$$

Now the familiar acceleration $\mathbf{g}$ of gravity as measured by any gravity measuring instrument is not, however, the acceleration due to $\mathbf{F}_g$ alone, but rather the acceleration due to the resultant of $\mathbf{F}_g$ and the centrifugal reaction $-m\boldsymbol{\omega} \times (\boldsymbol{\omega} \times \mathbf{r})$ so that

$$m\mathbf{g} = \mathbf{F}_g - m\boldsymbol{\omega} \times (\boldsymbol{\omega} \times \mathbf{r}) \qquad (1.247)$$

and substituting into Eq. (1.246), we obtain the useful relation

$$\mathbf{f} = \mathbf{g} - 2\boldsymbol{\omega} \times \mathbf{v} \qquad (1.248)$$

As this final equation does not involve the position vector of the particle, we can take the origin of coordinates at any convenient point instead of at the center of the earth.

Problem 1.13(a) Show that the angular acceleration of $\boldsymbol{\omega}$ is the same whether measured in the inertial or a rotating coordinate system.

Problem 1.13(b) Two spheres of equal mass m, connected by a taut inextensible string of length l, have holes drilled through their centers so that they can slide along a straight bar. The bar rotates about a fixed axis at right angles to its length with a constant angular velocity ω. Assuming the spheres to be initially at rest relative to the rotating bar at distances d and $d-l$ from the axis, find the speed with which they are sliding along the bar at time t, and determine their positions for no sliding, discussing the stability of the equilibrium configuration. Treat the spheres as particles.
Ans. $v = \tfrac{1}{2}\omega(d - \tfrac{1}{2}l)(e^{\omega t} - e^{-\omega t})$

References

Bowman, F. 1958. *Introduction to Bessel Functions*. New York: Dover.
Carslaw, H. S. 1930. *Introduction to the Theory of Fourier's Series and Integrals*. New York: Dover.
Gibbs, J. W., and E. B. Wilson. 1925. *Vector Analysis*. New Haven: Yale University Press.
Kellogg, O. D. 1929. *Foundations of Potential Theory*. New York: Dover.
Kraut, E. A. 1967. *Fundamentals of Mathematical Physics*. New York: McGraw-Hill.
Jeffreys, H., and B. S. Jeffreys. 1956. *Methods of Mathematical Physics*. Cambridge: Cambridge University Press.

Margenau, H., and G. M. Murphy. 1943. *The Mathematics of Physics and Chemistry.* New York: Van Nostrand.

Page L. 1935. *Introduction to Theoretical Physics.* New York: Van Nostrand.

Slater, J. C., and N. H. Frank. 1933. *Introduction to Theoretical Physics.* New York: McGraw-Hill.

Wangsness, R. K. 1963. *Introduction to Theoretical Physics.* New York: Wiley.

Whittaker, E. T., and G. N. Watson. 1948. *A Course of Modern Analysis.* New York: Macmillan.

Part Two

Thermodynamics and Hydrodynamics

CHAPTER 2

THERMODYNAMICS OF THE EARTH

2.1 Thermodynamics

We assume at the outset in this section that the reader has an understanding of the basic principles of thermodynamics. We shall simply restate some of these more basic relations.

For our purposes, the first and second laws of thermodynamics may be summarized as stating that the *intrinsic energy* U and the *entropy* S are exact differentials, where these quantities are defined by the relations

$$dU = \delta Q - \delta W \qquad (2.1)$$

and, for a reversible transformation,

$$dS = \frac{\delta Q}{T} \qquad (2.2)$$

and where Q is the heat taking in by a substance and W the work done by the substance, and where δ is used to designate a small quantity, which is not an exact differential. The summary statement of the combined first and second laws is then

$$dU = TdS - pdV \qquad (2.3)$$

or

$$du = Tds - pdv \qquad (2.4)$$

when referred to a unit mass of the substance, and where T is the temperature, p the pressure, u the intrinsic energy per unit mass, s the entropy per unit mass, and v the volume per unit mass of *specific volume*. We also have that we may at will choose any two of the five variables u, s, T, v, p as independent variables for the purpose of describing the state of a substance, expressing the remaining three as functions of these two. We note also that there are two types of reversible transformations that are of particular importance in developing certain thermodynamic relations—an *isothermal transformation* during which the temperature remains constant and an *adiabatic transformation* during which heat is neither taken in nor given out.

Since du is an exact differential, we may express it in the form

$$du = \left(\frac{\partial u}{\partial s}\right)_v ds + \left(\frac{\partial u}{\partial v}\right)_s dv \qquad (2.5)$$

from which we see that when u is expressed as a function of s and v that

$$T = \left(\frac{\partial u}{\partial s}\right)_v \qquad p = -\left(\frac{\partial u}{\partial v}\right)_s \qquad (2.6)$$

The subscript appended to the derivative indicates which variable is kept constant in the process of differentiation. Taking the second partial derivative of the first expression of Eqs. (2.6) with respect to v and that of the second with respect to s, we obtain the equality

$$\left(\frac{\partial T}{\partial v}\right)_s = -\left(\frac{\partial p}{\partial s}\right)_v \qquad (2.7)$$

There are three other expressions, similar to u, which are exact differentials. The *free energy* ψ is defined by

$$\psi = u - Ts \qquad (2.8)$$

for which we have

$$d\psi = du - Tds - sdT = -sdT - pdv \qquad (2.9)$$

and

$$\left(\frac{\partial s}{\partial v}\right)_T = \left(\frac{\partial p}{\partial T}\right)_v \qquad (2.10)$$

The *enthalpy* χ is defined by

$$\chi = u + pv \qquad (2.11)$$

for which we have

$$d\chi = du + pdv + vdp = Tds + vdp \qquad (2.12)$$

and

$$\left(\frac{\partial T}{\partial p}\right)_s = \left(\frac{\partial v}{\partial s}\right)_p \qquad (2.13)$$

The *thermodynamic potential* ζ is defined by

$$\zeta = u - Ts + pv \qquad (2.14)$$

for which we have

$$d\zeta = du - Tds - sdT + pdv + vdp = -sdT + vdp \qquad (2.15)$$

and

$$\left(\frac{\partial s}{\partial p}\right)_T = -\left(\frac{\partial v}{\partial T}\right)_p \qquad (2.16)$$

The functions ψ, χ, ζ are known as *Gibbs' functions* and the equalities (2.7), (2.10), (2.13), and (2.16) as *Maxwell's relations*.

Problem 2.1(a) The pressure on a perfect reflector due to radiation of energy u per unit volume is $p = \tfrac{1}{3}u$. Show that the energy density of radiation in equilibrium with matter at temperature T is proportional to the fourth power of the absolute temperature. This is known as the *Stefan-Boltzmann law*. *Hint*: The energy density u is a function of T only. Make use of the fact that ds is an exact differential.

Problem 2.1(b) A wire expands on being heated. Show that if a tension stress F is applied adiabatically, the temperature falls. *Hint*: Use the function $\chi = U - Fx$. Why?

2.2 Implicit Functions in Thermodynamics

We have that the state of a substance can be given by any two of the variables u, s, p, v, T acting as independent variables, the remaining three being given as functions of these two. We shall be interested in the implicit relations among the partial derivatives using two of the three variables p, v, T as independent.

In particular, there will be a functional relation among the three variables p, v, T, referred to as the *equation of state*, which we may express in the form

$$f(p, v, T) = 0 \qquad (2.17)$$

If we choose $v = v(p, T)$ as the dependent variable, or alternatively p or T as dependent variable, we may write successively

$$dv = \left(\frac{\partial v}{\partial p}\right)_T dp + \left(\frac{\partial v}{\partial T}\right)_p dT$$

$$dp = \left(\frac{\partial p}{\partial T}\right)_v dT + \left(\frac{\partial p}{\partial v}\right)_T dv \qquad (2.18)$$

$$dT = \left(\frac{\partial T}{\partial v}\right)_p dv + \left(\frac{\partial T}{\partial p}\right)_v dp$$

66 ¶ *Thermodynamics and Hydrodynamics*

If in the second of these equations $dp = 0$, we shall have

$$\left(\frac{\partial p}{\partial T}\right)_v dT + \left(\frac{\partial p}{\partial v}\right)_T dv = 0$$

$$\left(\frac{\partial T}{\partial v}\right)_p = \frac{1}{\left(\frac{\partial v}{\partial T}\right)_p} = -\frac{\left(\frac{\partial p}{\partial v}\right)_T}{\left(\frac{\partial p}{\partial T}\right)_v}$$

$$\left(\frac{\partial v}{\partial T}\right)_p \left(\frac{\partial p}{\partial v}\right)_T = -\left(\frac{\partial p}{\partial T}\right)_v \tag{2.19}$$

Taking the other two equations of Eqs. (2.18), similar reciprocal relations for the other two partial derivatives and the same final result, Eq. (2.19), would be obtained. Certain of these partial derivatives, or simple relations involving them, are conveniently measurable physical quantities. They are the *thermometric coefficients* α_v and α_p, being, respectively, the coefficient of change of pressure at constant volume and the coefficient of dilatation at constant pressure, defined by

$$\alpha_v = \frac{1}{p}\left(\frac{\partial p}{\partial T}\right)_v \tag{2.20}$$

and

$$\alpha_p = \frac{1}{v}\left(\frac{\partial v}{\partial T}\right)_p \tag{2.21}$$

and the *thermoelastic coefficients* k_T and k_s, being, respectively, the modulus of elasticity for an isothermal change and the modulus of elasticity for an adiabatic change, defined by

$$k_T = -v\left(\frac{\partial p}{\partial v}\right)_T \tag{2.22}$$

and

$$k_s = -v\left(\frac{\partial p}{\partial v}\right)_s \tag{2.23}$$

We see that Eqs. (2.22) and (2.23) are simply the bulk moduli defined for the thermodynamic condition under which the stress change has taken place. Substituting Eqs. (2.20), (2.21), and (2.22) into Eq. (2.19), we obtain the relation (2.19) in terms of these coefficients as

$$k_T \alpha_p = p \alpha_v \tag{2.24}$$

Taking alternate pairs of the three variables p, v, T as independent, we may write, for the differential heat,

Thermodynamics of the Earth ¶ 67

$$\delta q = \left(\frac{\delta q}{\partial p}\right)_v dp + \left(\frac{\delta q}{\partial v}\right)_p dv = \left(\frac{\delta q}{\partial T}\right)_p dT + \left(\frac{\delta q}{\partial p}\right)_T dp = \left(\frac{\delta q}{\partial T}\right)_v dT + \left(\frac{\delta q}{\partial v}\right)_T dv \quad (2.25)$$

Each of these six coefficients is a physically measurable and useful quantity. They are the *calorimetric coefficients* η_v and λ_p, being, respectively, the heat of compression at constant volume and the heat of expansion at constant pressure, given as

$$\eta_v = \left(\frac{\delta q}{\partial p}\right)_v \quad (2.26)$$

and

$$\lambda_p = \left(\frac{\delta q}{\partial v}\right)_p \quad (2.27)$$

c_v and c_p, being, respectively, the specific heat at constant volume and the specific heat at constant pressure, given as

$$c_v = \left(\frac{\delta q}{\partial T}\right)_v \quad (2.28)$$

and

$$c_p = \left(\frac{\delta q}{\partial T}\right)_p \quad (2.29)$$

h_T and l_T, being, respectively, the latent heat of change of pressure and the latent heat of change of volume, given as

$$h_T = \left(\frac{\delta q}{\partial p}\right)_T \quad (2.30)$$

and

$$l_T = \left(\frac{\delta q}{\partial v}\right)_T \quad (2.31)$$

We may then write Eq. (2.25) in terms of these coefficients as

$$\delta q = \eta_v dp + \lambda_p dv = c_p dT + h_T dp = c_v dT + l_T dv \quad (2.32)$$

Now we may consider separately the three equations given by the equalities of the last three members of Eq. (2.32). Only two of these will give independent equations. For each of these equations, we may successively take dp, dv, dT equal to zero to find the implicit relations among the calorimetric, the thermometric, and the thermoelastic coefficients. Again, only two of each of these three relations will be independent. We shall derive the following four independent relations among coefficients. Equating the third and fourth members of Eq. (2.32), we obtain

$$(c_p - c_v)dT + h_T dp - l_T dv = 0$$

68 ¶ *Thermodynamics and Hydrodynamics*

If $dp = 0$: we obtain from this expression

$$c_p - c_v - l_T \left(\frac{\partial v}{\partial T}\right)_p = 0$$

and then substituting from Eq. (2.21),

$$c_p = c_v + l_T v \alpha_p \tag{2.33}$$

If $dT = 0$, we obtain from this expression, using Eq. (2.22),

$$h_T - l_T \left(\frac{\partial v}{\partial p}\right)_T = 0$$

$$h_T = -\frac{v l_T}{k_T} \tag{2.34}$$

Equating the second and fourth members of Eq. (2.32), we obtain

$$(\lambda_p - l_T)dv + \eta_v dp - c_v dT = 0$$

If $dv = 0$, we obtain from this expression, using Eq. (2.20),

$$\eta_v - c_v \left(\frac{\partial T}{\partial p}\right)_v = 0$$

$$\eta_v = \frac{c_v}{p \alpha_v} \tag{2.35}$$

Equating the second and third members of Eq. (2.32), we obtain

$$(\eta_v - h_T)dp + \lambda_p dv - c_p dT = 0$$

If $dp = 0$, we obtain from this expression, using Eq. (2.21),

$$\lambda_p - c_p \left(\frac{\partial T}{\partial v}\right)_p = 0$$

$$\lambda_p = \frac{c_p}{v \alpha_p} \tag{2.36}$$

We obtain an additional relation involving the coefficient k_s. If in Eq. (2.32) $\delta q = 0$, we have from the first and second members

$$\eta_v + \lambda_p \left(\frac{\partial v}{\partial p}\right)_s = 0$$

Substituting from Eqs. (2.35), (2.24), (2.36), and (2.23) in this expression, we obtain

$$\frac{c_v}{k_T \alpha_p} - \frac{c_p v}{v \alpha_p k_s} = 0$$

or
$$\frac{k_s}{k_T} = \frac{c_p}{c_v} = \gamma \qquad (2.37)$$

where the symbol γ is used to denote the ratio of specific heats at constant pressure and constant volume.

It is of interest at this point to examine the relations among some of these quantities for a change of state of a substance. During a change of state from solid to liquid or liquid to gas, the temperature remains constant. Denoting by L the latent heat of change of state, that is, the heat necessary to change the state of a unit mass of the substance from solid to liquid or liquid to gas, we have from Eq. (2.31)

$$L = \delta q = l_T(v_2 - v_1)_T \qquad (2.38)$$

where v_1 is the specific volume before the change in state and v_2 that after. From Eq. (2.2) we may also expresss l_T under these conditions as

$$l_T = \left(\frac{\delta q}{\partial v}\right)_T = T\left(\frac{\partial s}{\partial v}\right)_T \qquad (2.39)$$

Substituting Eq. (2.39) and the Maxwell relation (2.10) into Eq. (2.38), we obtain

$$L = T\left(\frac{\partial p}{\partial T}\right)_v (v_2 - v_1)_T$$

or $\qquad (2.40)$

$$\frac{dT_0}{dp} = \left(\frac{\partial T}{\partial p}\right)_v = \frac{T_0}{L}(v_2 - v_1)$$

where the exact derivative can replace the partial derivative since the temperature at which a change of state occurs is a function of the pressure only, and not of the specific volume, where the subscript can be omitted from the expression $(v_2 - v_1)$ since the change of state necessarily takes place at constant temperature, and where T_0 is used instead of T to indicate the temperature of the change of state. Expression (2.40) is known as the *Clausius-Clapeyron equation*. It is an expression for the variation of the temperature at which a change of state occurs with respect to pressure given in terms of its absolute temperature, latent heat of change of state, and specific volumes before and after the change of state. We see that an increase in pressure raises the temperature at which the change in state takes place if $v_2 > v_1$, and lowers it if $v_2 < v_1$ since T and L are always positive. Thus, as ice contracts on melting, an increase in pressure lowers the freezing point of water. On the other hand, water expands on boiling so that an increase in pressure raises the boiling point.

Problem 2.2(a)   Follow through the derivations from the other two

70 ¶ *Thermodynamics and Hydrodynamics*

equations of Eqs. (2.18) to (2.19) and the reciprocal relations for the partial derivatives.

Problem 2.2(b) Derive the other five implicit relations from Eq. (2.32).

Problem 2.2(c) Derive these same relations from Eqs. (2.33), (2.34), (2.35), and (2.36).

Problem 2.2(d) For a mixture of water and steam at atmospheric pressure, $dp/dT = 2.68$ cm of mercury per °C, $L = 538.7$ cal/g, $T = 373°$ absolute, $v_1 = 1$ cm^3/g. Find the specific volume of steam.

Ans. 1686 cm^3/g

Problem 2.2(e) Find the depression of the freezing point of water for an increase in pressure of 1 atmosphere. $L = 80$ cal/g, $T = 273°$ absolute, $v_1 = 1.09$ cm^3/g, $v_2 = 1.00$ cm^3/g

Ans. 0.0073°C/atmosphere

2.3 Gravitational Adiabatic Equilibrium

It is of interest to derive an expression for the change in temperature as a function of depth, or elevation, under adiabatic conditions and under the influence of the earth's gravitational attraction. When a particle is lowered, or raised, from a given depth, it will undergo an increase, or decrease, in hydrostatic pressure due to the change in the weight of the overlying material. In general, it will also suffer a change in temperature. If this change in temperature is such that there is no exchange of heat, the system is said to be in adiabatic equilibrium. This temperature variation is the one we wish to derive. A system in adiabatic equilibrium is often a convenient reference system for geophysical problems of atmosphere, oceans, or the cooling of the earth.

For adiabatic equilibrium, we have from Eq. (2.32)

$$\delta q = c_p dT + h_T dp = 0 \tag{2.41}$$

From Eqs. (2.30), (2.2), (2.21) and the relation (2.16), we have

$$h_T = \left(\frac{\delta q}{\partial p}\right)_T = T\left(\frac{\partial s}{\partial p}\right)_T = -T\left(\frac{\partial v}{\partial T}\right)_p = -Tv\alpha_p \tag{2.42}$$

since the quantity T can be removed outside the partial derivative, which is kept constant with respect to T. Substituting Eq. (2.42) into Eq. (2.41), we obtain

$$\left(\frac{\partial T}{\partial p}\right)_s = \frac{Tv\alpha_p}{c_p} \tag{2.43}$$

If we take the depth z directed downward, we have that the change in pressure with respect to depth is simply

$$\frac{dp}{dz} = g\rho \qquad (2.44)$$

where g is the force per unit mass due to the earth's gravitational attraction and ρ the density. Since, by definition, the density is the reciprocal of the specific volume, we have finally

$$\frac{dT}{dz} = \left(\frac{\partial T}{\partial p}\right)_s \frac{dp}{dz} = \frac{g\alpha_p T}{c_p} \qquad (2.45)$$

Since all the quantities on the right-hand side of Eq. (2.45) are positive, we see that dT/dz will always be positive.

For the oceans, if there were complete mixing in a vertical direction, all temperature differences except those due to adiabatic effects would be eliminated. For such thermal equilibrium, the temperature at each depth would be fixed by purely adiabatic displacements of water, and for such equilibrium, the vertical distribution of temperature would remain invariable. Such a condition is sometimes referred to as *indifferent equilibrium*. The temperature of a water mass moved to the surface under these conditions is known as its *potential temperature*. If, instead, the vertical temperature gradient is greater than adiabatic, the equilibrium will be unstable in the vertical; for if a small water mass is displaced downward, it will remain colder, and heavier, than its surroundings in spite of the adiabatic heating and will be forced further downward. If, on the other hand, the vertical temperature gradient is less than adiabatic or negative, the equilibrium will be stable gravitationally in the vertical.

For the earth, the assumption is sometimes made that the earth was at some stage in its early history in a liquid state and has cooled to its present state since then. The further assumption is made that the convection of the high-temperature fluids was such as to obtain a nearly adiabatic condition. A comparison then of the Clausius-Clapeyron equation, (2.40), for the variation of melting temperature with pressure and Eq. (2.45) is of interest in indicating at what depths solidification will occur first and how it will proceed. We may put Eq. (2.40) into a form comparable with Eq. (2.45) through the use of Eq. (2.44), obtaining

$$\frac{dT_0}{dz} = \frac{\partial T_0}{\partial p}\frac{dp}{dz} = g\rho_2 \frac{T_0}{L}(v_2 - v_1)$$

$$= \frac{gT_0}{L}\left(1 - \frac{\rho_2}{\rho_1}\right) \qquad (2.46)$$

Problem 2.3(a) For a characteristic ocean situation, calculate the adiabatic gradient.

Ans. $\sim 0.1°C/km$

Problem 2.3(b) For silicate rocks at the earth's surface at their melting temperature, determine approximate values for Eqs. (2.45) and (2.46).

Ans. $\sim 0.3°C/km$, $\sim 3°C/km$

Problem 2.3(c) From the results of Problem 2.3(b) discuss how solidification would occur in the earth. Discuss the complications and uncertainties for extrapolating the results of Problem 2.3(b) directly in depth. Assuming the mantle to be silicate rock and the core to be nickel-iron, determine approximate values of the melting point for each at the core-mantle boundary and discuss solidification under these conditions.

2.4 Heat Conduction Equation

The fundamental physical fact from which the heat conduction equation is derived is that when there is a difference of temperature in a material body, heat will flow, and the rate of flow will be proportional to the difference in temperature. It is important to note that we are not considering heat transfer by radiation or heat transfer by flow of material of different temperatures —convection. The term *conduction* is used to refer to the transfer of heat under the above conditions. We may write this physical fact in mathematical terms by

$$w = -k \frac{\partial T}{\partial n} \tag{2.47}$$

where n is the coordinate measured in the direction of heat flow; k is the constant of proportionality; and w is the heat flow per unit time per unit cross section of area normal to the coordinate n. The quantity w is sometimes referred to as the *heat flux*. The constant k is referred to as the *thermal conductivity* of the material.

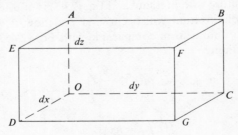

Fig. 2.1

Let us next consider the continuity of heat flow into and out of a small volume, for convenience taken as a small rectangular parallelopiped as shown in Fig. 2.1. Our derivation, which will equate the net heat flow into the volume to the increase in heat per unit time of the volume, follows the same reasoning as that used for the derivation of the equation of continuity for the mass flow into and the mass increase of a small volume. Our resultant equation, that is, *the heat conduction equation*, may be thought of then as simply the equation of continuity referred to heat flow. The heat flow per unit time through the face $OABC$, normal to the X coordinate, will be, from Eq. (2.47) and Fig. 2.1,

$$w_x dy\, dz = -k \frac{\partial T}{\partial x} dy\, dz$$

and that out through the face $DEFG$ will be

$$\left[w_x + \frac{\partial w}{\partial x} dx\right] dy\, dz = -\left[k \frac{\partial T}{\partial x} + \frac{\partial}{\partial x}\left(k \frac{\partial T}{\partial x}\right) dx\right] dy\, dz$$

The net heat flow into the volume from heat flow normal to the X coordinate will then be simply the difference of these two expressions, or

$$-\frac{\partial w_x}{\partial x} dx\, dy\, dz = \frac{\partial}{\partial x}\left(k \frac{\partial T}{\partial x}\right) dx\, dy\, dz \qquad (2.48)$$

Similarly, the net heat flow into the volume from heat flow normal to the Y and Z coordinates will be

$$-\frac{\partial w_y}{\partial y} dx\, dy\, dz = \frac{\partial}{\partial y}\left(k \frac{\partial T}{\partial y}\right) dx\, dy\, dz \qquad (2.49)$$

and

$$-\frac{\partial w_z}{\partial z} dx\, dy\, dz = \frac{\partial}{\partial z}\left(k \frac{\partial T}{\partial z}\right) dx\, dy\, dz \qquad (2.50)$$

From the definition of specific heat in Section 2.2, we may write for the increase of heat per unit time of our volume in terms of the increase in temperature per unit time as

$$c\rho\, dx\, dy\, dz\, \frac{\partial T}{\partial t} \qquad (2.51)$$

where ρ is the density and c the specific heat, making no distinction here, as is usually the case for solid materials, between the specific heat at constant volume and that at constant pressure. Equating the sum of Eqs. (2.48), (2.49), and (2.50) to Eq. (2.51), we obtain

$$-\nabla \cdot \mathbf{w} = c\rho \frac{\partial T}{\partial t} \qquad (2.52)$$

74 ¶ Thermodynamics and Hydrodynamics

If k is a constant with respect to the coordinates, this reduces to

$$k\nabla^2 T = c\rho \frac{\partial T}{\partial t} \tag{2.53}$$

or

$$\alpha \nabla^2 T = \frac{\partial T}{\partial t} \tag{2.54}$$

where

$$\alpha = \frac{k}{c\rho} \tag{2.55}$$

The constant α is referred to as the *thermal diffusivity*. The similarity of form of the heat conduction equation (2.54) with Laplace's equation of Section 1.7 and the wave equation of Section 1.10 should be noted. Some of the methods of solution used there will carry over to solutions of the heat conduction equation.

2.5 Periodic Flow of Heat

One of the simplest applications of the heat conduction equation is in the study of the periodic flow of heat from the diurnal or annual temperature cycle. We shall assume a daily or annual temperature cycle of the form

$$T = T_0 \sin \omega t \tag{2.56}$$

where T is now the temperature variation from the mean daily or annual temperature, T_0 half the total temperature range, and ω the circular frequency associated with a period of one day or one year.

We look for a solution of the one-dimensional form of Eq. (2.54), or

$$\alpha \frac{\partial^2 T}{\partial z^2} = \frac{\partial T}{\partial t} \tag{2.57}$$

where z is the depth, satisfying the initial condition Eq. (2.56). Let us apply the method of separation of variables and look for general solutions of the form

$$T = Z(z)\Theta(t) \tag{2.58}$$

Substituting Eq. (2.58) into Eq. (2.57), we obtain

$$\alpha Z''\Theta - Z\Theta' = 0$$

or

$$\frac{Z''}{Z} - \frac{\Theta'}{\alpha \Theta} = 0 \tag{2.59}$$

As before, if Eq. (2.59) is to have any solution at all, each of the terms on the left-hand side must separately be equal to a constant. We may then write

$$\frac{Z''}{Z} = -\gamma^2 \qquad \frac{\Theta'}{\alpha\Theta} = -\gamma^2 \qquad (2.60)$$

where the constant parameter γ^2 may have any value real, imaginary, or complex. Solutions of these two ordinary differential equations are simply

$$Z = a_1 e^{i\gamma z} + a_2 e^{-i\gamma z} \qquad (2.61)$$

and

$$\Theta = a_3 e^{-\alpha\gamma^2 t} \qquad (2.62)$$

To obtain a specific solution that will satisfy our initial condition, Eq. (2.56), we choose a value of γ such that

$$\gamma^2 = \frac{i\omega}{\alpha} \qquad (2.63)$$

or

$$\gamma = \sqrt{\frac{\omega}{2\alpha}}(1+i) \qquad (2.64)$$

remembering that

$$\sqrt{i} = e^{i(\pi/4)} = \frac{1+i}{\sqrt{2}}$$

Substituting Eqs. (2.64) and (2.63) into Eqs. (2.61) and (2.62), we obtain

$$Z = a_1 e^{(i-1)z\sqrt{\omega/2\alpha}} + a_2 e^{(-i+1)z\sqrt{\omega/2\alpha}} \qquad (2.65)$$

and

$$\Theta = a_3 e^{-i\omega t} \qquad (2.66)$$

In order to have a function decreasing toward infinity in the positive z direction, we take only the first term of Eq. (2.65), and to satisfy the initial condition, Eq. (2.56), we take the imaginary part of the product of Eqs. (2.65) and (2.66) and the coefficient $a_1 a_3$ to be T_0, from which we have our final solution

$$T = T_0 \, e^{-z\sqrt{\omega/2\alpha}} \sin\left(\omega t - z\sqrt{\frac{\omega}{2\alpha}}\right) \qquad (2.67)$$

We see that the temperature variation is represented by a wave motion whose amplitude decreases rapidly as a function of depth. We see, as we might expect, that the amplitude decreases more rapidly as a function of depth for short-period fluctuations, such as the diurnal temperature variation, than for long-period fluctuations, such as the annual temperature variation. From the discussion of Section 1.10, the velocity of propagation of the wave motion is seen to be $\sqrt{2\alpha\omega}$ so that the time lag for a particular phase that occurs at the surface will be at a depth z_1

$$t_1 = \frac{z_1}{\sqrt{2\alpha\omega}} \qquad (2.68)$$

The net heat flow across any surface at depth will be represented by the integral of Eq. (2.47) over a cycle. From Eq. (2.67) we see that the derivative dT/dz will only contain sine and cosine terms in the form of Eq. (2.67), and its integral over a cycle will be zero. As we might have expected, there is no net heat flow into the earth from a periodic temperature variation at the surface.

Problem 2.5(a) Consider a diurnal temperature variation of maximum and minimum values of 16°C and −4°C and a soil of diffusivity, $\alpha = 0.0049$ CGS. Find the temperature variation at depths of 30 cm and 1 m. What will the time lag be at a depth of 30 cm.

Ans. 1.4°C, 0.004°C, 9.7 hr

Problem 2.5(b) Consider an annual temperature variation of from 22°C to −8°C. Find the temperature variation at depths of 1 m and 10 m. How deep will a freezing temperature penetrate? At what depth will the seasons be reversed?

Ans. 19°C, 0.33°C, 170 cm, 7 m

Problem 2.5(c) Extend the solution [Eq. (2.67)] to an initial condition represented by a Fourier series. Discuss the characteristics of the harmonic terms in the solution.

2.6 Heat Flow

We shall be interested here in more general solutions of the heat conduction equation, but shall restrict ourselves to one-dimensional considerations. In one-dimensional form, the heat conduction equation reduces to Eq. (2.57), for which we may have solutions of the form of Eqs. (2.61) and (2.62), which we shall take for convenience in the form

$$\begin{aligned} T &= B e^{-\alpha \gamma^2 t} \cos(\gamma z) \\ T &= C e^{-\alpha \gamma^2 t} \sin(\gamma z) \end{aligned} \qquad (2.69)$$

These particular solutions may be generalized to an integral

$$T = \int_0^\infty [B \cos(\gamma z) + C \sin(\gamma z)] e^{-\alpha \gamma^2 t} \, d\gamma \qquad (2.70)$$

over the parameter γ.

Now if we should have an initial temperature distribution $f(z)$ at $t = 0$, we can evaluate the quantities B and C. At $t = 0$, Eq. (2.70) reduces to

$$T(z, 0) = f(z) = \int_0^\infty [B \cos(\gamma z) + C \sin(\gamma z)] \, d\gamma \qquad (2.71)$$

From the Fourier integral theorem [Eq. (1.161)], we shall then have that

$$B = \frac{1}{\pi} \int_{-\infty}^\infty f(\zeta) \cos(\gamma \zeta) \, d\zeta \qquad (2.72)$$

and

$$C = \frac{1}{\pi} \int_{-\infty}^\infty f(\zeta) \sin(\gamma \zeta) \, d\zeta \qquad (2.73)$$

Substituting Eqs. (2.72) and (2.73) back into Eq. (2.70), we obtain

$$T(z, t) = \frac{1}{\pi} \int_0^\infty e^{-\alpha \gamma^2 t} \, d\gamma \int_{-\infty}^\infty f(\zeta) \cos[\gamma(\zeta - z)] \, d\zeta \qquad (2.74)$$

We may evaluate one of these integrals, obtaining

$$T = \frac{1}{\pi} \int_{-\infty}^\infty f(\zeta) \, d\zeta \int_0^\infty e^{-\alpha \gamma^2 t} \cos[\gamma(\zeta - z)] \, d\gamma$$

$$= \frac{1}{\pi} \int_{-\infty}^\infty f(\zeta) \, d\zeta \left[\frac{\sqrt{\pi}}{2\sqrt{\alpha t}} e^{-(\zeta - z)^2/4\alpha t} \right]$$

$$= \frac{1}{2\sqrt{\pi \alpha t}} \int_{-\infty}^\infty f(\zeta) \, e^{-(\zeta - z)^2/4\alpha t} \, d\zeta$$

$$= \frac{1}{\sqrt{\pi}} \int_{-\infty}^\infty f\left(z + \frac{\beta}{\eta}\right) e^{-\beta^2} \, d\beta \qquad (2.75)$$

where, in the last equation, we have made the following substitution for the variable of integration

$$\beta = \frac{\zeta - z}{2\sqrt{\alpha t}} \qquad (2.76)$$

and have also substituted

$$\eta = \frac{1}{2\sqrt{\alpha t}} \qquad (2.77)$$

In many heat conduction problems, one is concerned with a semi-infinite solid, for which we shall have a given initial temperature distribution $f(z)$ in the positive z direction at $t = 0$ and also for which we shall have the surface kept at a constant temperature, taken for convenience to be zero, for all time. This second condition can simply be met by imposing a temperature distribution $-f(-z)$ for the negative z direction; under such a temperature distribution, there will be no heat flow across the plane $z = 0$, and its temperature will remain constant. From Eq. (2.75) we shall have

$$T = \frac{1}{2\sqrt{\pi\alpha t}} \left[\int_0^\infty f(\zeta) \, e^{-(\zeta-z)^2/4\alpha t} \, d\zeta + \int_{-\infty}^0 -f(-\zeta) \, e^{-(\zeta-z)^2/4\alpha t} \, d\zeta \right]$$

$$= \frac{1}{2\sqrt{\pi\alpha t}} \int_0^\infty f(\zeta) \, [e^{-(\zeta-z)^2/4\alpha t} - e^{-(\zeta+z)^2/4\alpha t}] \, d\zeta \qquad (2.78)$$

Making the change of variable to β defined, respectively, in this case for the first and second integrals,

$$\beta = \frac{\zeta - z}{2\sqrt{\alpha t}} \qquad \beta = \frac{\zeta + z}{2\sqrt{\alpha t}} \qquad (2.79)$$

we obtain the alternative form

$$T = \frac{1}{\sqrt{\pi}} \left[\int_{-z\eta}^\infty f\left(z + \frac{\beta}{\eta}\right) e^{-\beta^2} \, d\beta - \int_{z\eta}^\infty f\left(-z + \frac{\beta}{\eta}\right) e^{-\beta^2} \, d\beta \right] \qquad (2.80)$$

If the body were at a constant initial temperature, the solution [Eq. (2.80)] would reduce to

$$T = \frac{T_0}{\sqrt{\pi}} \left[\int_{-z\eta}^\infty e^{-\beta^2} \, d\beta - \int_{z\eta}^\infty e^{-\beta^2} \, d\beta \right]$$

$$= \frac{T_0}{\sqrt{\pi}} \int_{-z\eta}^{z\eta} e^{-\beta^2} \, d\beta$$

$$= \frac{2T_0}{\sqrt{\pi}} \int_0^{z\eta} e^{-\beta^2} \, d\beta$$

$$= T_0 Y(z\eta) \qquad (2.81)$$

where $Y(z\eta)$ is the *error function*. It is a conveniently tabulated function, and many heat conduction problems can be reduced to summations of error function values.

In other heat conduction problems, one may be interested in heat in a semi-infinite solid, for which there is a given temperature function $F(t)$ along the surface $z = 0$. Such a consideration arises in postglacial time calculations, where one can assume a given temperature distribution during the glacial period followed by a warmer temperature during the interglacial or postglacial times.

To arrive at this solution, we must first consider the effect of a heat source. Let us take an instantaneous source of heat of an amount Q per unit area acting at time $t = 0$ over a slab at $z = \lambda$ of thickness $\Delta\lambda$. Then from the definition of specific heat, the temperature of the slab will be raised by an amount $Q/c\rho\Delta\lambda$. In terms of the previous discussion, our initial condition $f(z)$ will then be

$$f(z) = \frac{Q}{c\rho\Delta\lambda} = \frac{S}{\Delta\lambda} \quad (\lambda < z < \lambda + \Delta\lambda) \tag{2.82}$$

and zero everywhere else. The quantity S defined by Eq. (2.82) is often referred to as the strength of the source. The integral [Eq. (2.75)] will reduce to

$$T = \frac{S}{2\Delta\lambda\sqrt{\pi\alpha t}} \int_\lambda^{\lambda+\Delta\lambda} e^{-(\zeta-z)^2/4\alpha t}\, d\zeta$$

$$= \frac{S}{2\sqrt{\pi\alpha t}} e^{-(\lambda-z)^2/4\alpha t} \tag{2.83}$$

in the limit as $\Delta\lambda \to 0$. Now if, instead, we have a constant source of strength S' per unit area per unit time located in the plane $z = \lambda$, which begins to liberate heat at $t = 0$, the total effect at some later time t will be the summation of each effect $S'd\tau$ that acted at a time $t-\tau$ previously, τ being the time variable with limits 0 and t. From Eq. (2.83) we shall then have

$$T = \frac{S'}{2\sqrt{\pi\alpha}} \int_0^t e^{-(\lambda-z)^2/4\alpha(t-\tau)} (t-\tau)^{-1/2}\, d\tau \tag{2.84}$$

For a source at the origin, this reduces to

$$T = \frac{S'}{2\sqrt{\pi\alpha}} \int_0^t e^{-z^2/4\alpha(t-\tau)} (t-\tau)^{-1/2}\, d\tau$$

$$= \frac{S'z}{2\alpha\sqrt{\pi}} \int_{z\eta}^\infty \frac{e^{-\beta^2}}{\beta^2}\, d\beta \tag{2.85}$$

where, in the last equation, we have substituted η and have made the following substitution for the variable of integration

$$\beta = \frac{z}{2\sqrt{\alpha(t-\tau)}} \tag{2.86}$$

To obtain a solution satisfying our desired initial condition $F(t)$, we must go through the subterfuge of considering a doublet source. If a source and sink, negative source, of equal strength S are made to approach each other while keeping constant the product of S and the distance $2b$ between them, the combination, in the limit, is referred to as a *doublet* of strength $S_d = 2bS$. From Eq. (2.83) we shall then have, for the combined effect of such a doublet source centered at the origin,

$$T = \frac{S}{2\sqrt{\pi\alpha t}}[e^{-(b-z)^2/4\alpha t} - e^{-(b+z)^2/4\alpha t}]$$

$$= \frac{S}{2\sqrt{\pi\alpha t}} e^{-(b^2+z^2)/4\alpha t}[e^{bz/2\alpha t} - e^{-bz/2\alpha t}]$$

$$= \frac{S_d}{4b\sqrt{\pi\alpha t}} e^{-(b^2+z^2)/4\alpha t}\left[\left(1 + \frac{bz}{2\alpha t} + \cdots\right) - \left(1 - \frac{bz}{2\alpha t} + \cdots\right)\right]$$

$$= \frac{S_d z}{4\sqrt{\pi\alpha^3 t^3}} e^{-z^2/4\alpha t} \tag{2.87}$$

in the limit as $b \to 0$. For a doublet source of strength $S'_d(t)$, we shall have from Eqs. (2.87) and (2.84)

$$T = \frac{z}{4\sqrt{\pi\alpha^3}} \int_0^t S'_d(\tau) e^{-z^2/4\alpha(t-\tau)} (t-\tau)^{-3/2} d\tau$$

$$= \frac{1}{\alpha\sqrt{\pi}} \int_{z\eta}^\infty S'_d\left(t - \frac{z^2}{4\alpha\beta^2}\right) e^{-\beta^2} d\beta \tag{2.88}$$

where we have made the same substitution for β as given in Eq. (2.86). Finally, if we now make the doublet source to be of strength $S'_d(t) = 2\alpha F(t)$, we obtain the desired result

$$T = \frac{2}{\sqrt{\pi}} \int_{z\eta}^\infty F\left(t - \frac{z^2}{4\alpha\beta^2}\right) e^{-\beta^2} d\beta \tag{2.89}$$

Expression (2.89) satisfies the differential equation (2.57), and at $z = 0$ reduces to

$$T(0, t) = \frac{2F(t)}{\sqrt{\pi}} \int_0^\infty e^{-\beta^2} d\beta = F(t)$$

our initial condition. From the method of derivation, it also satisfies the condition that $T = 0$ for $t = 0$.

If we were to have both an initial temperature distribution $f(z)$ at $t = 0$ and an initial temperature function $F(t)$ at $z = 0$ for a semi-infinite solid, the solution would simply be the sum of Eqs. (2.80) and (2.89). It should also be apparent from the derivations that the zero of the temperature scale is, like the zero of the Cartesian coordinate system, relative and not necessarily related to the zero of the centigrade or absolute temperature scales.

For heat flow in a problem with spherical symmetry, the Laplacian in spherical coordinates, Eq. (1.83), will reduce the heat conduction equation (2.54) to

$$\frac{\alpha}{r^2} \frac{\partial}{\partial r}\left(r^2 \frac{\partial T}{\partial r}\right) = \frac{\partial T}{\partial t}$$

$$\alpha \left(\frac{\partial^2 T}{\partial r^2} + \frac{2}{r} \frac{\partial T}{\partial r}\right) = \frac{\partial T}{\partial t}$$

$$\alpha \frac{\partial^2 (rT)}{\partial r^2} = \frac{\partial (rT)}{\partial t} \qquad (2.90)$$

similar to the reduction for Eq. (1.176). We are interested in a solution of Eq. (2.90), which will satisfy an initial condition $T = f(r)$ when $t = 0$. If we make the substitution

$$u = rT \qquad (2.91)$$

for the independent variable, the differential equation (2.90) becomes

$$\alpha \frac{\partial^2 u}{\partial r^2} = \frac{\partial u}{\partial t} \qquad (2.92)$$

and the initial condition becomes

$$u = rf(r) \qquad t = 0$$

plus
$$u = 0 \qquad r = 0 \qquad (2.93)$$

These conditions for u are the same as those for T in the derivation of Eq. (2.78), so that we may write down our solution immediately as

$$u = rT = \frac{1}{2\sqrt{\pi \alpha t}} \int_0^\infty \lambda f(\lambda) \left[e^{-(\lambda-r)^2/4\alpha t} - e^{-(\lambda+r)^2/4\alpha t}\right] d\lambda$$

or

$$T = \frac{1}{2r\sqrt{\pi \alpha t}} \int_0^\infty \lambda f(\lambda) \left[e^{-(\lambda-r)^2/4\alpha t} - e^{-(\lambda+r)^2/4\alpha t}\right] d\lambda \qquad (2.94)$$

Making the change of variable to β, defined, respectively, for the first and second integrals the same as Eqs. (2.79), we obtain the alternative form

$$T = \frac{1}{r\sqrt{\pi}} \left[\int_{-r\eta}^\infty \left(r + \frac{\beta}{\eta}\right) f\left(r + \frac{\beta}{\eta}\right) e^{-\beta^2} d\beta \right.$$
$$\left. - \int_{r\eta}^\infty \left(-r + \frac{\beta}{\eta}\right) f\left(-r + \frac{\beta}{\eta}\right) e^{-\beta^2} d\beta\right] \qquad (2.95)$$

If we had a spherical body of radius R at a constant initial temperature T_0, the solutions [Eqs. (2.94) and (2.95)] would reduce, respectively, to

$$T = \frac{T_0}{2r\sqrt{\pi \alpha t}} \int_0^R \lambda \left[e^{-(\lambda-r)^2/4\alpha t} - e^{-(\lambda+r)^2/4\alpha t}\right] d\lambda \qquad (2.96)$$

and

$$T = \frac{T_0}{r\sqrt{\pi}} \left[\int_{-r\eta}^{(R-r)\eta} \left(r + \frac{\beta}{\eta}\right) e^{-\beta^2} d\beta - \int_{r\eta}^{(R+r)\eta} \left(-r + \frac{\beta}{\eta}\right) e^{-\beta^2} d\beta \right] \quad (2.97)$$

Problem 2.6(a) Consider a layer 20 m thick of lava extruded at the earth's surface. Assuming the initial temperature of the lava to be 1000°C, that of the underlying rock to be zero, and that of the surface of the lava to remain at zero, calculate the temperature distribution as a function of depth after 1 and 9 yr. Use a value of $\alpha = 0.012$ CGS for both the hot lava and the underlying rock.

Problem 2.6(b) Consider an immense intrusion of roughly spherical size with a radius of 1000 m. Assuming the initial temperature of the material to be 1000°C and that of the surrounding rock to be zero, calculate the temperature distribution as a function of radial distance measured from the center of the intrusion after 100 and 10,000 yr. Use a value of $\alpha = 0.012$ CGS for both the intrusion and surrounding rock.

2.7 Internal Heat of the Earth

When we come to study the internal heat of the earth, we find unfortunately that there is only one directly measurable quantity that we have available. This is the outward heat flow through the crust, often referred to as the *geothermal flux*. Determinations of the geothermal flux require measurements of the vertical temperature gradients in the crust and of the thermal conductivities of the earth materials in which the gradient is measured. The geothermal flux is then calculated simply by Eq. (2.47). Proper consideration, of course, must be taken to correct for, or to measure at sufficiently great depths to eliminate, the effects of diurnal and annual periodic heat flow, to compensate for the thermal effects of the introduction of the heat probe measuring device into the earth materials, and to consider and correct, if necessary, for possible heat flow effects from nearby internal heat sources or from glacial and postglacial surface temperature differences.

In the derivations to be given in this section, we shall anticipate one of the results, namely, that the cooling of the earth by conduction has become considerable only down to depths of the order of 300 km, a distance small compared with the radius of the earth. We may then carry through our derivations with respect to the Cartesian coordinate z, rather than the spherical distance coordinate r. For the first derivation, let us make the gross assumption that when the earth initially reached a solid state, it was at a constant initial temperature T_0. The solution will then be that given by Eq. (2.81). Differentiating this for the temperature gradient, we obtain

$$\frac{\partial T}{\partial z} = \frac{2T_0}{\sqrt{\pi}} \frac{\partial}{\partial z}\left[\int_0^{z\eta} e^{-\beta^2}\,d\beta\right]$$

$$= \frac{2T_0}{\sqrt{\pi}} e^{-z^2\eta^2}\,\eta$$

$$= \frac{T_0}{\sqrt{\pi\alpha t}} e^{-z^2/4\alpha t} \tag{2.98}$$

For the temperature gradient at the earth's surface, we then have

$$\left(\frac{\partial T}{\partial z}\right)_0 = \frac{T_0}{\sqrt{\pi\alpha t}} \tag{2.99}$$

or

$$t = \frac{T_0^2}{\pi\alpha(\partial T/\partial z)_0^2} \tag{2.100}$$

From Eq. (2.100) we may obtain an age of the earth, using the observed value of the surface temperature gradient and under the assumptions of the derivation.

Let us next make the assumption that instead of a constant initial temperature, the temperature increases linearly with depth. This might be considered to be a first approximation to the melting point curve, Eq. (2.46), for the initial solidified earth. Our initial condition will then be

$$T(z, 0) = f(z) = mz + T_s \tag{2.101}$$

where T_s is the initial surface temperature. The solution satisfying the constant term initial condition T_s of Eq. (2.101) is given by Eq. (2.81). The solution satisfying the gradient term initial condition is simply $T = mz$, which is, in itself, a solution of the heat conduction equation; it satisfies the gradient portion of the initial condition Eq. (2.101) and satisfies the further subsidiary initial condition that $T = 0$ for $z = 0$. Our solution then is

$$T = ms + T_s Y(z\eta) \tag{2.102}$$

We could also have obtained Eq. (2.102) by substituting Eq. (2.101) directly into Eq. (2.80). The temperature gradient will then be

$$\frac{\partial T}{\partial z} = m + \frac{T_s}{\sqrt{\pi\alpha t}} e^{-z^2/4\alpha t} \tag{2.103}$$

For the temperature gradient at the earth's surface, we then have

$$\left(\frac{\partial T}{\partial z}\right)_0 = m + \frac{T_s}{\sqrt{\pi\alpha t}} \tag{2.104}$$

or

$$t = \frac{T_0{}^2}{\pi\alpha[(\partial T/\partial z)_0 - m]^2} \tag{2.105}$$

For the final derivation, let us consider the effect of internal heat sources within the earth. It is known from observations of surficial rocks that the content of the long-term radioactively decaying isotopes of uranium, thorium, and potassium is such that if these same rocks extended to any modest depths in the earth's crust with the same percentages of the radioactive materials, an appreciable fraction of the observed geothermal flux would be attributable to the heat generated by the radioactively disintegrating materials. We shall assume here that the radioactive heat generation may be approximated by an exponentially decreasing function from the surface down as a function of depth

$$\psi(z, t) = B e^{-bz} \tag{2.106}$$

where B is the quantity of heat generated per unit volume per second at the earth's surface. There is nothing particularly attractive in itself in taking an exponential function; it is simply a mathematical convenience. It is known, however, from the observed values of B that such materials need only extend down to a modest depth, 20 to 40 km, to account for all the measured geothermal flux. The exponential function does, at least, indicate such a decrease in radioactive heat generation as a function of depth. The total amount of heat generated per unit time, and necessarily for steady state in a closed system, escaping through the earth's surface per unit time will be

$$w_r = \int_0^\infty B e^{-bz}\, dz = \frac{B}{b} \tag{2.107}$$

If w_s is the total geothermal flux, we may write for the fraction n, due to radioactive heat generation,

$$n = \frac{w_r}{w_s} = \frac{B}{bw_s} \tag{2.108}$$

It is necessary for us to specify any two of the three quantities n, B, b, that is, fraction of geothermal flux attributable to radioactive heating, radioactive heat generated by surficial rocks, and effective depth to which radioactive materials extend, to be able to make any calculation.

Referring back to Section 2.4 for the derivation of the heat conduction equation, we see that if our small volume included heat generation ψ per unit volume per unit time, Eq. (2.53) would be altered to read

$$k\nabla^2 T + \psi = c\rho \frac{\partial T}{\partial t} \tag{2.109}$$

or, in one-dimensional form,

$$\alpha \frac{\partial^2 T}{\partial z^2} + \frac{\psi}{c\rho} = \frac{\partial T}{\partial t} \tag{2.110}$$

Substituting Eq. (2.106) into Eq. (2.110), we obtain

$$\frac{\partial T}{\partial t} = \alpha \frac{\partial^2 T}{\partial z^2} + Ce^{-bz} \tag{2.111}$$

where we have made the further substitution

$$C = \frac{B}{c\rho} \tag{2.112}$$

It will be convenient for us to reduce Eq. (2.111) down to the form of our original heat conduction equation. We may do this by making the following substitution for the independent variable T,

$$u = T + \frac{C}{b^2 \alpha} e^{-bz} \tag{2.113}$$

obtaining

$$\frac{\partial u}{\partial t} = \alpha \frac{\partial^2 u}{\partial z^2} \tag{2.114}$$

Taking our previous initial condition Eq. (2.101) and that $T = 0$ for $z = 0$, these will become in terms of u from Eq. (2.113)

$$u = mz + T_s + \frac{C}{b^2 \alpha} e^{-bz} \qquad t = 0 \tag{2.115}$$

and

$$u = \frac{C}{b^2 \alpha} \qquad z = 0 \tag{2.116}$$

Since our problem would be further simplified if the second condition for the dependent variable were to be zero at $z = 0$, we make the further substitution

$$v = u - \frac{C}{b^2 \alpha} \tag{2.117}$$

giving the same differential equation (2.114) in terms of v and the two conditions

$$v = f(z) = mz + T_s - \frac{C}{b^2 \alpha} + \frac{C}{b^2 \alpha} e^{-bz} \qquad t = 0 \tag{2.118}$$

and

$$v = 0 \qquad z = 0 \tag{2.119}$$

Our solution is then that given by Eq. (2.80). Substituting Eq. (2.118) into Eq. (2.80), we obtain, after some reduction,

$$v = mz + \left(T_s - \frac{C}{b^2\alpha}\right)\Upsilon(z\eta)$$

$$+ \frac{C}{b^2\alpha\sqrt{\pi}}\left[e^{(b^2/4\eta^2)-bz}\int_{(b/2\eta)-z\eta}^{\infty} e^{-\gamma^2}d\gamma - e^{(b^2/4\eta^2)+bz}\int_{(b/2\eta)+z\eta}^{\infty} e^{-\gamma^2}d\gamma\right] \quad (2.120)$$

or for T from Eqs. (2.113) and (2.117)

$$T = mz + T_s\Upsilon(z\eta) + \frac{C}{b^2\alpha}\left\{1 - \Upsilon(z\eta) - e^{-bz}\right.$$

$$\left. + \tfrac{1}{2}e^{(b^2/4\eta^2)-bz}\left[1 - \Upsilon\left(\frac{b}{2\eta} - z\eta\right)\right] - \tfrac{1}{2}e^{(b^2/4\eta^2)+bz}\left[1 + \Upsilon\left(\frac{b}{2\eta} + z\eta\right)\right]\right\} \quad (2.121)$$

For the temperature gradient at the earth's surface, we obtain, after some reduction,

$$\left(\frac{\partial T}{\partial z}\right)_0 = m + \frac{2T_s\eta}{\sqrt{\pi}} + \frac{C}{b\alpha}\left\{1 - e^{b^2/4\eta^2}\left[1 - \Upsilon\left(\frac{b}{2\eta}\right)\right]\right\} \quad (2.122)$$

When $b/2\eta$ is large, which is the case for the earth, this reduces to

$$\left(\frac{\partial T}{\partial z}\right)_0 = m + \frac{T_s}{\sqrt{\pi\alpha t}} + \frac{C}{b\alpha}\left(1 - \frac{1}{b\sqrt{\pi\alpha t}}\right) \quad (2.123)$$

or

$$t = \frac{\left(T_s - \dfrac{C}{b^2\alpha}\right)^2}{\pi\alpha[(\partial T/\partial z)_0 - m - C/b\alpha]^2} \quad (2.124)$$

For a more thorough examination of the internal temperature structure of the earth and of the earth's thermal history, it is necessary to reexamine some of the assumptions used in the above derivations. In particular, we have assumed a constant value for the thermal conductivity k and for the thermal diffusivity α; this is certainly not correct for the earth as a whole. Of more importance, we have assumed that heat flow has been entirely by conduction; this is probably not correct. Generally, the core is assumed to be liquid, and its thermal state is assumed to be adiabatic. Further, if there are any convection cells, or convective overturn, in the mantle on any geologic time scale, the heat transfer by such convection will be substantially larger than the corresponding heat transfer by conduction over the same time period for such zones. Still further, we have neglected heat transfer by radiation. We have, from the results of Problem 2.1(a), that the energy density of radiation is proportional to T^4, and we may anticipate that the associated heat flow may become important at the elevated temperatures of the mantle.

In addition, we have made a rather simple assumption as to the distribution of radioactive heat sources. Measurements of heat flow through continental and oceanic areas indicate that the geothermal flux is comparable in the two areas. From observations of surficial rocks, it is generally assumed that there is a greater quantity of radioactive heat source materials in the continental crustal rocks than in the oceanic crustal rocks. The geothermal flux measurements would then indicate that there might have been a differentiation of the upper mantle beneath the continental and oceanic crusts to produce a balancing lower quantity of radioactive heat source material in the continental upper mantle than in the oceanic upper mantle.

Problem 2.7(a)   Calculate an age of the earth from Eqs. (2.100), (2.105), and (2.124), assuming $T_0 = 4000°C$, $\alpha = 0.012$ CGS, $(\partial T/\partial z)_0 = 1°C$ for 40 m, $m = 5°C/km$, $w_s = 1.3 \times 10^{-6}$ CGS, $n = 0.25$, $1/b = 20$ km, and $k = 0.006$ CGS.

Ans. 216×10^6 yr, 337×10^6 yr, 634×10^6 yr

Problem 2.7(b)   Obtain Eq. (2.102) directly from Eq. (2.80).

Problem 2.7(c)   Calculate a temperature curve for the earth from Eq. (2.81) after cooling for (1) 50×10^6 yr and (2) 500×10^6 yr. Assume $\alpha = 0.012$ CGS. Discuss the significance of the results and their effect on the earth's thermal history.

Problem 2.7(d)   Carry through the reduction to Eq. (2.120).

Problem 2.7(e)   Carry through the reduction to Eqs. (2.122) and (2.123).

Problem 2.7(f)   Assume that the earth solidified from the core-mantle boundary outward and that the loss of internal heat since solidification has been by conduction. Taking various reasonable assumptions for the melting point curve variation in the earth, discuss the present thermal condition of the earth, zones where slow convection currents may exist, and the possible significance and effect of the seismic transition zone at 400 km.

References

Carslaw, H. S. 1921. *Mathematical Theory of the Conduction of Heat in Solids.* New York: Dover.

———, and J. C. Jaeger. 1947. *Conduction of Heat in Solids.* Oxford: Clarendon Press.

Garland, G. D. 1971. *Introduction to Geophysics.* Philadelphia: W. B. Saunders.
Ingersoll, L. R., O. J. Zobel, and A. C. Ingersoll. 1948. *Heat Conduction.* New York: McGraw-Hill.
Jeffreys, H. 1952. *The Earth.* Cambridge: Cambridge University Press.
Lee, W. H. K., ed. 1965. "Terrestrial heat flow." Washington, D.C.: American Geophysical Union, Geophysical Monograph 8.
Page, L. 1935. *Introduction to Theoretical Physics.* New York: Van Nostrand.
Stacey, F. D. 1969. *Physics of the Earth.* New York: Wiley.

CHAPTER 3

HYDRODYNAMICS

3.1 Kinematic Preliminaries

In hydrodynamics, we are dealing with a continuum, that is, a fluid, in motion in which there is relative motion of one part with respect to the next. This is to be contrasted with the problems in dynamics such as the motion of a particle or a small, finite number particles under the action of a given force field and interaction forces among particles for which there are a discrete number of equations, the solutions of which define the motion of the particles; or such as the motion of a rigid body in which distances between fixed points in the body remain constant; or such as elastic wave propagation for which the motions are small and are about an equilibrium position. We must develop a new formulation to describe hydrodynamic motion before we can develop the equations of motion and their solutions.

We may alternatively seek a solution that gives a description of the velocity of the fluid, its pressure, and its density at all points of space occupied by the fluid at all instants of time, or seek a solution that gives a description of the motion of each particle. The equations obtained by the two methods are referred to, respectively, as being in *Eulerian* or *Lagrangian* form. We shall generally seek a solution in terms of the first method; however, it will be necessary for the derivation of the equations of motions, that is, the application of Newton's second law for the forces acting on and the acceleration of a given particle, to use the concepts of the second method. For the solutions, our attention will be fixed on a particular point in the observer's inertial system, and the velocity of fluid, its pressure, and its density at this point, and all other points, will be determined as functions of time.

For the first five sections of this chapter, we shall be concerned with a perfect fluid. By a *perfect fluid* we mean one that cannot support a tangential stress. In the last section, we shall consider tangential stresses with the introduction of the concept of viscosity.

Let $\mathbf{v}$ denote the linear velocity of the fluid at a point $P(x, y, z)$ at the time t, and v_x, v_y, v_z the components of $\mathbf{v}$ along the three coordinate directions. To calculate the rate at which any function $F(x, y, z, t)$ varies for a moving particle of the fluid, we note that at the time $t+\delta t$, the particle that was originally at $P(x, y, z)$ will now be at $Q(x+\delta x, y+\delta y, z+\delta z)$, where Q is given in terms of the components of the linear velocity as $Q(x+v_x\delta t, y+v_y\delta t, z+v_z\delta t)$. Then the value of F at Q will be

$$F(x+v_x\delta t, y+v_y\delta t, z+v_z\delta t, t+\delta t)$$
$$= F + \frac{\partial F}{\partial x}v_x\delta t + \frac{\partial F}{\partial y}v_y\delta t + \frac{\partial F}{\partial z}v_z\delta t + \frac{\partial F}{\partial t}\delta t \quad (3.1)$$

If we introduce the symbol D/Dt to denote a differentiation following the motion of the fluid, the new value of F may also be expressed, by definition, as $F+(DF/Dt)\delta t$ so that from Eq. (3.1), we obtain

$$\frac{DF}{Dt} = \frac{\partial F}{\partial t} + v_x\frac{\partial F}{\partial x} + v_y\frac{\partial F}{\partial y} + v_z\frac{\partial F}{\partial z} \quad (3.2)$$

or

$$\frac{DF}{Dt} = \frac{\partial F}{\partial t} + \mathbf{v}\cdot\nabla F \quad (3.3)$$

For the acceleration of a particle of the fluid, we shall then have, substituting $\mathbf{v}$ for F,

$$\frac{D\mathbf{v}}{Dt} = \frac{\partial \mathbf{v}}{\partial t} + \mathbf{v}\cdot\nabla\mathbf{v} \quad (3.4)$$

Let us next consider the rotation of neighboring fluid particles about $P(x, y, z)$ at the instant t. Since the medium is not rigid, different neighboring particles may have different angular velocities about P, and we cannot avail ourselves of the relation stated in Section 1.1 between the angular velocity $\boldsymbol{\omega}$ and the curl of the linear velocity $\mathbf{v}$ for a rigid body, Eq. (1.32). Nevertheless, we can show that $\nabla\times\mathbf{v}$ at P represents both in magnitude and direction twice the *average* angular velocity about P of the fluid particles. Consider a small sphere with P as center as shown in Fig. 3.1. Then for an annular ring of radius p about the Z axis at a distance dz from P, we shall have, applying Stokes' theorem, Eq. (1.41), to the area bounded by the ring,

$$\int_\sigma \nabla\times\mathbf{v}\cdot d\boldsymbol{\sigma} = \oint \mathbf{v}\cdot d\boldsymbol{\lambda}$$

$$\pi p^2 \left[|\nabla\times\mathbf{v}|_z + \left|\frac{\partial}{\partial z}(\nabla\times\mathbf{v})\right|_z dz \right] = 2\pi p \bar{v}_\lambda$$

$$|\nabla\times\mathbf{v}|_z + \left|\frac{\partial}{\partial z}(\nabla\times\mathbf{v})\right|_z dz = 2\bar{\omega}_z$$

where $|\nabla\times\mathbf{v}|_z$ is the z component of $\nabla\times\mathbf{v}$, $\bar{v}_\lambda$ the average value of the tangential component of $\mathbf{v}$ around the circumference of the annular ring, and $\bar{\omega}_z$ the corresponding average angular velocity, given by $\bar{\omega}_z = \bar{v}_\lambda/p$. If we

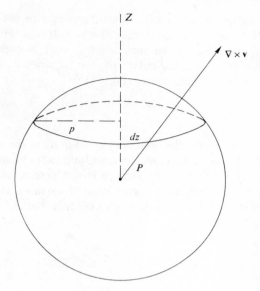

Fig. 3.1

average this over two annular rings equidistant from P on opposite sides, the term in dz will necessarily disappear. Hence, as the volume of the sphere can be divided into such pairs of annular rings, we get

$$|\nabla \times \mathbf{v}|_z = 2\bar{\omega}_z$$

for the entire sphere. Since similar expressions hold for the other components, the mean angular velocity of the particles contained in a small sphere with P as center will be

$$\bar{\boldsymbol{\omega}} = \tfrac{1}{2}\nabla \times \mathbf{v} \qquad (3.5)$$

If the mean angular velocity is everywhere zero, $\mathbf{v}$ by definition from Section 1.1. is an irrotational vector. In this case, the motion is said to be *irrotational*. Now it was stated in Section 1.1 that any irrotational vector may be expressed as the gradient of a scalar function of position in space. So if the motion is irrotational, we may write

$$\mathbf{v} = -\nabla \Phi \qquad (3.6)$$

The scalar function of position is known as the *velocity potential*, since the velocity is obtained from it by the same mathematical operation as that used to get the force from the potential energy in the case of a conservative dynamical system. For such a situation, we may seek a solution, generally more easily, in terms of the scalar function Φ rather than the vector function $\mathbf{v}$.

A surface over which Φ is a constant is known, similar to gravitational theory, as an *equipotential surface*. The *streamlines* in the fluid are curves having everywhere the direction of the velocity **v**. It was stated in Section 1.1 that $\nabla \Phi$ is perpendicular to the surface $\Phi(x, y, z) = $ constant. Therefore, the streamlines are perpendicular to the equipotential surfaces, the differential equation of these lines being

$$\frac{dx}{v_x} = \frac{dy}{v_y} = \frac{dz}{v_z} \tag{3.7}$$

Finally, let us consider the boundary conditions at the surface of a solid with which a fluid is in contact. As a perfect fluid cannot support a tangential stress, its velocity tangent to the surface is not determined by that of the solid. On the other hand, its velocity normal to the surface must be the same as that of the solid for it to maintain contact. Let AB in Fig. 3.2 repre-

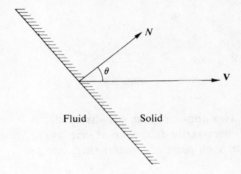

Fig. 3.2

sent the surface of a solid moving with velocity **V** in a direction making an angle θ with the normal N. Then if l, m, n are the direction cosines of the normal, the components of velocity of a fluid in contact with the solid must satisfy the boundary condition

$$lv_x + mv_y + nv_z = V \cos \theta \tag{3.8}$$

at the surface AB.

3.2 Conjugate Functions

If the potential is a function of x and y only, the problem under consideration reduces to one in two dimensions. In solving such a problem, the method of conjugate functions is often useful. Let z represent the complex quantity $x + iy$ so that $z = x + iy$, where $i = \sqrt{-1}$. Then any function $F(z)$ of z may be written

$$F(z) = F(x + iy) = \varphi(x, y) + i\psi(x, y) \tag{3.9}$$

where the real function $\varphi(x, y)$ and $\psi(x, y)$ are known as *conjugate functions*. Now we shall have

$$\frac{\partial F}{\partial x} = \frac{dF}{dz} \qquad \frac{\partial^2 F}{\partial x^2} = \frac{d^2 F}{dz^2}$$

and (3.10)

$$\frac{\partial F}{\partial y} = i\frac{dF}{dz} \qquad \frac{\partial^2 F}{\partial y^2} = -\frac{d^2 F}{dz^2}$$

so that

$$\frac{\partial^2 F}{\partial x^2} + \frac{\partial^2 F}{\partial y^2} = \frac{d^2 F}{dz^2} - \frac{d^2 F}{dz^2} = 0$$

showing that $F(z)$ satisfies the two-dimensional Laplace's equation. Consequently, the real part $\varphi(x, y)$ and the imaginary part $\psi(x, y)$ of *any* function $F(z)$ of the complex variable $x + iy$ are each solutions of Laplace's equation. Either of them represents the potential in a two-dimensional problem, for which the velocity potential is a solution of Laplace's equation, such as, for example, the conditions given by the case for Eq. (3.19).

From Eq. (3.10), if we replace F by $\varphi + i\psi$, we shall have

$$\frac{\partial \varphi}{\partial y} + i\frac{\partial \psi}{\partial y} = \frac{\partial F}{\partial y} = i\frac{\partial F}{\partial x} = i\left(\frac{\partial \varphi}{\partial x} + i\frac{\partial \psi}{\partial x}\right)$$

from which we get separating real and imaginary parts

$$\frac{\partial \varphi}{\partial y} = -\frac{\partial \psi}{\partial x} \qquad \frac{\partial \psi}{\partial y} = \frac{\partial \varphi}{\partial x} \qquad (3.11)$$

Consequently,

$$(\nabla \varphi) \cdot (\nabla \psi) = \frac{\partial \varphi}{\partial x}\frac{\partial \psi}{\partial x} + \frac{\partial \varphi}{\partial y}\frac{\partial \psi}{\partial y} = 0$$

We know that $\nabla \varphi$ is normal to the curve $\varphi =$ constant and that $\nabla \psi$ to the curve $\psi =$ constant. Hence, the two families of curves $\varphi(x, y) =$ constant and $\psi(x, y) =$ constant intersect orthogonally. If one family represents the traces on the plane of the equipotential surfaces in a hydrodynamic problem, the other represents the streamlines.

Problem 3.2(a)   Consider the function $F = Az^n$. Introducing polar coordinates, we shall have $\varphi = Ar^n \cos n\theta$, $\psi = Ar^n \sin n\theta$. Examine, discuss, and sketch the cases for $n = 1, 2,$ and -1.

Problem 3.2(b)   Examine and discuss the equipotential surfaces and streamlines represented by the function $F = -\mu \log z$.

Problem 3.2(c) Determine the potential and stream functions for $F = -z + i\beta e^{-ikz}$.

Ans. $\varphi = -x + \beta e^{ky} \sin(kx)$
$\psi = -y + \beta e^{ky} \cos(kx)$

3.3 Equation of Continuity

We have essentially derived the equation of continuity in Section 2.4. It is of interest to rederive it here using the concepts of Section 3.1. If Q be the volume of a small, moving element of the fluid, we have, on account of the constancy of mass, that

$$\frac{D}{Dt}(\rho Q) = 0$$

or

$$\frac{1}{Q}\frac{DQ}{Dt} + \frac{1}{\rho}\frac{D\rho}{Dt} = 0 \tag{3.12}$$

To calculate the value of the first term in Eq. (3.12), let the volume element in question at time t be a rectangular space, dx, dy, dz, having one corner at P and edges parallel to the three coordinate axes. Then at a later time $t + dt$, the same volume element will form an oblique parallelopiped. The side of original length dx will now be oriented, so its projections on the three coordinate axes will be

$$\left(1 + \frac{\partial v_x}{\partial x} dt\right) dx, \quad \frac{\partial v_y}{\partial x} dt dx, \quad \frac{\partial v_z}{\partial x} dt dx$$

A similar orientation will hold for the other two side, and we may show that the change in volume per unit volume will be

$$\frac{1}{Q}\frac{DQ}{Dt} = \frac{\partial v_x}{\partial x} + \frac{\partial v_y}{\partial y} + \frac{\partial v_z}{\partial z} \tag{3.13}$$

As we see, the expression (3.13) measures the rate of dilatation of the fluid and is sometimes referred to as the *expansion*. Substituting Eq. (3.13) into Eq. (3.12), we obtain

$$\frac{D\rho}{Dt} + \rho\left(\frac{\partial v_x}{\partial x} + \frac{\partial v_y}{\partial y} + \frac{\partial v_z}{\partial z}\right) = 0 \tag{3.14}$$

or

$$\frac{D\rho}{Dt} + \rho \nabla \cdot \mathbf{v} = 0 \tag{3.15}$$

which is the form of the equation of continuity in terms of the derivative (D/Dt) following the motion. Substituting Eq. (3.3) into Eq. (3.15), we obtain

$$\frac{\partial \rho}{\partial t} + \mathbf{v}\cdot\nabla\rho + \rho\nabla\cdot\mathbf{v} = 0$$

or

$$\frac{\partial \rho}{\partial t} + \nabla\cdot(\rho\mathbf{v}) = 0 \qquad (3.16)$$

which is the same as Eq. (1.28).

If the fluid is incompressible, which is often the assumption in hydrodynamics, ρ is a constant, and Eq. (3.16) becomes

$$\nabla\cdot\mathbf{v} = 0 \qquad (3.17)$$

or

$$\frac{\partial v_x}{\partial x} + \frac{\partial v_y}{\partial y} + \frac{\partial v_z}{\partial z} = 0 \qquad (3.18)$$

If, in addition, the motion is irrotational, we obtain, substituting Eq. (3.6) into Eq. (3.17),

$$\nabla\cdot\nabla\Phi = 0 \qquad (3.19)$$

The differential equation is Laplace's equation discussed in Section 1.7.

The equation of continuity is of fundamental importance in hydrodynamics. For a number of hydrodynamic problems, the method of attack consists simply in finding a solution of this differential equation that satisfies the bondary conditions.

We have examined here the application of the principle of continuity to the derivation of an equation of continuity for mass transport and, in Section 2.4, its application for heat flux to the derivation of the heat conduction equation. Let us now look at its application to other quantities occurring in hydrodynamics. We shall first consider the salinity of sea water. *Salinity s* is usually defined as a ratio expressing the number of grams of salt per kilogram of sea water. Returning to the derivation of Section 2.4, the flux of salt **S**, expressed in grams per kilogram, will then be

$$\mathbf{S} = \rho s \mathbf{v} \qquad (3.20)$$

Following through the same derivation as given there or simply using ρs for ρ in Section 1.1, we obtain

$$\frac{\partial}{\partial t}(\rho s) + \nabla\cdot(\rho s \mathbf{v}) = 0 \qquad (3.21)$$

which can be reduced, using Eq. (3.16), to

$$\rho\left[\frac{\partial s}{\partial t} + \mathbf{v}\cdot\nabla s\right] + s\left[\frac{\partial \rho}{\partial t} + \nabla\cdot(\rho\mathbf{v})\right] = 0$$

$$\frac{\partial s}{\partial t} + \mathbf{v}\cdot\nabla s = 0 \tag{3.22}$$

or from Eq. (3.3) in terms of D/Dt

$$\frac{Ds}{Dt} = 0 \tag{3.23}$$

We could have written Eq. (3.23) down directly. It is simply the direct statement that any particular quantity of water moves without change in total salt content. If the conditions are stationary, Eq. (3.22) reduces to

$$\mathbf{v}\cdot\nabla s = 0 \tag{3.24}$$

or

$$v_x \frac{\partial s}{\partial x} + v_y \frac{\partial s}{\partial y} + v_z \frac{\partial s}{\partial z} = 0 \tag{3.25}$$

We may extend this to any conservative quantity Q associated with fluid motion, that is, one that remains unchanged for any particular quantity of sea water, for which we shall have the same equation as Eq. (3.23)

$$\frac{DQ}{Dt} = 0 \tag{3.26}$$

or

$$\frac{\partial Q}{\partial t} + \mathbf{v}\cdot\nabla Q = 0 \tag{3.27}$$

Problem 3.3(a) Show that the change in the volume of the distorted parallelopiped is as given in Eq. (3.13).

3.4 Equation of Motion

Since we are dealing with a perfect fluid in which there are no tangential stresses, the stress relations reduce down simply to a hydrostatic pressure. Let us examine first the effect of such an internal hydrostatic pressure distribution on the equation of motion. Consider a rectangular parallelopiped as shown in Fig. 3.3 of volume $dxdydz$ fixed relative to the axes XYZ. If we denote by p the pressure at the center P of the parallelopiped, the force on the fluid inside the parallelopiped due to the hydrostatic pressure on the face

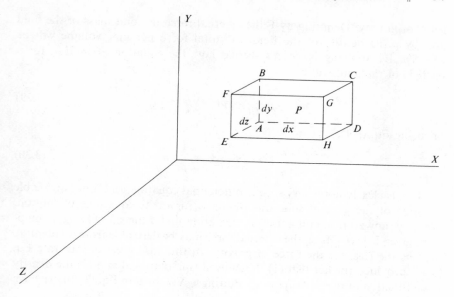

Fig. 3.3

$ABFE$ will be the pressure multiplied by the area of the face or

$$\left(p - \frac{\partial p}{\partial x}\frac{dx}{2}\right) dydz$$

acting in the positive X direction, and that on the face $DCGH$ will be

$$\left(p + \frac{\partial p}{\partial x}\frac{dx}{2}\right) dydz$$

acting in the negative X direction. The net force in the positive X direction due to the pressure on these two faces will then be simply

$$-\frac{\partial p}{\partial x} dxdydz$$

Similar expressions hold for the forces in the Y and Z directions for the faces perpendicular to these two directions. Dividing by the volume $dxdydz$ and then reducing the parallelopiped to zero, we obtain for the net force per unit volume at P

$$-\mathbf{i}\frac{\partial p}{\partial x} - \mathbf{j}\frac{\partial p}{\partial y} - \mathbf{k}\frac{\partial p}{\partial z} = -\nabla p \qquad (3.28)$$

In addition to the internal stress due to the pressure, there may be external, or body, forces acting on the elements of the fluid, such as the

98 ¶ Thermodynamics and Hydrodynamics

force of gravity. Denoting by **F** the external force per unit mass of the fluid and by ρ the density of the fluid, the total force per unit volume will be $\rho \mathbf{F} - \nabla p$. By applying Newton's second law, the equation of motion for a particle of the fluid will be

$$\rho \frac{D\mathbf{v}}{Dt} = \rho \mathbf{F} - \nabla p \tag{3.29}$$

or, using Eq. (3.4),

$$\frac{\partial \mathbf{v}}{\partial t} + \mathbf{v} \cdot \nabla \mathbf{v} = \mathbf{F} - \frac{1}{\rho} \nabla p \tag{3.30}$$

This is clearly not a very simple differential equation and is not amenable to some of the general solutions obtained for other equations of motion. We may, however, obtain a useful integration under the simplifying assumptions given below. Often, the external force may be derivable from a potential, such as the case for the force of gravity. In this case, we can integrate Eq. (3.30) provided further that (1) the motion is irrotational and (2) the density ρ is a function of the pressure only. Putting $-\nabla \Omega$ for **F** in Eq. (3.30), we have

$$\frac{\partial \mathbf{v}}{\partial t} + \mathbf{v} \cdot \nabla \mathbf{v} = -\nabla \Omega - \frac{1}{\rho} \nabla p \tag{3.31}$$

As the motion is irrotational, $\nabla \times \mathbf{v}$ is zero so that expanding by the triple vector product rule of Section 1.1,

$$(\nabla \times \mathbf{v}) \times \mathbf{v} = \mathbf{v} \cdot \nabla \mathbf{v} - (\nabla \mathbf{v}) \cdot \mathbf{v} = 0$$

or

$$\mathbf{v} \cdot \nabla \mathbf{v} = (\nabla \mathbf{v}) \cdot \mathbf{v} = \nabla(\tfrac{1}{2} \mathbf{v} \cdot \mathbf{v}) = \nabla(\tfrac{1}{2} v^2) \tag{3.32}$$

Substituting Eq. (3.32) into Eq. (3.31) and putting $-\nabla \Phi$ for **v**, we obtain

$$-\nabla \frac{\partial \Phi}{\partial t} + \nabla(\tfrac{1}{2} v^2) = -\nabla \Omega - \frac{1}{\rho} \nabla p \tag{3.33}$$

Taking the scalar product of Eq. (3.33) with an arbitrary position vector $d\boldsymbol{\lambda}$ and remembering the definition of the gradient, particularly Eq. (1.25), we obtain

$$-d\left(\frac{\partial \Phi}{\partial t}\right) + d(\tfrac{1}{2} v^2) = d\Omega - \frac{1}{\rho} dp \tag{3.34}$$

Integrating, we obtain

$$\int \frac{dp}{\rho} = \frac{\partial \Phi}{\partial t} - \Omega - \tfrac{1}{2} v^2 + G(t) \tag{3.35}$$

where the constant of integration may be a function of the time since our integration is with respect to the space coordinates only and where by hypothesis ρ is a function of p only. If the motion is steady, Φ is not a function of the time, and Eq. (3.35) reduces to

$$\int \frac{dp}{\rho} + \Omega + \tfrac{1}{2}v^2 = C \tag{3.36}$$

where C is a constant. Equation (3.36) is known as *Bernoulli's equation*. If, in addition, the fluid is incompressible, ρ is a constant, and Eq. (3.36) reduces to

$$\frac{p}{\rho} + \Omega + \tfrac{1}{2}v^2 = D \tag{3.37}$$

where D is a constant.

Problem 3.4(a)   Find the pressure at a distance r from the center of the earth due to gravitational attraction, assuming the earth to be an incompressible fluid of spherical form without rotation and of density 5.6 g/cm^3 throughout. What is the pressure at the center?

Ans. $p = \tfrac{1}{2}g\rho a[1-(r^2/a^2)]$, 13,000 tons/in.^{2}. Here ρ is the density and a the radius of the earth.

Problem 3.4(b)   Obtain the result of Eq. (3.32) by direct substitution of $\mathbf{v} = -\nabla \Phi$ into the left-hand member and reduction.

Problem 3.4(c)   Consider the flow of fluid of constant density in a smooth horizontal trough. Discuss the relation between pressure and velocity. Consider flow in the same trough in which there are variations in cross-sectional area normal to the flow. Discuss the relation between pressure and velocity.

Problem 3.4(d)   Consider a vertical tank of fluid with a small hole near the bottom. Determine the velocity v with which the fluid leaves the container.

Ans. $v^2 = 2gh$, where h is the height of the fluid surface above the hole. This result is known as *Torricelli's law*.

3.5 Kelvin's Circulation Theorem

We are interested in this section in showing that if the initial motion of the liquid is irrotational, it will remain irrotational forever provided that the external force is derivable from a potential. This result is known as *Kelvin's circulation theorem*.

Consider a closed curve λ lying in the fluid and moving along with the fluid motion, as shown in Fig. 3.4. The *circulation K* is defined as the line integral along this curve

$$K = \oint \mathbf{v} \cdot \delta \boldsymbol{\lambda} \tag{3.38}$$

where the differential symbol $\delta\boldsymbol{\lambda}$ is used to distinguish a differential around this loop, the loop itself moving with the fluid. Now it was stated in Section 1.1. that if the line integral of a vector vanishes about every closed path, the

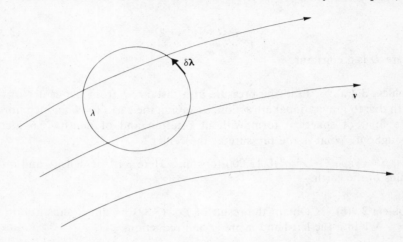

Fig. 3.4

vector is irrotational. Therefore, the condition for irrotational motion of a fluid is that the circulation shall vanish everywhere.

Let us calculate the time rate of change of the circulation, or

$$\frac{DK}{Dt} = \frac{D}{Dt}\oint \mathbf{v} \cdot \delta\boldsymbol{\lambda} = \oint \frac{D\mathbf{v}}{Dt} \cdot \delta\boldsymbol{\lambda} + \oint \mathbf{v} \cdot \frac{D(\delta\boldsymbol{\lambda})}{Dt} \tag{3.39}$$

For the second integral, we shall have, interchanging the order of differentiation,

$$\oint \mathbf{v} \cdot \delta\left(\frac{D\boldsymbol{\lambda}}{Dt}\right) = \oint \mathbf{v} \cdot \delta\mathbf{v} = \oint \delta(\tfrac{1}{2}\mathbf{v}\cdot\mathbf{v}) = \oint \delta(\tfrac{1}{2}v^2)$$

where the rate of change of $\boldsymbol{\lambda}$, moving along with the fluid, is simply the fluid velocity $\mathbf{v}$. The last expression is an integral of an exact differential of a single valued function about a closed loop, which is necessarily zero, since its value at the beginning and end points of the integration is the same. We then have for Eq. (3.39), substituting the equation of motion (3.29) into the first integral and using the assumed condition that the external force is derivable from a potential,

$$\frac{DK}{Dt} = -\oint \nabla\Omega \cdot \delta\lambda - \oint \frac{1}{\rho} \nabla p \cdot \delta\lambda$$

$$= -\oint \delta\Omega - \oint \frac{1}{\rho} \delta p \tag{3.40}$$

where we have again used the definition of the gradient, Eq. (1.25). The first integral of this expression will be zero for the same reason as stated above. If ρ is a constant or a singled valued function of the pressure, the integrand of the second integral will also be an exact differential and its integral around a closed loop zero. We then have

$$\frac{DK}{Dt} = 0 \tag{3.41}$$

This equation tells us that, as the closed curve around which we are integrating is carried through the fluid by the motion of the fluid, the line integral of the velocity around this curve remains unchanged in value. Thus, if the circulation is initially zero everywhere in the fluid, it will remain zero for all time. As zero circulation is equivalent to irrotational motion; irrotational motion once established, which was our original assumption, will persist indefinitely. These conclusions are subject to the condition that the external force is derivable from a potential and that the density is a constant or a function of the pressure only, and, of course, they are true only for a perfect fluid free from viscosity. If an external force not derivable from a potential acts on the fluid or if there are frictional forces between the layers of fluid and the walls of a containing vessel, a circulation may be set up in a fluid whose motion was initially irrotational.

3.6 Equation of Motion of a Viscous Fluid

The dynamics of a viscous fluid are very similar to those of an elastic solid, restated in Section 6.1, the main point of difference being that the shearing stress in a fluid is proportional to the time rate of change of strain instead of to the strain itself. In particular, the *rate of strain dyadic* $\nabla \mathbf{v}$ will be similar to the strain dyadic of Section 6.1 and may be considered as the sum of a pure strain dyadic Φ, which is symmetric, and a rotation dyadic Θ, which is skew-symmetric, and where both are given by

$$\Phi = \tfrac{1}{2}[\nabla \mathbf{v} + (\nabla \mathbf{v})_c] \tag{3.42}$$

and

$$\Theta = \tfrac{1}{2}[\nabla \mathbf{v} - (\nabla \mathbf{v})_c] \tag{3.43}$$

In particular, it is found by experiment, similar to Hooke's law, that the excess stress in the fluid over the hydrostatic stress is a homogeneous linear function of the rate of pure strain. We may then write down in principal axis form, referring again to Section. 6.1,

$$P+p = \lambda \nabla \cdot \mathbf{v} + 2\mu \frac{\partial v_x}{\partial x}$$

$$Q+p = \lambda \nabla \cdot \mathbf{v} + 2\mu \frac{\partial v_y}{\partial y}$$

$$R+p = \lambda \nabla \cdot \mathbf{v} + 2\mu \frac{\partial v_z}{\partial z}$$

where P, Q, R are the principal axis tensions and where the mean pressure p is given simply by

$$P+Q+R = -3p$$

Adding the three above equations, we then have

$$\lambda = -\tfrac{2}{3}\mu$$

so that the excess stress rate of pure strain relation in general form, similar to the stress-strain relation of Section 6.1, is simply

$$\Psi = -(p+\tfrac{2}{3}\mu \nabla \cdot \mathbf{v})I + 2\mu \Phi \tag{3.44}$$

where Ψ is the stress dyadic and I the idemfactor. In general, Eq. (3.44) is equivalent to nine scalar equations, of which three are redundant on account of the symmetry of the dyadics involved. The coefficient μ is known as the *coefficient of viscosity*.

To obtain the equation of motion for a viscous fluid, we have shown in Section 6.1 in the derivation of the equation of motion for an elastic solid that the force per unit volume due to the internal stresses is $\nabla \cdot \Psi$. From Eq. (3.44) we shall have

$$\nabla \cdot \Psi = -\nabla p - \tfrac{2}{3}\mu \nabla \nabla \cdot \mathbf{v} + 2\mu \nabla \cdot \Phi \tag{3.45}$$

since $\nabla \cdot I = \nabla$. To express $\nabla \cdot \Phi$ in terms of $\mathbf{v}$, we have from Eq. (3.42)

$$\nabla \cdot \Phi = \tfrac{1}{2}[\nabla \cdot \nabla \mathbf{v} + \nabla \cdot (\nabla \mathbf{v})_c]$$

$$= \tfrac{1}{2}[\nabla \cdot \nabla \mathbf{v} + \nabla \nabla \cdot \mathbf{v}] \tag{3.46}$$

Substituting Eq. (3.46) into Eq. (3.45), we have

$$\nabla \cdot \Psi = -\nabla p + \tfrac{1}{3}\mu \nabla \nabla \cdot \mathbf{v} + \mu \nabla \cdot \nabla \mathbf{v} \tag{3.47}$$

If in addition to the internal stresses, there is an external force $\mathbf{F}$ per unit mass, the equation of motion, similar to Eq. (3.29) for a perfect fluid, becomes

$$\rho \frac{D\mathbf{v}}{Dt} = \rho \mathbf{F} - \nabla p + \tfrac{1}{3}\mu \nabla \nabla \cdot \mathbf{v} + \mu \nabla \cdot \nabla \mathbf{v} \tag{3.48}$$

This equation obviously reduces to Eq. (3.29) when μ is put equal to zero. Using Eq. (3.4), we obtain the usual form

$$\rho \frac{\partial \mathbf{v}}{\partial t} + \rho \mathbf{v} \cdot \nabla \mathbf{v} = \rho \mathbf{F} - \nabla p + \tfrac{1}{3}\mu \nabla \nabla \cdot \mathbf{v} + \mu \nabla \cdot \nabla \mathbf{v} \tag{3.49}$$

Equation (3.49) is referred to as the *Navier-Stokes' equation*.

If the fluid is incompressible, the equation of continuity requires that $\nabla \cdot \mathbf{v}$ vanish so that Eq. (3.49) reduces to

$$\rho \frac{\partial \mathbf{v}}{\partial t} + \rho \mathbf{v} \cdot \nabla \mathbf{v} = \rho \mathbf{F} - \nabla p + \mu \nabla \cdot \nabla \mathbf{v} \tag{3.50}$$

If, in addition, the velocity of the fluid elements is small, we can neglect the second term on the left of Eq. (3.50), reducing it still further to

$$\rho \frac{\partial \mathbf{v}}{\partial t} = \rho \mathbf{F} - \nabla p + \mu \nabla \cdot \nabla \mathbf{v} \tag{3.51}$$

References

Lamb, H. 1932. *Hydrodynamics*. New York: Dover.
Page, L. 1935. *Introduction to Theoretical Physics*. New York: Van Nostrand.
Proudman, J. 1953. *Dynamical Oceanography*. New York: Wiley.

CHAPTER 4

PHYSICAL OCEANOGRAPHY—CIRCULATION

4.1 Thermodynamic Circulation

In this chapter, we shall be concerned with the application of some of the results from the previous two chapters on thermodynamics and hydrodynamics to a description of circulation in the ocean. We shall consider the general categories of thermodynamic circulation, mixing and diffusion, steady currents without friction, and driven currents with friction.

We shall consider thermodynamic circulation only in a rather general and qualitative manner to obtain general principles rather than specific results. Let us consider a vertical section in the ocean, for which we shall arbitrarily assume a circulation around the rectangle $ABCD$ as shown in Fig. 4.1. We shall assume that the circulation is stationary, that the driving

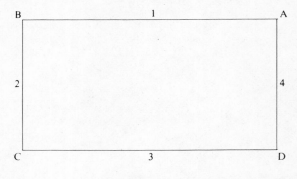

Fig. 4.1

energy for the circulation is thermodynamic in nature, that is, caused by heat sources and sinks along the circulation, and that the driving energy is balanced by frictional energy losses of the water circulation itself to create the stationary conditions.

Let us consider first a heat source along CD and a heat sink along AB. Let us suppose that the circulation is in the direction $ABCD$. The thermometric coefficient α_p of Eq. (2.21) is positive for sea water so that for an increase in temperature T, there will be a corresponding increase in specific volume v, or what is the same a decrease in its reciprocal, the density ρ. Along CD, T will increase and ρ decrease; along AB, T will decrease and ρ increase. Along the arms BC and DA, both T and ρ will be constant. Therefore, $\rho_{BC} > \rho_{DA}$, and the forces of gravity will maintain the circulation in the direction $ABCD$. If we had initially assumed a circulation in the direction $DCBA$, the same conclusion would be obtained with $\rho_{AD} > \rho_{BC}$ and the circulation maintained in the direction $DCBA$.

If we next consider a heat sink along CD and a heat source along AB, the opposite conclusion will be reached. The gravitational forces along the arms BC and DA will act in the same direction as the frictional forces to impede, rather than maintain, the circulation. Stationary circulation is not possible. Both of these conclusions should, of course, be considered within the limits of gravitational adibatic equilibrium as discussed in Section 2.3.

Let us next consider a heat source along DA and a heat sink along AB. Then T will be a maximum at A and constant along BC and CD. The average value of T along DA will be greater than that along BC so that $\rho_{BC} > \rho_{DA}$, and the circulation will be maintained by the gravity forces. If we had initially assumed a circulation $DCBA$, T would be a minimum at A so that $\rho_{AD} > \rho_{BC}$, and the circulation would again be maintained by the gravity forces in the direction $DCBA$.

Finally, let us consider a heat source along the first half of AB and a heat sink along the second half. T will be a maximum at the midpoint of AB and constant and of the same value along BC, CD, and DA. The forces of gravity would not then maintain the circulation. Under such conditions, we must then consider the transfer of heat by the slower process of heat conduction. Heat conduction will cause a rise in T in the arm DA and a lowering of T in the arm BC. Then, $\rho_{BC} > \rho_{DA}$ so that gravity can maintain a circulation in the direction $ABCD$.

For the stationary circulation of Fig. 4.1 where the fluid velocity is constant, the equations of motion (3.30) for each of the arms AB, BC, CD, and DA reduce to

$$0 = -\frac{1}{\rho}\frac{dp}{dx} - R$$

$$0 = g - \frac{1}{\rho}\frac{dp}{dx} - R \qquad (4.1)$$

$$0 = -\frac{1}{\rho}\frac{dp}{dx} - R$$

$$0 = -g - \frac{1}{\rho}\frac{dp}{dx} - R$$

where **F** is **g**, the force of gravity, and **R** is the friction per unit mass and where the coordinate direction x is taken in the direction of circulation $ABCD$. On multiplying Eqs. (4.1) by ρ and integrating for one complete turn around the circulation loop, we obtain

$$g \int_B^C \rho\, dx - g \int_D^A \rho\, dx = \oint R\rho\, dx \tag{4.2}$$

Since R is positive, we see that, as before, the density in the arm BC must be greater than that in the arm DA for the circulation to be maintained in the direction $ABCD$. If ρ is a known function of the circulation loop and R is a known function of fluid velocity, Eq. (4.2) can be used to obtain the stationary circulation velocity.

On integrating Eqs. (4.1), as they stand, for one complete turn around the circulation loop, we obtain

$$-\oint v\, dp = \oint R\, dx \tag{4.3}$$

where we have substituted for the density ρ its reciprocal the specific volume v. Now, since

$$\oint v\, dp + \oint p\, dv = \oint d(pv) = 0 \tag{4.4}$$

we can write Eq. (4.3) as

$$\oint p\, dv = \oint R\, dx \tag{4.5}$$

which is simply an expression for the work done around the loop. We could have obtained Eq. (4.5) directly from Section 3.5 if we had included a frictional resistance force and had not assumed ρ to be a function of p only. Since the integral on the right-hand-side of Eq. (4.5) must be positive, we see that for a circulation to exist, the expansions, which correspond to positive values of dv, take place, in general, at higher pressures than the contractions.

From the first of the implicit thermodynamic relations [Eq. (2.18)] with the use of the definitions [Eqs. (2.21) and (2.22)], we may write

$$dv = v\alpha_p dT - \frac{v}{k_T} dp \tag{4.6}$$

Integrating Eq. (4.6) around the circulation loop, using Eq. (4.4) and a similar relation for $d(Tv)$ and assuming the coefficients to be constant, we obtain

$$0 = \alpha_p \oint v\, dT - \frac{1}{k_T} \oint v\, dp$$

$$0 = \alpha_p \oint T\, dv - \frac{1}{k_T} \oint p\, dv$$

or

$$\oint p\, dv = k_T \alpha_p \oint T\, dv \tag{4.7}$$

Since the integral on the left-hand side is positive and the coefficients k_T and α_p are positive, we see that for a circulation to exist, the expansions take place, in general, at higher temperatures than the contractions. Combining this conclusion with that of the previous paragraph, it follows that for a circulation to exist, the greater values of temperature are, in general, associated with the greater values of pressure.

4.2 Continuity, Mixing, and Diffusion

As was mentioned in the previous chapter, a number of physical oceanographic problems can be solved simply by the application of considerations of continuity along with the associated boundary conditions. We shall start, in this section, with a few simple examples and then go in the latter part of this section and in the following two sections to somewhat more complex examples.

Consider a body of water, which has one major inflow section and one major outflow section, as shown in Fig. 4.2. Let A_1 and A_2 be the cross-

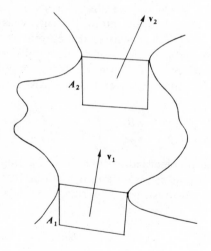

Fig. 4.2

sectional areas at the entrance and exit, and v_1 and v_2 the average water velocities normal to these two sections. Let r denote the mean rainfall plus fresh water river flow over evaporation per unit area between the two cross sections, and let B denote the surface area between the two sections. Then assuming the density ρ to be constant and assuming that there is complete mixing of the salt water entering A_1 and the fresh water represented by rB

and that the conditions are stationary, we shall have, for continuity of the water mass,

$$A_1 v_1 + rB = A_2 v_2 \tag{4.8}$$

and for the continuity of the salt mass

$$A_1 v_1 s_1 = A_2 v_2 s_2 \tag{4.9}$$

where s_1 and s_2 are the average salinities of the sea water at sections A_1 and A_2, respectively. Solving Eqs. (4.8) and (4.9) for $A_1 v_1$ and $A_2 v_2$, we obtain

$$A_1 v_1 = \frac{rB}{\dfrac{s_1}{s_2} - 1} \tag{4.10}$$

and

$$A_2 v_2 = \frac{rB}{1 - \dfrac{s_2}{s_1}} \tag{4.11}$$

From Eqs. (4.10) and (4.11) we see that when r is positive, that is, the influx of fresh water exceeds evaporation, the current is in the direction of decreasing salinity; when r is negative, the current is in the direction of increasing salinity, as expected. Now, if we denote by h the mean depth of water beneath the surface area B, we see from Eq. (4.11) that the time necessary to empty the volume between the two cross sections, or the *flushing time*, is given by

$$t = \frac{hB}{A_2 v_2} = \left(1 - \frac{s_2}{s_1}\right)\frac{h}{r} \tag{4.12}$$

If A, v, and s can be regarded as functions of a single space coordinate under the same conditions as above and if r can be regarded as a constant, for example, no influx of fresh water by river flow, Eqs. (4.8) and (4.9) can be generalized to

$$\frac{\partial}{\partial x}(Av) = br \tag{4.13}$$

and

$$\frac{\partial}{\partial x}(Avs) = 0 \tag{4.14}$$

where $b = b(x)$ is the width of the area B normal to x. Combining these two equations, we get

$$Av\frac{\partial s}{\partial x} + brs = 0 \tag{4.15}$$

Let us consider next the similar problem of a partially enclosed body with inflow to and outflow from the body through one principal opening, channel, or the like. Let us presume that there is a stratified flow of the incoming and outgoing currents as shown in Fig. 4.3. Then Eqs. (4.8) to (4.12) apply.

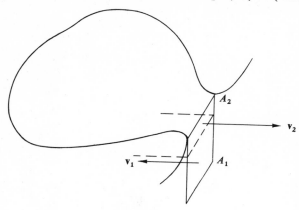

Fig. 4.3

So far we have assumed that there has been complete mixing. We shall now be interested in examining the mixing processes themselves. We shall first look at a phenomenon known as diffusion. For our purposes, *diffusion* is defined as the dispersion of a property of the fluid, such as salinity, without any net mass transfer of the fluid itself. In physical oceanography, we can recognize diffusive transfer due to molecular motion of the water particles, in which case it is referred to as *molecular diffusion*, and due to the turbulence of the water masses, in which case it is referred to as *eddy diffusion*. The corresponding dispersion of a property of the fluid due to the mass transfer of the fluid is known as *advection*.

Consider the diffusion across a unit cross section normal to the Z axis as shown in Fig. 4.4. Due to the turbulence or molecular motion, we shall have a mass of fluid $m_d(z)$ passing down through the cross section per unit time for fluid above the plane $z = 0$, and a corresponding mass of fluid $m_u(z)$ passing up through the cross section per unit time for fluid below the plane $z = 0$. Since there is no net transfer of fluid, we can write

$$\int_- m_u(z)\,dz = \int_+ m_d(z)\,dz \qquad (4.16)$$

where the first integral is taken for negative values of z and the second for positive values of z. The salt flux S or salt mass per unit area per unit time progressing in the positive z direction will then be

$$S = \int_- m_u(z)s(z)\,dz - \int_+ m_d(z)s(z)\,dz \qquad (4.17)$$

110 ¶ *Thermodynamics and Hydrodynamics*

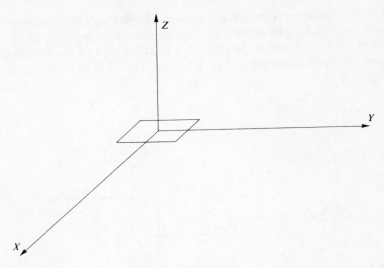

Fig. 4.4

Since these integrals will have appreciable values only in the neighborhood of the plane $z = 0$, we can write for $s(z)$

$$s(z) = s_0 + \frac{\partial s}{\partial z} z \tag{4.18}$$

Substituting Eq. (4.18) into Eq. (4.17) with the use of Eq. (4.16) and assuming $\partial s/\partial z$ to be constant over the range of integration, we obtain

$$S = -\left[\int_{-} z m_u(z)\, dz + \int_{+} z m_d(z)\, dz\right] \frac{\partial s}{\partial z}$$

$$S = -\eta \frac{\partial s}{\partial z} \tag{4.19}$$

where we have made the substitution η for the expression in the brackets. The quantity η is known as the *coefficient of diffusion*. Given a particular $m_u(z)$ and $m_d(z)$ for molecular or turbulent motion, the coefficient of diffusion can be evaluated in terms of other physical quantities. Equation (4.19) states that the diffusive salt flux in a given direction is proportional to the salinity gradient in that direction. In general, the flux of any conservative quantity will be proportional to the concentration gradient in the same direction.

If the coefficient of diffusion η is a constant with respect to the space coordinates, which is generally the case for molecular diffusion but not for eddy diffusion, Eq. (4.19) may be generalized to

$$\mathbf{S} = -\eta \nabla s \tag{4.20}$$

If η is not a constant, it is generally more convenient to use the three scalar relations

$$S_x = -\eta_x \frac{\partial s}{\partial x} \quad S_y = -\eta_y \frac{\partial s}{\partial y} \quad S_z = -\eta_z \frac{\partial s}{\partial z} \tag{4.21}$$

rather than considering η in its tensor form, since usually for such cases the eddy coefficient of diffusion is appreciable for only one coordinate direction and can be neglected for the other two.

As in Section 3.3 we can express the diffusive flux of salt in terms of a diffusion velocity **w** as

$$\mathbf{S} = \rho s \mathbf{w} \tag{4.22}$$

similar to Eq. (3.20). The equation of continuity applies equally well to salt diffusion so that, if ρ is a constant, we have from Eq. (3.21)

$$\frac{\partial s}{\partial t} = -\nabla \cdot (s\mathbf{w}) \tag{4.23}$$

or

$$\rho \frac{\partial s}{\partial t} = -\nabla \cdot (\mathbf{S}) \tag{4.24}$$

If Eq. (4.20) applies, we shall then have from Eq. (4.24)

$$\frac{\partial s}{\partial t} = \frac{\eta}{\rho} \nabla^2 s \tag{4.25}$$

which is the same as the heat conduction equation. If Eq. (4.20) does not apply, we shall have from Eqs. (4.21)

$$\frac{\partial s}{\partial t} = \frac{1}{\rho} \left[\frac{\partial}{\partial x} \left(\eta_x \frac{\partial s}{\partial x} \right) + \frac{\partial}{\partial y} \left(\eta_y \frac{\partial s}{\partial y} \right) + \frac{\partial}{\partial z} \left(\eta_z \frac{\partial s}{\partial z} \right) \right] \tag{4.26}$$

For most problems in physical oceanography, the coefficient of eddy diffusion is much larger than the coefficient of molecular diffusion, so that we are usually concerned with an equation of the form of Eq. (4.26). If we were dealing with a nonconservative quantity, instead of the salinity s, or with a problem involving sources or sinks for either a conservative or nonconservative quantity, we should have to alter the equation of continuity, and consequently Eq. (4.25) or (4.26) for such spatial or time variations, much as was done in Section 2.7. The coefficients of eddy diffusion η_x, η_y, η_z are sometimes referred to as the *exchange coefficients*. Since, to the first order, ρ is equal to unity, it is often omitted from Eqs. (4.25) and (4.26) in physical oceanographic problems.

We shall now derive the equation for eddy diffusion through an alternative method, which will provide us with a different relation for the eddy diffusion coefficients. Consider a medium in which there is small turbulence.

The instantaneous velocity **v** may be taken as equal to a mean velocity $\bar{\mathbf{v}}$ and a velocity deviation **v**'. We shall assume that the velocity deviation is sufficiently well behaved across its zero position, so that the mean values of the derivatives with respect to the independent variables defining it are also zero. We may then write

$$v_x = \bar{v}_x + v'_x \quad v_y = \bar{v}_y + v'_y \quad v_z = \bar{v}_z + v'_z \qquad (4.27)$$

$$\overline{v'_x} = \overline{v'_y} = \overline{v'_z} = 0$$

and

$$\overline{\frac{\partial v'_x}{\partial t}} = \overline{\frac{\partial v'_x}{\partial x}} = \overline{\frac{\partial v'_x}{\partial y}} = \overline{\frac{\partial v'_x}{\partial z}} = 0$$

and similarly for the partial derivatives with respect to v'_y and v'_z. Substituting the relations (4.27) into the equation of continuity (3.18) for ρ a constant and taking the mean value, we have

$$\frac{\partial \bar{v}_x}{\partial x} + \frac{\partial \bar{v}_y}{\partial y} + \frac{\partial \bar{v}_z}{\partial z} = 0$$

or

$$\nabla \cdot \bar{\mathbf{v}} = 0 \qquad (4.28)$$

From Eqs. (4.28) and (3.18) we then have also

$$\frac{\partial v'_x}{\partial x} + \frac{\partial v'_y}{\partial y} + \frac{\partial v'_z}{\partial z} = 0$$

or

$$\nabla \cdot \mathbf{v}' = 0 \qquad (4.29)$$

Equations (4.28) and (4.29) may be considered the equations of continuity for the mean velocity and for the velocity deviation, respectively. We may take in the same way the instantaneous salinity s to be equal to a mean salinity $\bar{s}$ and a salinity deviation s' with similar relations to Eqs. (4.27). Substituting into the equation of continuity for salt mass [Eq. (3.22)] and taking the mean value we have, using the relation (4.29),

$$\frac{\partial \bar{s}}{\partial t} + \mathbf{v} \cdot \nabla \bar{s} + \overline{\mathbf{v}' \cdot \nabla s'} = 0$$

$$\frac{\partial \bar{s}}{\partial t} + \bar{\mathbf{v}} \cdot \nabla \bar{s} + \nabla \cdot \overline{(s'\mathbf{v}')} = 0 \qquad (4.30)$$

or

$$\frac{\partial \bar{s}}{\partial t} + \bar{v}_x \frac{\partial \bar{s}}{\partial x} + \bar{v}_y \frac{\partial \bar{s}}{\partial y} + \bar{v}_z \frac{\partial \bar{s}}{\partial z} + \frac{\partial}{\partial x} \overline{(s'v'_x)} + \frac{\partial}{\partial y} \overline{(s'v'_y)} + \frac{\partial}{\partial z} \overline{(s'v'_z)} = 0 \qquad (4.31)$$

If we now define the functions η_x, η_y, η_z by

$$\eta_x = -\frac{\overline{\rho s' v'_x}}{\frac{\partial \bar{s}}{\partial x}} \quad \eta_y = -\frac{\overline{\rho s' v'_y}}{\frac{\partial \bar{s}}{\partial y}} \quad \eta_z = -\frac{\overline{\rho s' v'_z}}{\frac{\partial \bar{s}}{\partial z}} \quad (4.32)$$

Eq. (4.31) becomes

$$\frac{\partial \bar{s}}{\partial t} = \frac{1}{\rho}\left[\frac{\partial}{\partial x}\left(\eta_x \frac{\partial \bar{s}}{\partial x}\right) + \frac{\partial}{\partial y}\left(\eta_y \frac{\partial \bar{s}}{\partial y}\right) + \frac{\partial}{\partial z}\left(\eta_z \frac{\partial \bar{s}}{\partial z}\right)\right]$$
$$-\bar{v}_x \frac{\partial \bar{s}}{\partial x} - \bar{v}_y \frac{\partial \bar{s}}{\partial y} - \bar{v}_z \frac{\partial \bar{s}}{\partial z} \quad (4.33)$$

which is the same as Eq. (4.26) with the addition of the advection term $\bar{\mathbf{v}} \cdot \nabla \bar{s}$. Equation (4.33) states that the local time rate of change of salinity $\partial \bar{s}/\partial t$ within a small volume is equal to the diffusion into the volume minus the advection out of the volume.

To summarize, the diffusion equation, Eq. (4.25) or (4.26), is the same form as the heat conduction equation. When for a given physical oceanographic problem the coefficient of eddy diffusion can be taken as a constant, which is sometimes the case of one-dimensional diffusion, the solutions of Section 2.6 for the heat conduction equation apply directly. Further, when the equation including the advection term, Eq. (4.33), under stationary conditions with the coefficient of eddy diffusion again a constant applies, the one-dimensional solutions of the heat conduction equation of Section 2.6 again apply. The derivations in the preceding sections have been taken for the salinity, but they apply equally well to any other conservative quantity such as the concentration of some other dissolved substance, biological organisms, and temperature for such problems where these quantities can be considered conservative. We could also have derived a coefficient of eddy viscosity in analogy to the coefficient of molecular viscosity discussed in Section 3.6. In the mixing process, there is a change of momentum just as there is a change in concentration. The rate of change of momentum, or stress terms, will be $\overline{\rho v'^2_x}$, $\overline{\rho v'_x v'_y}$, $\overline{\rho v'_x v'_z}$ for the X, Y, Z stress acting on a surface normal to the X axis. By analogy with Section 3.6, equating these stress terms to the time rate of change of strain will give the apparent coefficients of eddy viscosity.

Problem 4.2(a) Northerly winds produce a water transport offshore of 10 cm/sec from an east-west oriented coast in a surface layer 100 m thick out to a distance of 100 km offshore. What must be the net vertical velocity to maintain continuity?

Ans. 0.01 cm/sec

Thermodynamics and Hydrodynamics

Problem 4.2(b) For the Irish Sea, the average salinity at the southern entrance is 34.83‰ and at the nothern exit 34.33‰. If the average depth and surface area of the Irish Sea are 60 m and 18,600 km², the per unit area excess of rainfall and fresh water inflow over evaporation 61 cm/yr, and the cross-sectional area of the northern exit 7.14 km², determine the flushing time and the volume transport and velocity across the exit section.

Ans. 1.4 yr, 790 km³/yr, 0.35 cm/sec

Problem 4.2(c) For the Arctic Sea, the principal salt water inflow is through the Faroe-Shetland channel to the east of Iceland and the major outflow through the Denmark strait to the west of Iceland. The mean salinity of the entering salt water is 35.3‰ and that of the exiting salt water 32.5‰. If the rate of volume transport through the Faroe-Shetland channel is 3×10^6 m³/sec, determine the volume transport through the Denmark strait and the net rate of influx of fresh water to the Arctic Sea.

Ans. 3.26×10^6 m³/sec, 2.6×10^5 m³/sec

Problem 4.2(d) The mean depth and surface area of the Mediterranean Sea are 1400 m and 2,500,000 km². The water flowing into the Mediterranean from the Atlantic has a mean salinity of 36.25‰ and that flowing out a mean salinity of 37.75‰. The rate of volume transport of water from the Atlantic to the Mediterranean through the Straits of Gibraltar is estimated to be 1.75×10^6 m³/sec. Determine the rate of volume transport out of the Mediterranean, the rate of influx of fresh water to the Mediterranean, and the flushing time.

Ans. 1.68×10^6 m³/sec, -7×10^4 m³/sec, 66 yr

Problem 4.2(e) The mean depth and surface area of the Black Sea are 1200 m and 420,000 km². The rate of volume transport into the Black Sea through the Bosporus is estimated to be 6100 m³/sec and that out through the Bosporus from the Black Sea to be 12,600 m²/sec. Determine the rate of influx of fresh water to the Black Sea and the flushing time.

Ans. 6.5×10^3 m³/sec, 1300 yr

Problem 4.2(f) Compare and discuss the results of Problems 4.2(d) and 4.2(e).

4.3 Ocean Mixing

The investigations of ocean mixing phenomena have been largely descriptive and empirical. We shall follow that lead in the theoretical discussion presented here.

One can recognize several major current systems throughout the deep ocean areas of the world, such as the subtropical lower water, subantarctic intermediate water, North Atlantic deep water, and so forth. These current systems appear to be both permanent and stationary. Let us look at the mixing by diffusion between currents in a direction normal to the current flow, that is, essentially in the vertical Z direction. Since the currents are stationary, we shall presume that each current at a particular location can be characterized by a particular temperature and salinity. We shall also presume that the diffusion itself is sufficiently well behaved that the coefficients of diffusion for both the temperature and salinity mixing are the same. These, of course, are rather gross assumptions. Then for the mixed water we shall have

$$T = q_1 T_1 + q_2 T_2 \tag{4.34}$$

and

$$s = q_1 s_1 + q_2 s_2 \tag{4.35}$$

where

$$q_1 + q_2 = 1 \tag{4.36}$$

and where T_1, s_1, and T_2, s_2 are the temperatures and salinities of the two original masses of water, respectively, and q_1 and q_2 are the mixing ratios. Combining these three equations, we obtain

$$T = \frac{T_1 - T_2}{s_1 - s_2} s + \frac{T_2 s_1 - T_1 s_2}{s_1 - s_2} \tag{4.37}$$

We see then that under these assumptions, temperature and salinity are not independent quantities and that a plot of T versus s will provide a linear relation with end points T_1, s_1 and T_2, s_2 irregardless of the degree of mixing. For two original water masses with temperatures and salinities as shown in Fig. 4.5(a) and (b) and with possible degrees of mixing as shown by the dotted lines, the $T-s$ plot will be the same line as shown in Fig 4.5(c) for any degree of mixing. If there were three masses of water involved in the

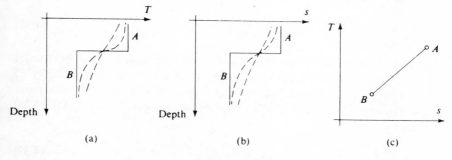

Fig. 4.5

mixing, the relations would be as shown in Fig. 4.6(a), (b) and (c). In this latter case, if the degree of mixing has progressed sufficiently far to effect the core of the intermediate water mass, the acute angle at B on the $T-s$ plot would be rounded off. It is found that within a given area when the temperatures and corresponding salinities are plotted against each other, the points generally fall on a well-defined curve, which can be approximated by a series of straight lines, and further that repeated measurements for a

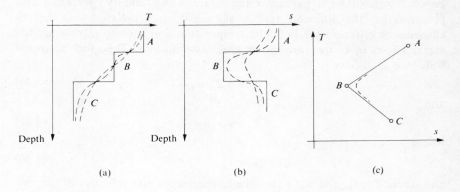

Fig. 4.6

particular station location produce the same curve. These results indicate that our simple assumptions as to the similarity of the temperature and salinity mixing and the stationary nature of the currents are valid. The $T-s$ plot has proved to be a useful empirical method for identifying and delineating the characteristics of major current systems of the deep ocean areas.

Let us now look at the diffusion process in somewhat more detail. We shall assume that the conditions are stationary, that the diffusion is in a vertical direction only, and that the advection due to the current flow is in a horizontal direction only. Equation (4.33) then reduces to

$$\rho v_x \frac{\partial s}{\partial x} = \frac{\partial}{\partial z}\left(\eta_z \frac{\partial s}{\partial z}\right) \tag{4.38}$$

where we have omitted the bar for the average salinity and horizontal velocity. Equation (4.38) again is simply a statement of the continuity of salt mass that the diffusion in a vertical direction is equal to the advection in a horizontal direction. If either $\partial \eta_z/\partial z$ or $\partial s/\partial z$ is equal to zero, Eq. (4.38) reduces to

$$\rho v_x \frac{\partial s}{\partial x} = \eta_z \frac{\partial^2 s}{\partial z^2} \tag{4.39}$$

or

$$\frac{\eta_z}{v_x} = \frac{\rho \frac{\partial s}{\partial x}}{\frac{\partial^2 s}{\partial z^2}} \qquad (4.40)$$

which permits us to calculate η_z/v_x from vertical measurements at selected stations in the direction of the current. Since, in general, $\partial \eta_z/\partial z$ is not zero, we are restricted in the application of Eq. (4.39) to depths in the vicinity of salinity maxima or minima. From Fig. 4.6 we see that such maxima and minima occur at depths corresponding to the vertices on the $T-s$ plot.

Although it is perhaps superfluous to mention and not necessarily informative, we see that for those cases in which there is diffusion in only one direction, either normal to, transverse to, or along the current direction, and the coefficient of diffusion is constant, Eq. (4.39) will apply, which is exactly the one-dimensional heat conduction equation whose solutions are given in Section 2.6. We see from Eq. (2.69) that the form of variation of salinity will be exponential along the current direction and sinusoidal in the direction of diffusion.

Problem 4.3(a) Construct the $T-s$ diagram for an ocean section, where there are the five following currents—subtropical lower water, subantarctic intermediate water, North Atlantic deep water, North Atlantic bottom water, and Antarctic bottom water—whose characteristic temperatures and salinities are, respectively, 18.0°C, 35.93‰; 3.25°C, 34.15‰; 4.0°C, 35.00‰; 2.5°C, 34.90‰; and 0.4°C, 34.67‰.

Problem 4.3(b) Compute the quantity η_z/v_x for the salinity minimum of the Antarctic intermediate current at succeeding pairs of stations 1000 km apart in the direction of the current for which the values in the vicinity of the minimum are as given in the table below.

Station	A	B	C	D	E
$z_0 - 200$ m	34.20‰	34.27	34.40	34.45	34.51
	34.18	34.21	34.29	34.34	34.38
$z_0 + 200$ m	34.22	34.28	34.44	34.49	34.54

4.4 Estuary Mixing

Mixing phenomena in an estuary generally tend to be more complex than those discussed in the preceding section because of the addition of two parameters. One, there is usually a substantial influx of fresh water to the estuary from the major river draining into it. Two, there is substantial longitudinal mixing due to the tidal flow into and out of the estuary. The salinity then will vary in a longitudinal direction from zero at the entrance of the fresh water to the estuary to sea water salinity at the ocean end of the estuary. It may also be a function of depth if, for example, the salt water movement is principally shoreward at depth and the fresh water movement principally seaward at the surface. It may also be a function of the coordinate transverse to the estuary if there are differences in circulation and mixing from one side of the estuary to the other. The same sort of variation conditions will also apply to a chemical pollutant introduced into the estuary. We shall consider here a couple of simplified examples of estuary mixing that are amenable to direct solution.

Consider an estuary with rectangular sides as shown in Fig. 4.7 with the

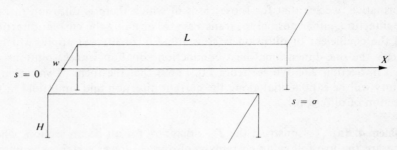

Fig. 4.7

x coordinate longitudinal to the estuary. We shall assume that the conditions are stationary and the mixing is complete for any section normal to x. Then the salinity will be a function of x only, $s = s(x)$. We shall take, as our boundary conditions at the entrance of the river water to the estuary,

$$s = 0$$
$$\frac{ds}{dx} = 0 \quad (x = 0) \tag{4.41}$$

and at the seaward end of the estuary

$$s = \sigma \quad (x = h) \tag{4.42}$$

where σ is the salinity of the sea water. If D is the volume discharge of the river, then the mean velocity of the water in the channel, v_x, due to the river is

$$v_x = \frac{D}{wH} \tag{4.43}$$

where w and H are the width and depth of the estuary, respectively. If the length of the channel is small compared with the wavelength of the tide, which is the usual case, we may consider the tide to be simultaneous and uniform over the entire estuary and the tidal height above mean sea level to be given by

$$h = h_0 \cos \omega t \qquad (4.44)$$

where ω is the angular frequency of the tide. From consideration of mass continuity across two adjacent sections normal to x, as shown in Fig. 4.8,

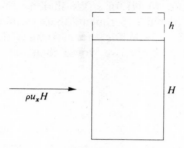

Fig. 4.8

we see that the tidal current u_x is given in terms of the tidal height by

$$\frac{\partial(\rho h)}{\partial t} = - \frac{\partial(\rho u_x H)}{\partial x}$$

or

$$\frac{\partial h}{\partial t} = -H \frac{\partial u_x}{\partial x} \qquad (4.45)$$

where we have taken the density ρ to be a constant. Substituting Eq. (4.44) into Eq. (4.45) and integrating, we obtain

$$u_x = \frac{\omega h_0 x}{H} \sin \omega t \qquad (4.46)$$

The total displacement ξ can then be obtained by integrating Eq. (4.46) with respect to time, giving

$$\xi = - \frac{h_0 x}{H} \cos \omega t \qquad (4.47)$$

The constants of integration for both Eqs. (4.46) and (4.47) are zero since, when the tidal height is at a maximum, the tidal velocity is zero and the tidal displacement a maximum in the negative x direction.

For steady state and a density of unity, the equation of continuity of salt mass, Eq. (4.33), reduces to

$$v_x \frac{\partial s}{\partial x} = \frac{\partial}{\partial x} \left(\eta_x \frac{\partial s}{\partial x} \right) \qquad (4.48)$$

120 ¶ *Thermodynamics and Hydrodynamics*

Since s is a function of x only, this may be integrated to

$$v_x s = \eta_x \frac{ds}{dx} \qquad (4.49)$$

where from the boundary conditions [Eqs. (4.41)], the constant of integration is zero. We are now interested in arriving at an approximate value for η_x so that we may solve Eq. (4.49) for s. We shall assume that the tidal mixing is such that in an isolated experiment, there would be complete mixing over a tidal period $T = 2\pi/\omega$ for the maximum tidal displacement $2\xi_0$. For an original linear gradient, we would then have from Eq. (4.20) and Fig. 4.9

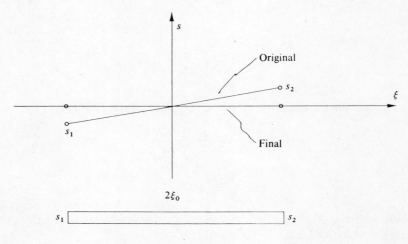

Fig. 4.9

$$S = -\eta_x \frac{ds}{dx}$$

$$-\frac{2}{T} \int_0^{\xi_0} \frac{s_2 - s_1}{2\xi_0} x \, dx = -\eta_x \frac{s_2 - s_1}{2\xi_0}$$

$$\frac{\xi_0^2}{T} = \eta_x$$

or from Eq. (4.47)

$$\eta_x = \frac{\omega h_0^2 x^2}{2\pi H^2} \qquad (4.50)$$

Substituting Eqs. (4.50) into (4.49) and integrating, we obtain

$$v_x s = \frac{\omega h_0^2}{2\pi H^2} x^2 \frac{ds}{dx}$$

$$\frac{ds}{s} = \frac{2\pi v_x H^2}{\omega h_0^2} \frac{dx}{x^2}$$

$$\log s = -\frac{2\pi v_x H^2}{\omega h_0^2} \frac{1}{x} + C$$

Substituting in the boundary condition [Eq. (4.42)], we obtain for C

$$C = \log \sigma + \frac{2\pi v_x H^2}{\omega h_0^2 L}$$

and finally for s

$$s = \sigma e^{F(1 - 1/\lambda)} \tag{4.51}$$

where we have made the substitution of the dimensionless parameter

$$\lambda = \frac{x}{L} \tag{4.52}$$

and where

$$F = \frac{2\pi v_x H^2}{\omega h_0^2 L} \tag{4.53}$$

Equation (4.51) defines a set of curves that are asymptotic to $s = 0$ for large values of F and asymptotic to $s = \sigma$ for small values of F.

Understandably, the simplification of Fig. 4.9 is often inappropriate to a particular problem. In such cases, the measured values of $s = s(x)$ can be used to determine numerically the values of $\eta_x = \eta_x(x)$. Multiplying Eq. (4.49) by the area $A = A(x)$, where in this case the area need not be a constant, we shall have

$$Ds = A\eta_x \frac{ds}{dx}$$

or

$$\eta_x = \frac{Ds}{A \frac{ds}{dx}} \tag{4.54}$$

under the conditions to which Eq. (4.49) applies. From a knowledge of the volume discharge of the river and the cross section area of the estuary, the measured values of salinity, and the determined values of the horizontal salinity gradient, the value of the horizontal diffusion coefficient η_x can be determined sequentially down the estuary from Eq. (4.54).

Let us presume that η_x has been determined by some such a procedure as Eq. (4.54). Then we may determine, in a simple manner, the horizontal distribution of a pollutant in the estuary provided that its coefficient of diffusion is the same as that for salt, which is a reasonable assumption under most conditions. Let us also consider that the pollutant is nonconservative and that in an isolated state, it will decrease exponentially in the fashion $c = c_0 e^{-\alpha t}$. Then its time derivative will be $dc/dt = -\alpha c$ so that the equation of continuity for pollutant following the particle motion, instead of being equal to zero as given in Eq. (3.23), will be given by

$$\frac{Dc}{Dt} = -\alpha c \qquad (4.55)$$

The total seaward flux of pollutant $C(x)$ past any cross section $A(x)$ of the estuary will be the sum of the horizontal advection and diffusion terms or from Eq. (4.20)

$$C = A v_x c - A \eta_x \frac{dc}{dx}$$

$$= Dc - A\eta_x \frac{dc}{dx} \qquad (4.56)$$

If the pollutant were conservative and if there were no tidal diffusion, C would be a constant downstream of the outfall of the pollutant and zero above it. For a nonconservative pollutant with tidal diffusion under steady-state conditions, taking ρ equal to unity, we shall have from Eqs. (4.55) and (4.56)

$$\frac{dC}{dx} = -\alpha c$$

or

$$\frac{d}{dx}\left(Dc - A\eta_x \frac{dc}{dx}\right) + \alpha c = 0 \qquad (4.57)$$

At the outfall, Eq. (4.57) will be replaced by

$$\frac{d}{dx}\left(Dc - A\eta_x \frac{dc}{dx}\right) + \alpha c = \psi \qquad (4.58)$$

where ψ is the rate of supply of pollutant at the outfall. From Eqs. (4.57) and (4.58), the distribution of the pollutant $c(x)$ can be determined numerically.

In more complex cases in which there may be stratified longitudinal flow with both horizontal and vertical advection and both horizontal and vertical diffusion, the same concepts of continuity of water mass and salt mass can usually be used in conjunction with observations of currents and salinity to distinguish the magnitudes of each of the distribution factors.

4.5 Equation of Motion and Circulation Theorem

For physical oceanographic problems, the water motion is subject to the force of gravity and the effects due to the rotation of the earth. From Eq. (1.248) we have that the force per unit mass for motion relative to the earth is given by

$$\mathbf{f} = \mathbf{g} - 2\boldsymbol{\omega} \times \mathbf{v} \tag{4.59}$$

where $\mathbf{g}$ is the gravitational force per unit mass and $-2\boldsymbol{\omega} \times \mathbf{v}$ the Coriolis force per unit mass, $\boldsymbol{\omega}$ being the angular velocity of the earth's rotation and $\mathbf{v}$ the water particle velocity. Substituting Eq. (4.59) into Eq. (3.30), we have

$$\frac{\partial \mathbf{v}}{\partial t} + \mathbf{v} \cdot \nabla \mathbf{v} = \mathbf{g} - 2\boldsymbol{\omega} \times \mathbf{v} - \frac{1}{\rho} \nabla p \tag{4.60}$$

Equation (4.60) is the hydrodynamic equation for a water particle relative to the earth. For some problems we may also want to add a term for a driving force and for a frictional resistance.

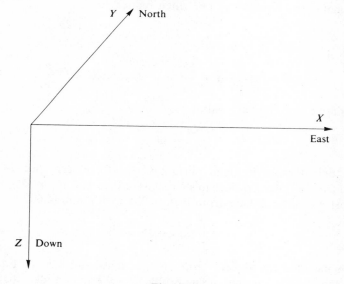

Fig. 4.10

The custom in physical oceanographic problems has unfortunately been to use a left-handed coordinate set with X axis to the east, the Y axis to the north, and the Z axis down as shown in Fig. 4.10. We shall follow this custom. The net effect of this change is to alter the signs of the components [Eq. (1.12)] of the vector product of two vectors so that the sign of the vector product, when expanded into component form with respect to a left-handed coordinate system, is the negative of what it would be for a right-handed coordinate system. With reference to Fig. 4,10, the components of **ω** will be

$$\omega_x = 0 \quad \omega_y = \omega \cos \varphi \quad \omega_z = -\omega \sin \varphi \tag{4.61}$$

where φ is the latitude. From Eqs. (4.61) and for the coordinate system of Fig. 4.10, we obtain for Eq. (4.60)

$$\frac{\partial v_x}{\partial t} + v_x \frac{\partial v_x}{\partial x} + v_y \frac{\partial v_x}{\partial y} + v_z \frac{\partial v_x}{\partial z} = 2\omega v_y \sin \varphi + 2\omega v_z \cos \varphi - \frac{1}{\rho} \frac{\partial p}{\partial x}$$

$$\frac{\partial v_y}{\partial t} + v_x \frac{\partial v_y}{\partial x} + v_y \frac{\partial v_y}{\partial y} + v_z \frac{\partial v_y}{\partial y} = -2\omega v_x \sin \varphi - \frac{1}{\rho} \frac{\partial p}{\partial y} \tag{4.62}$$

$$\frac{\partial v_z}{\partial t} + v_x \frac{\partial v_z}{\partial x} + v_y \frac{\partial v_z}{\partial y} + v_z \frac{\partial v_z}{\partial z} = g - 2\omega v_x \cos \varphi - \frac{1}{\rho} \frac{\partial p}{\partial z}$$

For most problems we are concerned only with horizontal motion. Further, the Coriolis term in the third of Eq. (4.62) is always much smaller than the gravitational term. Equations (4.62) then reduce to

$$\frac{\partial v_x}{\partial t} + v_x \frac{\partial v_x}{\partial x} + v_y \frac{\partial v_x}{\partial y} = 2\omega v_y \sin \varphi - \frac{1}{\rho} \frac{\partial p}{\partial x}$$

$$\frac{\partial v_y}{\partial t} + v_x \frac{\partial v_y}{\partial x} + v_y \frac{\partial v_y}{\partial y} = -2\omega v_x \sin \varphi - \frac{1}{\rho} \frac{\partial p}{\partial y} \tag{4.63}$$

and the simple and familiar relation

$$\frac{\partial p}{\partial z} = \rho g \tag{4.64}$$

Since only the gravitational portion of the force term [Eq. (4.59)] is derivable from a potential, Kelvin's circulation theorem, Eq. (3.41), will not apply. Instead, we shall have from Eqs. (3.39), (3.40), and (4.60)

$$\frac{DK}{Dt} = -\oint 2\boldsymbol{\omega} \times \mathbf{v} \cdot \delta \boldsymbol{\lambda} - \oint \frac{1}{\rho} \delta p \tag{4.65}$$

We have included the second term on the right-hand side of Eq. (3.40) since, in the physical oceanographic problems we shall consider, the density is not a single function of the pressure. In other words, the surfaces of constant pressure, *isobaric* surfaces, do not coincide with the surfaces of constant density, *isoteric* or *isopycnal* surfaces. For horizontal motion where **v** is given by

$$\mathbf{v} = \mathbf{i} v_x + \mathbf{j} v_y \tag{4.66}$$

and the effective Coriolis force terms by Eqs. (4.63) or

$$-2\boldsymbol{\omega} \times \mathbf{v} = 2\omega \sin \varphi (\mathbf{i} v_y - \mathbf{j} v_x) \tag{4.67}$$

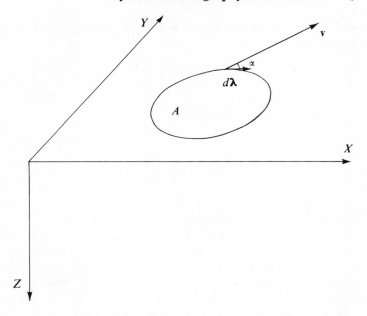

Fig. 4.11

Eq. (4.65) becomes, taking the circulation loop in a horizontal plane,

$$\frac{DK}{Dt} = 2\omega \sin \varphi \oint (v_y dx - v_x dy) - \oint \frac{1}{\rho} dp$$

$$= 2\omega \sin \varphi \oint |\mathbf{v} \times d\boldsymbol{\lambda}| - \oint \frac{1}{\rho} dp \qquad (4.68)$$

remembering the cross product relations for a left-handed coordinate system. From Fig. 4.11 we see that the integrand of the first term on the right hand side is $v \sin \alpha \, d\lambda$ and the integral the rate at which the area A of the circulation loop is increasing, giving

$$\frac{DK}{Dt} = 2\omega \sin \varphi \frac{dA}{dt} - \oint \frac{1}{\rho} dp \qquad (4.69)$$

Equation (4.69) is referred to as the *Bjerknes circulation theorem*.

4.6 Pressure Gradient and Geostrophic Effects

We shall examine, in this section, the water motion under the conditions of steady state when following this motion, that is, $D\mathbf{v}/Dt = 0$. We may consider that this is equivalent to steady-state local time conditions,

$\partial \mathbf{v}/\partial t = 0$, and no expansion effects, $\mathbf{v}\cdot\nabla\mathbf{v} = 0$. The equations of motion (4.63) then reduce to

$$2\omega v_y \sin \varphi = \frac{1}{\rho}\frac{\partial p}{\partial x}$$
$$-2\omega v_x \sin \varphi = \frac{1}{\rho}\frac{\partial p}{\partial y}$$
(4.70)

or

$$2\omega v \sin \varphi = \frac{1}{\rho}\frac{dp}{dn}$$
(4.71)

where v is the magnitude of the current, and n is a coordinate direction at right angles to $\mathbf{v}$ with its positive direction to the right of $\mathbf{v}$.

We see that under these conditions, the force term related to the Coriolis effect is balanced by the force term related to the horizontal pressure gradient. We also note that neither of these terms represents the driving force. We have included no driving force term nor frictional resistance term in Eq. (4.63). We have assumed frictionless motion. The Coriolis force is the result of the constant water velocity motion. The pressure gradient term is a necessary resultant from the equations of motion to preserve the assumed steady-state conditions. In physical oceanography, the Coriolis acceleration, or force per unit mass, is referred to as the *geostrophic* acceleration, or force per unit mass.

If we consider such a current extending to the sea surface, we see that the sea surface itself will not be a level surface, coincident with the horizontal surface of our coordinate system, for this case of a dynamic system of an ocean current, as it is for the static case of gravity determinations. We may show these relations diagrammatically in Fig. 4.12 for the northern hemisphere

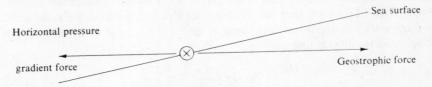

Fig. 4.12

for a current directed into the page with the geostrophic force to the right and the horizontal pressure gradient force to the left. For the southern hemisphere, the force relations would be reversed. Currents of this type are known as *geostrophic currents*.

If we further consider that the atmospheric pressure at the sea surface is constant, we may obtain from Fig. 4.13 the following relation for the

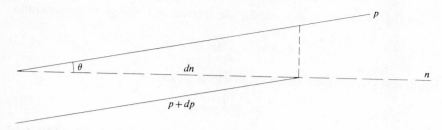

Fig. 4.13

horizontal pressure gradient in terms of the inclination of the sea surface, or any other isobaric surface,

$$\frac{dp}{dn} = g\rho \tan \theta \tag{4.72}$$

Substituting Eq. (4.72) into Eq. (4.71), we obtain for the current velocity

$$v = \frac{g \tan \theta}{2\omega \sin \varphi} \tag{4.73}$$

Equation (4.73) provides us with a convenient means of determining the velocities of the major ocean current systems from physically measurable quantities. Let us take two oceanographic stations a distance L apart at right angles across a current as shown in Fig. 4.14. We shall assume that the isobaric surfaces produce linear traces on a vertical section between these two stations and shall be interested in computing the heights of these isobaric surfaces above a level surface of no motion at some depth. We cannot state a priori at what depth no motion will occur, so that we shall actually be computing current velocities relative to the velocity at this depth. From Fig. 4.14 we shall have

$$g \tan \theta = \frac{g(h_B - h_A)}{L} \tag{4.74}$$

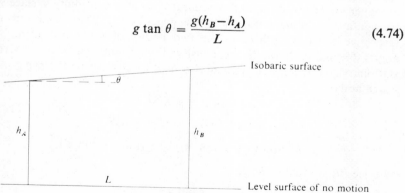

Fig. 4.14

From Eq. (4.64) we shall have, for the incremental change in geopotential,

$$g\Delta h = -\frac{1}{\rho}\Delta p = -\alpha \Delta p \qquad (4.75)$$

where α is the specific volume, from which we obtain

$$gh = \int_p^{p_0} \alpha\, dp \qquad (4.76)$$

where p_0 is the pressure at the depth of no motion and p the pressure at the depth at which the current is to be determined. Substituting Eqs. (4.74) and (4.76) into Eq. (4.73), we obtain finally

$$v = \frac{1}{2\omega L \sin \varphi} \int_p^{p_0} (\alpha_B - \alpha_A)\, dp \qquad (4.77)$$

Equation (4.77) is referred to as the *Helland-Hansen formula*. The quantity gh of Eq. (4.76) is referred to as the dynamic height. In the application of Eq. (4.77), the integral is determined numerically from values of α determined empirically from measurements of salinity and temperature at each oceanographic station as a function of depth. The incremental pressure intervals Δp are taken to a sufficient approximation to be given by $g\Delta z$, where Δz is the depth interval. It is apparent that a plot of dynamic heights for a number of stations in a given area will, in itself, be a measure of the current structure, the direction of the current being parallel to the dynamic height contours and its magnitude being proportional to the gradient of the contours.

For many of the deep ocean current systems, the current transport will bring waters of different physical properties in contact, and a density discontinuity surface, or a zone of rapid change in density, will occur between the overlying current water mass and the underlying still water mass. We shall be interested in determining the relation between the slope of this discontinuity surface and the current velocity. In Fig. 4.15 we have two bodies of water of homogeneous densities ρ' and ρ, separated by a discontinuity surface whose trace in a vertical plane normal to it is given by PQ. The inclinations of the respective isobaric surfaces in the two bodies of water are given by the angles θ' and θ. Now, if the density discontinuity surface is to remain stationary, the increase in pressure from P to Q must be the same in both water bodies,

$$\rho g SQ = \rho' g PS'$$

$$\rho(SR + RQ) = \rho'(PR' + R'S')$$

$$\rho PR(\tan \theta + \tan \gamma) = \rho' R'Q(\tan \gamma + \tan \theta')$$

or since PR is equal to $R'Q$,

$$\tan \gamma = \frac{\rho \tan \theta - \rho' \tan \theta'}{\rho' - \rho} \qquad (4.78)$$

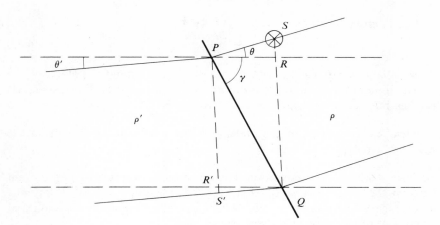

Fig. 4.15

In the usual case, $\rho' > \rho$ and $\rho \tan \theta > \rho' \tan \theta'$, so that γ will be positive. This means that when the isobaric surfaces slope upward, the density discontinuity surface slopes downward. If in the left-hand body of water the isobaric surfaces are level, that is, still water, we shall have

$$\tan \gamma = \frac{\rho \tan \theta}{\rho' - \rho} \qquad (4.79)$$

For the northern hemisphere, for a current into the page for the right-hand body of water, the density discontinuity surface will be as shown in Fig. 4.15. For the southern hemisphere, the same would be true for a current out of the page. Considering Eq. (4.78) in the limit, for a zone of rapid change in velocity, we shall have

$$\tan \gamma = -\frac{\partial}{\partial \rho}(\rho \tan \theta) \qquad (4.80)$$

where the trace PQ now represents isopycnal surfaces. For most physical oceanographic problems, the quantity $\rho/(\rho' - \rho)$ will be large of the order of 10^3 and the quantity $\tan \theta$ very small of the order of 10^{-6}.

Substituting Eq. (4.73) into Eq. (4.78), we obtain

$$\tan \gamma = \frac{2\omega \sin \varphi}{g}\left(\frac{\rho v - \rho' v'}{\rho' - \rho}\right) \qquad (4.81)$$

Corresponding to Eqs. (4.79) and (4.80), we obtain, respectively,

$$\tan \gamma = \frac{2\omega \sin \varphi}{g}\left(\frac{\rho v}{\rho' - \rho}\right) \qquad (4.82)$$

and
$$\tan \gamma = -\frac{2\omega \sin \varphi}{g} \frac{\partial}{\partial \rho}(\rho v) \qquad (4.83)$$

These relations provide us with a means of determining the slope of the density discontinuity surface in terms of the current velocity. Equation (4.81) is often referred to as *Margules' equation*.

Problem 4.6(a) Obtain the Helland-Hansen formula directly from the Bjerknes' circulation theorem.

Problem 4.6(b) The equatorial counter current is an east-going current lying in the northern hemisphere. At 120°E longitude and 10°N latitude, its average velocity in August is 24 nmi/day. Determine the direction and magnitude of the slope of the isobaric surfaces.

Ans. Slope upward toward the south, 0.133×10^{-5}

Problem 4.6(c) Determine the direction of inclination and magnitude of the slope of the lower boundary of the East Greenland current at 73°N latitude. Assume the current to consist of water of density 1.0271 g/cm^3 flowing southerly with a velocity of 20 cm/sec overlying still water of density 1.0281 g/cm^3.

Ans. Inclination is downward to the west; slope is 2.92 m/km.

Problem 4.6(d) For the gulf stream, ρ increases and ρv decreases in a downward direction. The changes in density are largely determined by changes in temperature, increasing with decreasing temperature. On the left boundary of the gulf stream, the changes in density are so great as to approach a discontinuity surface. Sketch on a section normal to the gulf stream the isopycnal and isothermal contours and discuss.

4.7 Inertia Effects

A rather simple type of current is one in which there are no frictional effects and no pressure gradient effects. The equations of motion (4.63) reduce to

$$\frac{Dv_x}{Dt} = 2\omega v_y \sin \varphi$$

$$\frac{Dv_y}{Dt} = -2\omega v_x \sin \varphi \qquad (4.84)$$

or

$$\frac{D\mathbf{v}}{Dt} = \mathbf{n}_1 2\omega v \sin \varphi \qquad (4.85)$$

where $\mathbf{n}_1$ is a unit vector normal to $\mathbf{v}$, with its positive sense to the right of $\mathbf{v}$. Resolving components of the acceleration $D\mathbf{v}/Dt$ normal to and tangential to the velocity, as given, for example, by Eq. (6.6), we have

$$\mathbf{t}_1 \frac{Dv}{Dt} + \mathbf{n}_1 \frac{v^2}{r} = \mathbf{n}_1 2\omega v \sin \varphi \qquad (4.86)$$

or

$$r = \frac{v}{2\omega \sin \varphi} \qquad (4.87)$$

The motion will be circular with a radius r given by Eq. (4.87). The time T for one complete revolution of the circle will be

$$T = \frac{2\pi r}{v} = \frac{\pi}{\omega \sin \varphi} \qquad (4.88)$$

In the northern hemisphere, the circulation motion will be clockwise, and in the southern hemisphere, counterclockwise. Currents of this type are known as *inertia currents*.

Problem 4.7(a) Determine the period for a circle of inertia at the poles, latitude 30°, and at the equator.

Ans. 12 hr, 24 hr, ∞

Problem 4.7(b) Determine the radius of an inertia current of velocity 15 cm/sec and period 14 hr.

Ans. 1.2 km

4.8 Equation of Motion with Internal Friction

In Section 3.6 we derived an expression, the Navier-Stokes equation, for the motion of a fluid under the effects of a viscous, or internal frictional, resistance. For most oceanographic problems, the ocean may be considered sufficiently incompressible that the first viscous resistance term of Eq. (3.49) may be considered small with respect to the second, We may then write for Eq. (3.50)

$$\frac{D\mathbf{v}}{Dt} = \frac{\partial \mathbf{v}}{\partial t} + \mathbf{v} \cdot \nabla \mathbf{v} = \mathbf{F} - \frac{1}{\rho} \nabla p + \frac{\mu}{\rho} \nabla \cdot \nabla \mathbf{v} \qquad (4.89)$$

where the internal frictional resistance term is

$$\mathbf{R} = \frac{\mu}{\rho} \nabla \cdot \nabla \mathbf{v} \qquad (4.90)$$

132 ¶ Thermodynamics and Hydrodynamics

The frictional resistance **R** will be numerically a negative quantity. Following the derivation of Section 4.5, the equation of motion for physical oceanographic circulation under the influence of viscosity will then be

$$\frac{\partial \mathbf{v}}{\partial t} + \mathbf{v} \cdot \nabla \mathbf{v} = \mathbf{g} - 2\boldsymbol{\omega} \times \mathbf{v} - \frac{1}{\rho}\nabla p + \frac{\mu}{\rho}\nabla \cdot \nabla \mathbf{v} \tag{4.91}$$

We must now, however, make a distinction between *molecular viscosity* and *eddy viscosity*, similar to the distinction that was made in Section 4.2 in the discussion of diffusion. The coefficient of molecular viscosity will be small and can be neglected with respect to the coefficient of eddy viscosity. The coefficient of eddy viscosity will not be a constant but a tensor, depend on the particular current system considered, and, in general, be a function of the coordinates themselves. We shall restrict ourselves here to problems in which the motion is essentially horizontal, and the vertical velocity gradients in the direction of motion are small. From Eq. (4.90) and Section 3.6 we may then write, for the frictional resistance components,

$$\begin{aligned} R_x &= \frac{1}{\rho}\frac{\partial}{\partial y}\left(\mu_t \frac{\partial v_x}{\partial y}\right) + \frac{1}{\rho}\frac{\partial}{\partial z}\left(\mu_z \frac{\partial v_x}{\partial z}\right) \\ R_y &= \frac{1}{\rho}\frac{\partial}{\partial x}\left(\mu_t \frac{\partial v_y}{\partial x}\right) + \frac{1}{\rho}\frac{\partial}{\partial z}\left(\mu_z \frac{\partial v_y}{\partial z}\right) \\ R_z &= 0 \end{aligned} \tag{4.92}$$

where μ_t and μ_z are the eddy coefficients of lateral and vertical viscosity. In many problems we can neglect the eddy coefficient of lateral viscosity, Eqs. (4.92) becoming

$$\begin{aligned} R_x &= \frac{1}{\rho}\frac{\partial}{\partial z}\left(\mu_z \frac{\partial v_x}{\partial z}\right) \\ R_y &= \frac{1}{\rho}\frac{\partial}{\partial z}\left(\mu_z \frac{\partial v_y}{\partial z}\right) \\ R_z &= 0 \end{aligned} \tag{4.93}$$

If the coefficient μ_z may be considered a constant with respect to the z coordinate, Eqs. (4.93) reduce further to

$$\begin{aligned} R_x &= \frac{\mu_z}{\rho}\frac{\partial^2 v_x}{\partial z^2} \\ R_y &= \frac{\mu_z}{\rho}\frac{\partial^2 v_y}{\partial z^2} \\ R_z &= 0 \end{aligned} \tag{4.94}$$

Under the conditions for which the relations (4.94) apply, Eqs. (4.91) reduce to Eq. (4.64) and to

$$\frac{\partial v_x}{\partial t} + v_x \frac{\partial v_x}{\partial x} + v_y \frac{\partial v_x}{\partial y} = 2\omega v_y \sin \varphi - \frac{1}{\rho}\frac{\partial p}{\partial x} + \frac{\mu_z}{\rho}\frac{\partial^2 v_x}{\partial z^2}$$
$$\frac{\partial v_y}{\partial t} + v_x \frac{\partial v_y}{\partial x} + v_y \frac{\partial v_y}{\partial y} = -2\omega v_x \sin \varphi - \frac{1}{\rho}\frac{\partial p}{\partial y} + \frac{\mu_z}{\rho}\frac{\partial^2 v_y}{\partial z^2}$$
(4.95)

similar to Eqs. (4.63) with the addition of the frictional resistance term. For steady-state conditions $D\mathbf{v}/Dt = 0$, Eqs. (4.95) reduce to

$$2\omega v_y \sin \varphi - \frac{1}{\rho}\frac{\partial p}{\partial x} + \frac{\mu_z}{\rho}\frac{\partial^2 v_x}{\partial z^2} = 0$$
$$-2\omega v_x \sin \varphi - \frac{1}{\rho}\frac{\partial p}{\partial y} + \frac{\mu_z}{\rho}\frac{\partial^2 v_y}{\partial z^2} = 0$$
(4.96)

Currents that do not include pressure gradient effects but only geostrophic and friction effects, the first and third terms of Eqs. (4.96), are referred to as *drift currents*. Currents that include pressure gradient effects as well as frictional and geostrophic effects, all three terms of Eqs. (4.96), are referred to as *gradient currents*.

4.9 Friction and Geostrophic Effects

For a drift current under steady-state conditions, the equations of motion (4.96) reduce to

$$\frac{\partial^2 v_x}{\partial z^2} = -\frac{2\rho\omega \sin \varphi}{\mu_z} v_y$$
$$\frac{\partial^2 v_y}{\partial z^2} = \frac{2\rho\omega \sin \varphi}{\mu_z} v_x$$
(4.97)

If we assume that v_x and v_y are functions of z only and let the complex variable w be defined by

$$w = v_x + iv_y \qquad (4.98)$$

we obtain, multiplying the second of Eqs. (4.97) by i and adding,

$$\frac{d^2 w}{dz^2} = \frac{2i\rho\omega \sin \varphi}{\mu_z} w \qquad (4.99)$$

The form of solution of this ordinary differential equation is simply

$$w = e^{\pm(1+i)\alpha z} \qquad (4.100)$$

Thermodynamics and Hydrodynamics

where

$$\alpha^2 = \frac{\rho\omega \sin\varphi}{\mu_z} \tag{4.101}$$

Since we shall want the velocity components to vanish for large values of z, the desired solution will be

$$w = Ae^{-(1+i)\alpha z} \tag{4.102}$$

We shall assume that the steady-state drift current is maintained by a wind directed in the positive Y direction at the sea surface. Such a wind will produce a tangential stress T in the ocean at the sea surface given by

$$T = -\mu_z \frac{\partial v_y}{\partial z}$$
$$\quad (z = 0) \tag{4.103}$$
$$0 = -\mu_z \frac{\partial v_x}{\partial z}$$

In terms of w, this boundary condition becomes

$$-iT = \mu_z \frac{dw}{dz} \quad (z = 0) \tag{4.104}$$

Substituting Eq. (4.104) into Eq. (4.102), we obtain

$$A = \frac{iT}{(1+i)\alpha\mu_z} = \frac{(1+i)T}{2\alpha\mu_z} \tag{4.105}$$

or

$$w = \frac{(1+i)T}{2\alpha\mu_z} e^{-(1+i)\alpha z} \tag{4.106}$$

Separating into real and imaginary parts, we then have

$$v_x = \frac{T}{2\alpha\mu_z} e^{-\alpha z} (\cos \alpha z + \sin \alpha z)$$
$$= \frac{T}{\sqrt{2}\,\alpha\mu_z} e^{-\alpha z} \cos\left(\frac{\pi}{4} - \alpha z\right)$$
$$= v_0 e^{-\alpha z} \cos\left(\frac{\pi}{4} - \alpha z\right) \tag{4.107}$$

and

$$v_y = \frac{T}{2\alpha\mu_z} e^{-\alpha z} (\cos \alpha z - \sin \alpha z)$$
$$= \frac{T}{\sqrt{2}\,\alpha\mu_z} e^{-\alpha z} \sin\left(\frac{\pi}{4} - \alpha z\right)$$
$$= v_0 e^{-\alpha z} \sin\left(\frac{\pi}{4} - \alpha z\right) \tag{4.108}$$

where v_0 is given by

$$v_0 = \frac{T}{\sqrt{2}\,\alpha\mu_z} = \frac{T}{(2\rho\mu_z\omega\sin\varphi)^{1/2}} \tag{4.109}$$

We see then that at the sea surface, the velocity of a drift current has a magnitude v_0 and is directed at 45° to the right of the wind direction in the northern hemisphere. With increasing depth, the angle of deflection increases, and the magniude of the velocity decreases. We may show this diagrammatically in Fig. 4.16, which, when drawn to scale, is known as the *Ekman spiral*.

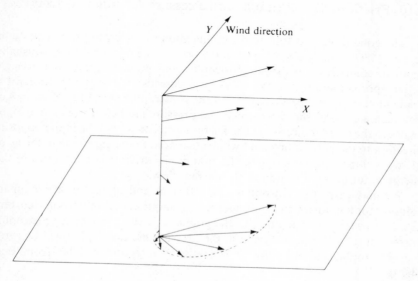

Fig. 4.16

It is sometimes useful to write the results [Eqs. (4.107) and (4.108)] in terms of the parameter D rather than α, where D is defined by

$$D = \frac{\pi}{\alpha} = \pi\left(\frac{\mu_z}{\rho\omega\sin\varphi}\right)^{1/2} \tag{4.110}$$

The results [Eqs. (4.107) and (4.108)] become

$$v_x = v_0 e^{-\pi z/D}\cos\left(\frac{\pi}{4} - \frac{\pi z}{D}\right) \tag{4.111}$$

and

$$v_y = v_0 e^{-\pi z/D}\sin\left(\frac{\pi}{4} - \frac{\pi z}{D}\right) \tag{4.112}$$

At the depth, $z = D$, the current is in a direction π to the current at the surface and of a magnitude $e^{-\pi} = \frac{1}{23}$ of its surface value. The quantity D, then, is a convenient index of the effective depth of a drift current and is referred to as the *frictional depth*.

Problem 4.9(a) Show that the total mass transport per unit column of a drift current is to the right of the wind direction in the northern hemisphere and is given by $T/2\omega \sin \varphi$ per unit area of a vertical column.

4.10 Friction, Geostrophic, and Pressure Gradient Effects

The conditions that lead to a simplification of the steady-state equations to Eqs. (4.97) usually do not exist in the ocean. The frictional force is usually small in comparison with the geostrophic and pressure gradient forces. A better approximation in many cases is to consider the frictional force of a gradient current as a perturbation on the geostrophic current of Section 4.6. For such a current, the geostrophic and pressure gradient forces are at right angles to the current direction. The frictional resistance will be in the opposite direction to the current flow and will be balanced in the aggregate in terms of the work done in overcoming the friction by the driving force in terms of the energy input in the direction of the current flow.

We may carry this reasoning a step further and obtain an approximate solution for a gradient current under the assumption of a geostrophic current as a first-order approximation to the gradient current. Taking a geostrophic current in the positive X direction, the pressure gradient $\partial p/\partial x$ will be zero; the geostrophic current velocity U in the X direction will be, from Eqs. (4.70),

$$U = \frac{-1}{2\rho\omega \sin \varphi} \frac{\partial p}{\partial y} \qquad (4.113)$$

which will be assumed to be a constant. Substituting Eq. (4.113) into the equations of motion (4.96), we obtain

$$\frac{\partial^2 v_x}{\partial z^2} = -\frac{2\rho\omega \sin \varphi}{\mu_z} v_y$$

$$\frac{\partial^2 v_y}{\partial z^2} = \frac{2\rho\omega \sin \varphi}{\mu_z} (v_x - U) \qquad (4.114)$$

Making the same change of variable to w of Eq. (4.98), we obtain

$$\frac{d^2 w}{dz^2} = \frac{2i\rho\omega \sin \varphi}{\mu_z} (w - U) \qquad (4.115)$$

whose solution is simply

$$w - U = Ae^{(1+i)\alpha z} + Be^{-(1+i)\alpha z} \tag{4.116}$$

where α is as given before by Eq. (4.101).

For this example, we shall take boundary conditions that there is no wind stress at the ocean surface, or

$$\frac{\partial v_x}{\partial z} = \frac{\partial v_y}{\partial z} = \frac{dw}{dz} = 0 \quad (z = 0) \tag{4.117}$$

and that in consideration of the internal frictional effects, there is no motion at the bottom, or

$$v_x = v_y = w = 0 \quad (z = d) \tag{4.118}$$

Substituting Eq. (4.116) into the first boundary condition, we obtain

$$A = B$$

and into the second boundary condition,

$$A = \frac{-U}{e^{(1+i)\alpha d} + e^{-(1+i)\alpha d}}$$

from which the final solution will be

$$w = U\left(1 - \frac{e^{(1+i)\alpha z} + e^{-(1+i)\alpha z}}{e^{(1+i)\alpha d} + e^{-(1+i)\alpha d}}\right) \tag{4.119}$$

Separating into real and imaginary parts, and after some reduction, we have for v_x and v_y,

$$v_x = (1 - \varphi)U \tag{4.120}$$

and

$$v_y = \psi U \tag{4.121}$$

where

$$\varphi = \frac{\cosh\frac{\pi}{D}(d+z)\cos\frac{\pi}{D}(d-z) + \cosh\frac{\pi}{D}(d-z)\cos\frac{\pi}{D}(d+z)}{\cosh\frac{2\pi}{D}d + \cos\frac{2\pi}{D}d} \tag{4.122}$$

and where ψ is the same as φ with cosh and cos replaced by sinh and sin in the numerator and where D is the same frictional depth as defined by Eq. (4.110).

From Eq. (4.122) we see that for values of d substantially greater than D that φ and ψ will have appreciable values only in the range from $z = d$ to $z = d - D$. At distances from the bottom greater than D, there is practically

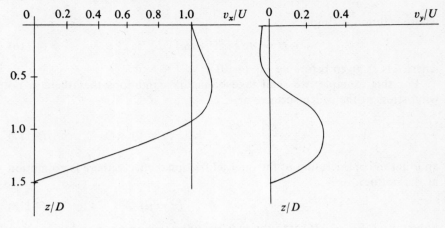

Fig. 4.17

a uniform velocity at right angles to the pressure gradient. This result, of course, is entirely to be expected on the basis of the assumptions made in the solution. For a value of $d/D = 1.5$, the plots of v_x and v_y versus depth are as shown in Fig. 4.17. In terms of an Ekman spiral diagram similar to Fig. 4.16, the current velocity will remain essentially constant in magnitude and direction down to a depth $z = d - D$ and then turn around in a counterclockwise direction, in the northern hemisphere, with decreasing amplitude as the bottom is approached.

In a general way, we may think of the current system in a homogeneous ocean as comprised of three parts, as illustrated in Fig. 4.18. The *deep current* is the geostrophic portion of the gradient current. The *bottom current* is the portion of the gradient current affected by internal friction under the assumption of no motion at the bottom. The *surface current* is the resultant of the deep current and the drift current generated by the wind. If the geostrophic current velocity of the deep current and the net wind stress are in the same direction, the vector relations of the current will be as shown diagram-

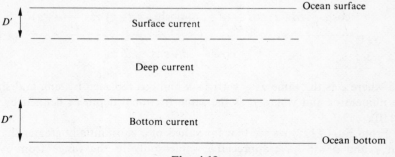

Fig. 4.18

Surface current   Deep current   Bottom current

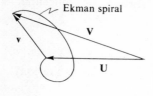
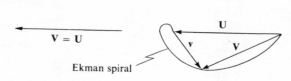

Fig. 4.19

matically in Fig. 4.19. If, in addition, we have the condition that the net mass transport is in the direction of the deep current, the mass transport of the surface current to the right of the deep current will be balanced by the mass transport of the bottom current to the left of the deep current.

Problem 4.10(a)   Carry through the reduction of Eq. (4.119) to Eqs. (4.120) and (4.121).

Problem 4.10(b)   Show that the combined mass transport per unit column in the positive X direction, east, for an eastward-moving geostrophic current extending down to a depth $z = d$, at which the current falls to zero and a surface drift current is

$$M_x = \frac{1}{2\omega \sin \varphi} \left(T_y - \frac{\partial P}{\partial y} \right)$$

where T_y is the wind stress in the Y direction, and P is given by the integral

$$P = \int_0^d p \, dz$$

4.11 Wind-driven Ocean Circulation

The wind system over the oceans is generally considered to be the driving force for the major surface and near-surface ocean currents. It is of interest to derive some simple and approximate relations between the wind stress field and the oceanographically measurable quantities of current mass transport and pressure gradients.

For steady-state conditions, we may write Eqs. (4.96) as

$$2\rho v_y \omega \sin \varphi - \frac{\partial p}{\partial x} + \frac{\partial}{\partial z}\left(\mu_z \frac{\partial v_x}{\partial z}\right) = 0$$

$$-2\rho v_x \omega \sin \varphi - \frac{\partial p}{\partial y} + \frac{\partial}{\partial z}\left(\mu_z \frac{\partial v_y}{\partial z}\right) = 0 \tag{4.123}$$

We shall want to integrate these equations from the surface down to a depth d at which the current falls to zero. At the surface, the tangential stress components due to the wind are given by

$$T_x = -\mu_z \frac{\partial v_x}{\partial z}$$
$$T_y = -\mu_z \frac{\partial v_y}{\partial z} \qquad (z = 0) \qquad (4.124)$$

similar to Eqs. (4.103), and at the depth d, these stress components vanish. Integrating Eqs. (4.123), we thus obtain

$$2\omega M_y \sin \varphi - \frac{\partial P}{\partial x} + T_x = 0$$
$$-2\omega M_x \sin \varphi - \frac{\partial P}{\partial y} + T_y = 0 \qquad (4.125)$$

where M_x and M_y are the mass transports per unit column of the current in the X and Y directions, respectively, given by

$$M_x = \int_0^d \rho v_x \, dz$$
$$M_y = \int_0^d \rho v_y \, dz \qquad (4.126)$$

and where P is given by

$$P = \int_0^d p \, dz \qquad (4.127)$$

We also have, from the equation of continuity (3.14) under steady-state conditions, the relation

$$\frac{\partial M_x}{\partial x} + \frac{\partial M_y}{\partial y} = 0 \qquad (4.128)$$

We have three equations, Eqs. (4.125) and (4.128), for the three oceanographic variables M_x, M_y, P in terms of the surface wind stress components T_x, T_y.

Let us apply these equations to the equatorial currents of the Pacific Ocean. We may eliminate P by differentiating the first of Eqs. (4.125) with respect to y and the second of these equations with respect to x and subtracting. Remembering that the latitude φ is a function of y only and using Eq. (4.128), we obtain

$$\frac{\partial T_x}{\partial y} - \frac{\partial T_y}{\partial x} + \frac{2\omega}{\lambda} M_y \cos \varphi = 0 \qquad (4.129)$$

where λ is a constant given by

$$\lambda = \frac{dy}{d\varphi} \tag{4.130}$$

the variation of the north coordinate distance with respect to latitude. For the trade wind belt of the equatorial currents, it is possible to put $\partial T_y/\partial x = 0$ reducing Eq. (4.129) to

$$M_y = \frac{-\lambda}{2\omega \cos \varphi} \frac{\partial T_x}{\partial y} \tag{4.131}$$

From Eqs. (4.128) and (4.131) we also have

$$\frac{\partial M_x}{\partial x} = \frac{1}{2\omega \cos \varphi} \left(\frac{\partial T_x}{\partial y} \tan \varphi + \lambda \frac{\partial^2 T_x}{\partial y^2} \right) \tag{4.132}$$

If we consider some distance $x = x_0$ at the eastern end of the equatorial currents where the currents themselves are negligible so that we have $M_x = 0$, we may integrate Eq. (4.132) with respect to x, obtaining approximately

$$M_x = \frac{x_1 - x_0}{2\omega \cos \varphi} \left(\frac{\partial T_x}{\partial y} \tan \varphi + \lambda \frac{\partial^2 T_x}{\partial y^2} \right) \tag{4.133}$$

From Eqs. (4.125) and (4.131) we have

$$\frac{\overline{\partial P}}{\partial x} = \overline{T}_x - \frac{\overline{\partial T_x}}{\partial y} \lambda \tan \varphi \tag{4.134}$$

Equations (4.133) and (4.134) are relations for M_x and $\partial P/\partial x$ as a function of latitude, or y coordinate distance, in terms of the wind stress field.

Using the average wind stress distribution obtained from the wind field data given in climatological charts, Eq. (4.133) correctly predicts the westward-flowing north and south equatorial currents and the eastward-flowing counter equatorial current, located between the two westward-flowing currents. The theoretical calculations of Eqs. (4.133) and (4.134) agree well with the experimentally observed oceanographic quantities of $(\overline{\partial P/\partial x})$ and M_x, where M_x is calculated from the result of Problem 4.10(b).

As a final example, let us look, in a general way, at the effect of the variation of the Coriolis parameter, $2\omega \sin \varphi$, with latitude on large-scale ocean circulation. We shall want to include the effects of internal friction transverse and longitudinal to the current. Including these terms from Section 4.8 in Eq. (4.123), we shall obtain similar to Eqs. (4.125)

$$\begin{aligned} 2\omega M_y \sin \varphi - \frac{\partial P}{\partial x} + T_x + H_x = 0 \\ -2\omega M_x \sin \varphi - \frac{\partial P}{\partial y} + T_y + H_y = 0 \end{aligned} \tag{4.135}$$

where H_x and H_y are the integrals of the horizontal friction terms. Equation (4.129) will then become

$$\left(\frac{\partial T_y}{\partial x} - \frac{\partial T_x}{\partial y}\right) - \beta M_y + \left(\frac{\partial H_y}{\partial x} - \frac{\partial H_x}{\partial y}\right) = 0 \qquad (4.136)$$

where β is simply

$$\beta = \frac{2\omega}{\lambda} \cos \varphi \qquad (4.137)$$

Since the wind stress vector **T** has components T_x, T_y only and is not a function of z, we see that the first term of Eq. (4.136) is the scalar magnitude of the curl of the vector **T**, or

$$\boldsymbol{\tau} = \nabla \times \mathbf{T} = \mathbf{k}\left(\frac{\partial T_y}{\partial x} - \frac{\partial T_x}{\partial y}\right) \qquad (4.138)$$

The vector $\boldsymbol{\tau}$ is referred to as the *vortex* vector of the wind stress. A similar relation holds for the frictional vector **H**. Equation (4.136) is referred to as a *vorticity* equation and the second term as the *planetary vorticity*.

Let us now consider a large ocean area, such as the North Atlantic. We shall consider a clockwise anticyclonic wind system over the whole ocean of constant vorticity. We see immediately that we cannot have a uniform current vortex with horizontal frictional resistance **H** unless the second term of Eq. (4.136) is zero, which is not the case. For a right-handed co-ordinate system with the X axis to the east and the Y axis north, the wind stress vorticity is a negative number and the frictional resistance vorticity a positive number. On the western side of the ocean, the transport will be to the north with βM_y positive, and on the eastern side of the ocean, to the south with βM_y negative. Designating the wind, resistance, and planetary vorticities by the numerically positive quantities a, b, c, respectively, we shall have from Eq. (4.136), for the western side of the ocean,

$$c = b - a$$

and for the eastern side of the ocean

$$c = a - b$$

Since we have assumed a to be constant, we see that if these relations are to hold, b must be larger than a on the western side; or the current, and consequent frictional resistance effect, must be intensified on the western side of the ocean with respect to the eastern side. These general considerations correctly predict the westward intensification to the formation of the gulf stream in the North Atlantic and the Kuroshio current in the North Pacific.

References

Arons, A. B., and H. Stommel. 1951. "A mixing-length theory of tidal flushing," *Am. Geophys. Un., Trans.*, vol. 32, pp. 419–421.
Defant, A. 1961. *Physical Oceanography*, vol. 1. New York: Pergamon.
Eckart, C. 1960. *Hydrodynamics of Oceans and Atmospheres.* New York: Pergamon.
Hill, M. N., ed. 1962. *The Sea*, vol. 1, *Physical Oceanography.* New York: Wiley.
Neumann, G. 1968. *Ocean Currents.* Amsterdam: Elsevier.
Proudman, J. 1953. *Dynamical Oceanography.* New York: Wiley.
Sverdrup, H. U., M. W. Johnson, and R. H. Fleming. 1942. *The Oceans.* Englewood Cliffs, N.J.: Prentice-Hall.

CHAPTER 5

PHYSICAL OCEANOGRAPHY—WAVES AND TIDES

5.1 Tidal Waves

In this chapter, we are concerned principally with the motion of the free surface of the ocean and with one of the main forces causing such motion —that of the gravitational attraction of the moon and the sun. It is convenient theoretically to divide the discussion of the motion of the ocean surface into two parts: that for which the wavelength of the motion is large compared with the ocean depth and that for which it is not. The former motion is variously referred to as *tidal* or *long* waves and the latter as *surface*, *gravity*, or *short* waves.

The description of tidal waves, as the name implies, is applicable to the wave motion produced by the tide generating forces of the gravitational attraction of the moon and the sun. The basic assumption in the theory of tidal waves is that the wavelengths are so long in comparison with the ocean depth that the water particle motion is mainly horizontal and is essentially the same for all particles in a given vertical plane. The vertical accelerations may then be neglected, and the pressure at any depth taken to be the hydrostatic pressure. We shall assume that the fluid is incompressible and for the first derivation that the particle motion is sufficiently small that the time derivative D/Dt may be replaced by $\partial/\partial t$.

Let us consider one-dimensional wave motion with the X axis in the direction of wave motion, the Z axis vertical upward, and the origin at the bottom of the ocean as shown in Fig. 5.1. Let the displacements of the ocean surface away from its horizontal, neutral position be denoted by ξ, η, respectively, in the X and Z directions, and let the depth of the ocean be h. Then the hydrostatic pressure at any depth will be directly proportional to the overlying column of water, and the horizontal pressure gradient will be given by

$$\frac{\partial p}{\partial x} = \rho g \frac{\partial \eta}{\partial x} \tag{5.1}$$

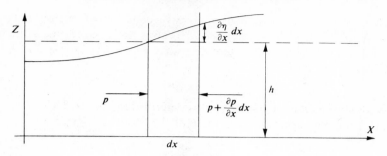

Fig. 5.1

The horizontal force per unit volume in the positive X direction will be the negative of Eq. (5.1) so that the equation of motion in the X direction will be simply

$$\rho \frac{\partial^2 \xi}{\partial t^2} = -\rho g \frac{\partial \eta}{\partial x} \tag{5.2}$$

The continuity conditions may be obtained by considering the net volume of water that has entered the column shown in Fig. 5.2 in a time t. The net displacements ξ and η in the X and Z directions will be the integrals of their respective velocity components so that we shall have, for continuity of an incompressible fluid,

$$\eta = -h \frac{\partial \xi}{\partial x} \tag{5.3}$$

Combining Eqs. (5.2) and (5.3), we obtain

$$\frac{\partial^2 \xi}{\partial t^2} = gh \frac{\partial^2 \xi}{\partial x^2} \tag{5.4}$$

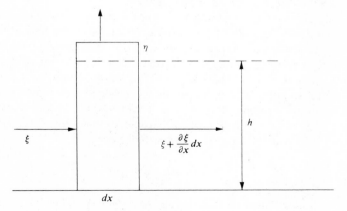

Fig. 5.2

which is simply the one-dimensional wave equation of section 1.10. The elimination of ξ from Eqs. (5.2) and (5.3) gives, as expected, the same wave equation for η. We see then that the wave motion is propagated with a velocity c, given by

$$c = \sqrt{gh} \tag{5.5}$$

and with all the other characteristics as described in Section 1.10. If we were to take a simple harmonic wave for ξ of the form

$$\xi = A \cos \frac{2\pi}{\lambda}(x-ct) \tag{5.6}$$

we would have from Eq. (5.3) for η,

$$\eta = \frac{2\pi h}{\lambda} A \sin \frac{2\pi}{\lambda}(x-ct) \tag{5.7}$$

showing that when h is much smaller than λ, the displacement, velocity, and acceleration in the vertical direction are small compared with those in the horizontal direction. In general, the solution to the wave equation (5.4) for a wave propagating in the positive X direction will be of the form

$$\xi = F(x-ct) \tag{5.8}$$

From Eqs. (5.8) and (5.3) we then have, for the ratio of the water particle velocity $\dot{\xi}$ to the wave velocity c,

$$\frac{\dot{\xi}}{c} = \frac{\eta}{h} \tag{5.9}$$

The ratio $\dot{\xi}$ to c is the same as the ratio of the wave height η to the water depth h. At a wave crest, the horizontal particle velocity is in the direction of wave propagation; at a wave trough, the horizontal particle velocity is opposite to the direction of wave propagation.

We may arrive at these same conclusions with regard to the value of the wave velocity and the ratio of the particle velocity to wave velocity through an interesting alternative approach, which we shall refer to as the *Rayleigh method*. Let us consider a tidal wave propagating in the negative X direction and impose on it a mass movement of the whole ocean with a constant velocity equal and opposite to that of the wave velocity, as shown in Fig. 5.3. Then the motion becomes steady. The ocean surface becomes stationary, while the forces acting on the water particles remain the same as before. Equation (3.37) will then apply, and we shall have

$$\frac{p}{\rho} = D - g(h+\eta) - \tfrac{1}{2}v^2 \tag{5.10}$$

where D is a constant, and v is the particle velocity, taken as before to be principally in the X direction. Our continuity condition now becomes that

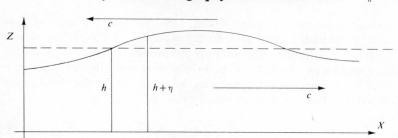

Fig. 5.3

the volume transport of water $v(h+\eta)$ through any vertical section of height $(h+\eta)$ must be a constant. Considering the wave motion height gradually reduced to zero, the volume transport through a section of vertical height h will be ch. The equation of continuity then becomes

$$v(h+\eta) = ch \tag{5.11}$$

Substituting Eq. (5.11) into Eq. (5.10), we obtain

$$\frac{1}{\rho}p(\eta) = D - g(h+\eta) - \tfrac{1}{2}\frac{c^2 h^2}{(h+\eta)^2} \tag{5.12}$$

If the ocean surface is to remain stationary, the pressure along this surface must be a constant. This will be achieved if $p'(\eta)$ is zero. From Eq. (5.12) we shall then have

$$\frac{1}{\rho}p'(\eta) = -g + \frac{c^2 h^2}{(h+\eta)^3} = 0$$

or

$$c^2 = gh\left(1 + \frac{\eta}{h}\right)^3 \tag{5.13}$$

To a first-order approximation for infinitesimal values of η, this will be simply

$$c = c_0 = \sqrt{gh} \tag{5.14}$$

the same as Eq. (5.5). To a second-order approximation for small values of η, this will be

$$c = c_0\left(1 + \frac{3}{2}\frac{\eta}{h}\right) \tag{5.15}$$

In this second approximation, we see that the wave velocity is not a constant. A wave of this type, with finite amplitudes of wave height, cannot be propagated without a change in profile since the wave velocity is a function of the wave height. From Eq. (5.11) we also have a first approximation that

$$v = c\left(1 - \frac{\eta}{h}\right) \tag{5.16}$$

Subtracting the imposed mass velocity c, we see that the ratio of particle velocity in the undisturbed state to the wave velocity is the same as that given by Eq. (5.9) in the direction of propagation.

We should now like to examine briefly, in a more formal manner, tidal waves of finite amplitude. In this case, we cannot make the substitution of $\partial/\partial t$ for D/Dt. The equation of motion then is

$$\frac{\partial v_x}{\partial t} + v_x \frac{\partial v_x}{\partial x} = -g \frac{\partial \eta}{\partial x} \tag{5.17}$$

where the symbol v_x is used for the particle velocity in the X direction instead of $\dot{\xi}$. To this next order of approximation, we see from Fig. 5.4 that equating

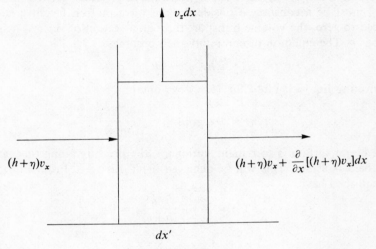

Fig. 5.4

the rate of mass transport into and out of the vertical column for an incompressible fluid gives, for the continuity relation,

$$\frac{\partial}{\partial x}\left[(h+\eta)v_x\right] = -v_z = -\frac{\partial \eta}{\partial t} \tag{5.18}$$

This may be rewritten as

$$\frac{\partial \eta}{\partial t} + v_x \frac{\partial \eta}{\partial x} = -(h+\eta)\frac{\partial v_x}{\partial x} \tag{5.19}$$

Let us now make a change of the independent variable from η to ψ, where ψ is defined by

$$\psi(\eta) = 2c_0\left[\left(1+\frac{\eta}{h}\right)^{1/2} - 1\right] \tag{5.20}$$

c_0 being given by Eq. (5.14). We shall then have, for the partial derivative of ψ with respect to x in terms of the partial derivative of η with respect to x,

$$\frac{\partial \psi}{\partial x} = \frac{\partial \eta}{\partial x} \frac{d\psi}{d\eta} = \left(\frac{g}{h+\eta}\right)^{1/2} \frac{\partial \eta}{\partial x} \tag{5.21}$$

Substituting Eq. (5.21) and the similar relation with respect to t into Eqs. (5.17) and (5.19), we obtain

$$\frac{\partial v_x}{\partial t} + v_x \frac{\partial v_x}{\partial x} = -a \frac{\partial \psi}{\partial x} \tag{5.22}$$

and

$$\frac{\partial \psi}{\partial t} + v_x \frac{\partial \psi}{\partial x} = -a \frac{\partial v_x}{\partial x} \tag{5.23}$$

where a is given by

$$a = [g(h+\eta)]^{1/2} = c_0 \left(1 + \frac{\eta}{h}\right)^{1/2} \tag{5.24}$$

Adding and subtracting Eq. (5.22) from Eq. (5.23), we get

$$\left[\frac{\partial}{\partial t} + (v_x + a) \frac{\partial}{\partial x}\right](\psi + v_x) = 0 \tag{5.25}$$

and

$$\left[\frac{\partial}{\partial t} + (v_x - a) \frac{\partial}{\partial x}\right](\psi - v_x) = 0 \tag{5.26}$$

The differential in the brackets of Eqs. (5.25) and (5.26) is in the form of the total time differential D/Dt of Eq. (3.3), following the motion of the fluid. Thus, Eq. (5.25) states that the quantity $(\psi + v_x)$ is constant for a point moving with the velocity $(v_x + a)$. Similarly, Eq. (5.26) states that the quantity $(\psi - v_x)$ is constant for a point moving with the velocity $(v_x - a)$.

These equations enable us to understand, in a general way, the nature of the motion. Consider an initial disturbance confined between the planes $x = x_1$ and $x = x_2$ so that for $x < x_1$ and $x > x_2$, both ψ and v_x are zero. The region within which $(\psi + v_x)$ is variable will advance in the positive direction, and the quantity $(\psi - v_x)$ will recede in the opposite direction. After a time, these regions will separate, leaving between a region for which both quantities are zero so that $\psi = 0$ and $\eta = 0$, or $v_x = 0$, that is, the fluid at rest and at its normal elevation. The original disturbance has been split into two progressive waves travelling in opposite directions. In the advancing wave, $\psi - v_x = 0$ so that $\psi = v_x$, and we have from Eq. (5.25) that the disturbance is propagated with the velocity

$$c = v_x + a = \psi + a = c_0 \left[3\left(1 + \frac{\eta}{h}\right)^{1/2} - 2\right] \tag{5.27}$$

where we have substituted for ψ and a from Eqs. (5.20) and (5.24). In the receding wave, we have $\psi + v_x = 0$ so that $\psi = -v_x$, and we obtain the same result from Eq. (5.26) for a wave propagating in the negative x direction with the velocity

$$c = v_x - a = -(\psi + a) \tag{5.28}$$

The first-order approximation for Eq. (5.27) gives the same result, [Eq. (5.15)], as we had obtained previously. Since the wave velocity increases with the vertical displacement, it appears that in a progressive wave, the slopes will become continually steeper in front and more gradual behind until a state is reached in which we are no longer justified in neglecting vertical accelerations.

Problem 5.1(a)   Obtain Eq. (5.2) directly from the hydrodynamic equations of motion.

Problem 5.1(b)   Obtain Eq. (5.3) directly from the hydrodynamic equation of continuity.

Problem 5.1(c)   Derive the expressions for ξ and η in simple harmonic form as a function of depth. Show that the water particle motion is elliptical with the vertical displacement decreasing with depth so that the motion along the bottom is linear horizontally.

Problem 5.1(d)   Using the concepts of wave fronts and rays, discuss the refraction of tidal waves propagating into a shoaling region.

5.2 Driven Tidal Waves

We shall now consider some of the characteristics of driven tidal wave motion. To do this, we shall want to include in the equation of motion an external force per unit mass X. In particular, we shall choose X to be of a form $f(x - Ut)$ corresponding to the gravitational attractive, tide-producing force of the moon. To a first approximation for an equatorial canal, the velocity U will be simply the circumference of the earth divided by one lunar day. In this discussion, we shall become aware of the distinction, common in the discussion of the motion of many mechanical and electrical systems, between *free* waves or oscillations, in which the propagation velocity is that determined by the system itself, c of Eq. (5.5), and *forced* waves or oscillations, in which the propagation velocity is that of the driving force U.

From Eqs. (5.2), (5.3), and (5.5) we shall have for the equation of motion

$$\frac{\partial^2 \xi}{\partial t^2} = c^2 \frac{\partial^2 \xi}{\partial x^2} + X \tag{5.29}$$

or from the hydrodynamic equations of motion

$$\frac{\partial^2 \xi}{\partial t^2} = c^2 \frac{\partial^2 \xi}{\partial x^2} - \frac{1}{\rho}\frac{\partial p}{\partial x} \tag{5.30}$$

where p is here the pressure on the ocean surface of the driving force X. Let us consider our solution ξ to be composed of two parts given as

$$\xi = \xi_1 + \xi_2 \tag{5.31}$$

where ξ_1 is a general solution of the differential equation (5.4) of the form of Eq. (5.8), consisting of two arbitrary functions representing waves propagating in the positive and negative x directions with the velocity c, and where ξ_2 is a particular solution of Eq. (5.29). Adding the two differential equations for ξ_1 and ξ_2 will give Eq. (5.29) for ξ. We may then be sure that such a solution is a complete solution and may now proceed to look for a particular solution of ξ_2 satisfying Eq. (5.29). The solution ξ_1 represents the free waves, and the solution ξ_2 represents the forced waves.

Anticipating the resultant form of the tide-producing force, we shall take X to be given by a simple harmonic function

$$X = -A \sin [k(x - Ut)] \tag{5.32}$$

We shall look for a particular solution for ξ_2 corresponding to this of the form

$$\xi_2 = a \sin [k(x - Ut)] \tag{5.33}$$

Substituting Eqs. (5.32) and (5.33) into Eq. (5.29), we obtain

$$ak^2 U^2 = ak^2 c^2 + A$$

or

$$a = \frac{A}{k^2(U^2 - c^2)}$$

so that

$$\xi_2 = \frac{A}{k^2(U^2 - c^2)} \sin [k(x - Ut)] \tag{5.34}$$

From Eqs. (5.34) and (5.3) we then have for the surface elevation

$$\frac{\eta_2}{h} = \frac{A}{k(c^2 - U^2)} \cos [k(x - Ut)]$$

and from Eqs. (5.29), (5.30), and (5.32) for the surface pressure

$$\frac{p}{\rho} = -\frac{A}{k} \cos [k(x - Ut)]$$

so that

$$-\frac{\eta_2}{h} = \frac{p}{\rho(c^2 - U^2)} \tag{5.35}$$

We see then that for a positive surface pressure of the driving force, the forced wave will have a depression, negative η_2, if $c > U$. The driving surface pressure and the resultant surface motion are said to be *in phase*. However, if $c < U$, then a positive surface pressure corresponds to a surface elevation, and the two are said to be *out of phase*.

For the tidal motion produced by the moon's gravitational attraction, the tides are said to be *direct* when high tide is in phase with the moon, and *inverted* when low tide is in phase with the moon. For the actual relative motion of the moon about the earth and ordinary ocean depths, c will be small compared with U, so that the normal tides will be inverted for an equatorial canal.

Added to this particular solution of the equation of motion, which represents the forced waves due to the attraction of the moon, we may also have solutions representing free waves such as were discussed in the preceding section. If we had included a frictional resistance, damping, term in the equation of motion, we would have found that the free wave solution contained a time-dependent damping factor. They would represent an initial, transient effect which, in comparison with the steady-state motion produced by the moon attraction, could be neglected.

5.3 Seiches

One of the simpler examples of free tidal wave motion is that of seiches. *Seiches* are oscillations of a body of water, such as a lake, that depend only on the dimensions of that body of water and that, when once set in motion by some external force, will continue for some time. The external forces causing such oscillations are usually either a change in atmospheric pressure or the relaxation resulting from a piling of water mass at one end of the body of water caused by the wind. The force is a transient phenomenon; the oscillations are a free wave, resonant motion.

Let us consider the one-dimensional oscillations of a rectangular body of water closed at both ends. The boundary conditions are that $\xi = 0$ at $x = 0$ and at $x = l$. This problem is exactly the same as the vibrating string problem of Section 7.1. We may then take a solution of the wave equation (5.4) in the form of Eq. (7.3),

$$\xi = [A \sin(kx) + B \cos(kx)] \cos(kct) \tag{5.36}$$

where c is given by Eq. (5.5). The first boundary condition gives $B = 0$, and the second gives the relation

$$kl = n\pi$$

or

$$P = \frac{2l}{nc} = \frac{2l}{n\sqrt{gh}} \qquad (5.37)$$

where n is an integer corresponding to the first, second, and so on, modes of oscillation, and P is the period. The expression Eq. (5.37) is sometimes referred to as *Merian's formula* in physical oceanography. Since the vertical displacement η is given by Eq. (5.3) in terms of ξ, we see that whereas each end of the body of water is a node for ξ, it is an antinode, or maximum displacement, for η. For the fundamental, or first, mode, the nodal line for η will be at $x = l/2$.

We may also have resonant oscillations for a body of water open at one end, such as a bay. In this case, the open end is approximated as a free boundary, or a maximum for ξ. The secondary boundary condition then gives

$$kl = (2n-1)\frac{\pi}{2}$$

or

$$P = \frac{4l}{(2n-1)c} = \frac{4l}{(2n-1)\sqrt{gh}} \qquad (5.38)$$

The open end of the body of water is here a nodal line for η.

Let us consider next the related problem of forced oscillation in a rectangular body of water open at one end, such as a bay or gulf. In this case, the generating force is that of the tide. The forced wave motion within the bay or the gulf is referred to as a *tidal cooscillation*. This problem is the analogy of the vibrating string driven at one end. We may solve it simply by taking as our boundary condition at the open end $x = l$ the tide amplitude

$$\eta = a \cos(\omega t) = a \cos\left(\frac{2\pi t}{T}\right) \qquad (5.39)$$

where T is the period of the tide. From Eqs. (5.36) and (5.3) we shall have for η in the bay

$$\eta = [-khA \cos(kx) + khB \sin(kx)] \cos(kct) \qquad (5.40)$$

The boundary condition at $x = 0$ gives $B = 0$ as before. The boundary condition at $x = l$ gives

$$-khA \cos(kl) \cos(kct) = a \cos\left(\frac{2\pi t}{T}\right) \qquad (5.41)$$

For this to hold for all values of time, we see that k must be given by

$$k = \frac{2\pi}{cT}$$

154 ¶ *Thermodynamics and Hydrodynamics*

as expected, from which we then get

$$A = \frac{-a}{kh \cos(kl)}$$

or

$$\eta = a \frac{\cos(kx)}{\cos(kl)} \cos\left(\frac{2\pi t}{T}\right) \tag{5.42}$$

We see that there will be large amplitudes, or resonance, in the bay under the conditions for which the denominator of Eq. (5.42) approaches zero. This gives

$$kl = (2n-1)\frac{\pi}{2} \tag{5.41}$$

or from Eq. (5.41)

$$l = \frac{(2n-1)cT}{4} \tag{5.43}$$

We see, as expected, that resonance occurs when the length of the bay is equal to one quarter the natural wavelength cT. For $l < \frac{1}{4}cT$, the vertical displacements in the bay will all be in phase. For $l > \frac{1}{4}cT$, there will be a nodal line in the bay.

Problem 5.3(a) For Loch Earn in Scotland, the average length is 10 km and depth 60 m. Calculate the fundamental period of the seiches.

Ans. 14 min

Problem 5.3(b) For Lake Baikal in Siberia, the average length is 665 km and depth 680 m. Calculate the fundamental period of the seiches.

Ans. 4.5 hr

Problem 5.3(c) The tides in the Bay of Fundy exhibit resonance. The length of the bay is 270 km and average depth 75 m. Calculate the ratio of the vertical displacement amplitude at the head of the bay to that at the bay entrance.

Ans. 6.4

5.4 Geostrophic Effects on Tidal Waves

Let us now look at the effect of the Coriolis force on tidal wave motion. From Eqs. (4.63) and (5.1) the two-dimensional equations of motion including the geostrophic term will be

$$\frac{\partial v_x}{\partial t} - 2\omega_0 v_y \sin \varphi = -g \frac{\partial \eta}{\partial x}$$

$$\frac{\partial v_y}{\partial t} + 2\omega_0 v_x \sin \varphi = -g \frac{\partial \eta}{\partial y} \tag{5.44}$$

where, as before, we have assumed that the motion is sufficiently small that the differential D/Dt may be replaced by $\partial/\partial t$ and where, for convenience, we have used ω_0 for the angular frequency of the earth's rotation to distinguish it from ω representing the angular frequency of the wave motion. The two-dimensional equation of continuity in terms of the velocity components, corresponding to Eq. (5.3), will be

$$\frac{\partial \eta}{\partial t} = -h \left(\frac{\partial v_x}{\partial x} + \frac{\partial v_y}{\partial y} \right) \tag{5.45}$$

For the case of simple harmonic wave motion, the time factor being given by $e^{-i\omega t}$, the vertical displacement η may be represented in the form

$$\eta(x, y, t) = \eta(x, y) e^{-i\omega t} \tag{5.46}$$

and similarly for the velocity components v_x and v_y. Equations (5.44) and (5.45) then become

$$i\omega v_x + 2\omega_0 v_y \sin \varphi = g \frac{\partial \eta}{\partial x}$$

$$i\omega v_y - 2\omega_0 v_x \sin \varphi = g \frac{\partial \eta}{\partial y} \tag{5.47}$$

and

$$i\omega \eta = h \left(\frac{\partial v_x}{\partial x} + \frac{\partial v_y}{\partial y} \right) \tag{5.48}$$

Solving Eqs. (5.47) for v_x and v_y, we obtain

$$v_x = \frac{-g}{\omega^2 - 4\omega_0^2 \sin^2 \varphi} \left(i\omega \frac{\partial \eta}{\partial x} - 2\omega_0 \sin \varphi \frac{\partial \eta}{\partial y} \right)$$

$$v_y = \frac{-g}{\omega^2 - 4\omega_0^2 \sin^2 \varphi} \left(i\omega \frac{\partial \eta}{\partial y} + 2\omega_0 \sin \varphi \frac{\partial \eta}{\partial x} \right) \tag{5.49}$$

Substituting these values of v_x and v_y into Eq. (5.48), we then obtain the differential equation for η

$$\frac{\partial^2 \eta}{\partial x^2} + \frac{\partial^2 \eta}{\partial y^2} + \frac{\omega^2 - 4\omega_0^2 \sin^2 \varphi}{gh} \eta = 0$$

or
$$\nabla^2 \eta + \frac{\omega^2 - 4\omega_0^2 \sin^2 \varphi}{gh} \eta = 0 \tag{5.50}$$

where ∇^2 is the two-dimensional Laplacian operator.

We shall next consider a few rather simplified applications of these equations to gain some understanding of the geostrophic effect on tidal wave motion. Let us take first the case of wave motion in which the velocity component in the Y direction, v_y, is zero. This may be considered to correspond to the case of an infinitely long, straight canal. Equations (5.47) and (5.48) reduce to

$$i\omega v_x = g \frac{\partial \eta}{\partial x}$$

$$-2\omega_0 v_x \sin \varphi = g \frac{\partial \eta}{\partial y} \tag{5.51}$$

$$i\omega \eta = h \frac{\partial v_x}{\partial x}$$

From the second of these equations, we see that the y dependent term will be in the form of an exponential, so that we can correctly assume a solution of the form

$$\eta = a e^{i(kx - ct) + my} \tag{5.52}$$

Substituting Eq. (5.52) into the first of Eqs. (5.51), we obtain

$$v_x = \frac{g}{c} \eta \tag{5.53}$$

from which we then obtain from the second and third equations

$$m = -\frac{2\omega_0 \sin \varphi}{c} \tag{5.54}$$

and the familiar relation of Eq. (5.5)

$$c^2 = gh \tag{5.55}$$

Our solution for η is then expressed in real form,

$$\eta = a e^{-(2\omega_0 \sin \varphi / c) y} \cos [k(x - ct)] \tag{5.56}$$

The wave velocity is unaffected by the earth's rotation, but the wave height is not everywhere the same in a section normal to the wave motion. The variation in wave amplitude, and consequently from Eq. (5.53) the horizontal particle velocity, is greater to the right of the wave motion than to the left for the northern hemisphere. In a section normal to the direction of propagation, the wave height slopes down from right to left facing the direction of propagation at a wave crest, or high water, and slopes down from left to right at a wave trough, or low water. These results are in accordance with what would be expected for the geostrophic effect from the discussion of Section 4.6. As given by Eq. (5.53) and illustrated in Fig. 5.5, the water particle velocity is in

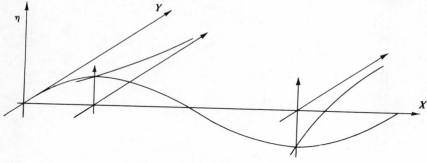

Fig. 5.5

the direction of wave motion at a wave crest and opposite to the direction of wave motion at a wave trough. Such geostrophic wave motions are sometimes referred to in physical oceanography as *Kelvin waves*.

Consider now the interference effect of two simple harmonic Kelvin waves of equal amplitude progressing in opposite directions. From Eq. (5.56) their combined effect may be represented by

$$\eta = ae^{-my} \cos [k(x-ct)] - ae^{my} \cos [k(x+ct)] \qquad (5.57)$$

where we have taken the origin such that the Y axis would have been a nodal line for two such interfering waves without the geostrophic term. As we watch this steady-state wave motion as a function of time, the crests of the waves, or high water, are defined by the relation $\partial \eta / \partial t = 0$, which from Eq. (5.57) gives

$$e^{-my} \sin [k(x-ct)] + e^{my} \sin [k(x+ct)] = 0$$

$$\sinh (my) \cos (kx) \sin (\omega t) + \cosh (my) \sin (kx) \cos (\omega t) = 0$$

$$\frac{\tanh (my)}{\tan (kx)} = -\cot (\omega t) \qquad (5.58)$$

For locations near the origin, this reduces to

$$y = -\frac{kx}{m} \cot (\omega t) \qquad (5.59)$$

Equation (5.59) defines a straight line centered at the origin, which rotates counterclockwise about the origin with the angular frequency ω. Such lines of equal tidal phase, in this case high tide, are referred to as *cotidal lines*. If

158 ¶ *Thermodynamics and Hydrodynamics*

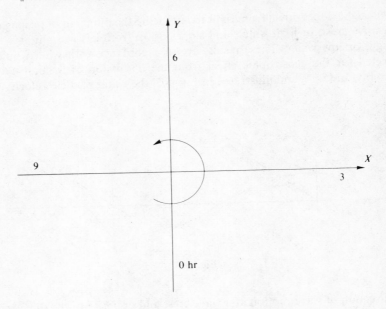

Fig. 5.6

instead of free tidal waves, the interference effect is that of a tidal cooscillation, the period of the rotation will be that of the lunar dirunal tide, or approximately 12 hr. For such a case, the time of high tide will progress counterclockwise in a regular manner around the bay as illustrated in Fig. 5.6. Instead of having nodal lines of no motion, we are now reduced to a central region, referred to as an *amphidromic region*, of no motion at the origin. The interference effect of two geostrophically controlled simple harmonic waves is to change the observed wave pattern from that of a linear standing wave to a rotary wave.

As a final example, we shall consider the somewhat fictitious one in which the quantities η, v_x, v_y are independent of y. In some manner the wave crests and troughs are maintained as horizontal lines in the Y direction. The equations of motion (5.44) and the equation of continuity (5.45) reduce to

$$\frac{\partial v_x}{\partial t} - 2\omega_0 v_y \sin \varphi = -g \frac{\partial \eta}{\partial x}$$
$$\frac{\partial v_y}{\partial t} + 2\omega_0 v_x \sin \varphi = 0$$
(5.60)

and

$$\frac{\partial \eta}{\partial t} + h \frac{\partial v_x}{\partial x} = 0 \qquad (5.61)$$

Assuming a simple harmonic wave propagating in the positive X direction, we will have η of the form

$$\eta = a \cos [k(x-ct)] \tag{5.62}$$

From the second of Eqs. (5.60) and from Eq. (5.61) we then obtain, respectively for v_x and v_y,

$$v_x = \frac{ac}{h} \cos [k(x-ct)] \tag{5.63}$$

and

$$v_y = \frac{2\omega_0 ac \sin \varphi}{\omega h} \sin [k(x-ct)] \tag{5.64}$$

Substituting Eqs. (5.62), (5.63), and (5.64) into the first of Eqs. (5.60), we then obtain, for the conditional relation defining the wave velocity

$$\frac{\omega c}{h} - \frac{4\omega_0^2 c \sin^2 \varphi}{\omega h} = \frac{\omega g}{c}$$

or

$$c^2 = \frac{gh}{1 - \dfrac{4\omega_0^2 \sin^2 \varphi}{\omega^2}} \tag{5.65}$$

We see from Eqs. (5.63) and (5.64) that the wave motion is such that the particle velocities, or currents, rotate in a clockwise sense in the northern hemisphere with a ratio of the maximum v_y to maximum v_x components of $2\omega_0 \sin \varphi/\omega$. If the tidal wave motion is that of the diurnal tide, ω is approximately equal to $2\omega_0$, and this ratio reduces to $\sin \varphi$. For this latter condition, the velocity of wave propagation of Eq. (5.65) becomes

$$c = (gh)^{1/2} \sec \varphi \tag{5.66}$$

5.5 Internal Tidal Waves

We should next like to consider some of the simpler aspects associated with tidal wave motion in two superposed liquids of constant, but different densities in which the density of the upper liquid is less than that of the lower liquid. Two solutions will result, one simply a direct extension of the surface tidal wave motion of Section 5.1, the other a tidal wave motion associated with and dependent on the internal boundary. This latter motion is referred to as an internal seiche, free tidal wave, or forced tidal wave, as the case may be.

Let η represent the vertical displacement of the upper, free surface and η' that of the internal boundary, as shown in Fig. 5.7. Let ρ and ρ' be their respective densities, h the thickness of the upper layer, h' the thickness of the lower

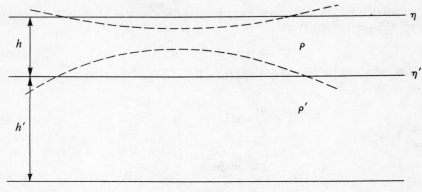

Fig. 5.7

layer measured from the internal boundary to the bottom, z the depth measured from the undisturbed free surface in the upper layer, and z' the depth measured from the undisturbed internal boundary in the lower layer. From Eq. (3.30) the one-dimensional equation of motion, assuming no external forces and that the displacements are sufficiently small that D/Dt may be replaced by $\partial/\partial t$ and neglecting the geostrophic term, will be

$$\frac{\partial v_x}{\partial t} = -\frac{1}{\rho}\frac{\partial p}{\partial x} \tag{5.67}$$

The hydrostatic pressures in the upper and lower fluids will be given, respectively, by

$$p = p_a + g\rho(\eta+z) \tag{5.68}$$

and

$$p' = p_a + g\rho(\eta+z-\eta') + g\rho'(\eta'+z') \tag{5.69}$$

where p_a is the atmospheric pressure, from which we obtain

$$\frac{\partial p}{\partial x} = g\rho \frac{\partial \eta}{\partial x} \tag{5.70}$$

and

$$\frac{\partial p'}{\partial x} = g\rho \frac{\partial \eta}{\partial x} + g\rho'\left(1 - \frac{\rho}{\rho'}\right)\frac{\partial \eta'}{\partial x} \tag{5.71}$$

Substituting these values into Eq. (5.67), we then have for the equations of motion in the upper and lower fluids, respectively,

$$\frac{\partial v_x}{\partial t} = -g\frac{\partial \eta}{\partial x} \tag{5.72}$$

$$\frac{\partial v'_x}{\partial t} = -g\frac{\rho}{\rho'}\frac{\partial \eta}{\partial x} - g\left(1 - \frac{\rho}{\rho'}\right)\frac{\partial \eta'}{\partial x} \tag{5.73}$$

With reference to Fig. 5.2, we obtain for the equation of continuity in the upper and lower fluids, respectively,

$$h\frac{\partial v_x}{\partial x} + \frac{\partial}{\partial t}(\eta-\eta') = 0 \tag{5.74}$$

$$h'\frac{\partial v'_x}{\partial x} + \frac{\partial \eta'}{\partial t} = 0 \tag{5.75}$$

similar to Eq. (5.3).

We shall look for a simple harmonic solution of the usual form for η as

$$\eta = A \cos [k(x-ct)] \tag{5.76}$$

From Eq. (5.72) we then have for v_x

$$v_x = \frac{gA}{c} \cos [k(x-ct)] \tag{5.77}$$

from Eq. (5.74) for η'

$$\eta' = \left(1 - \frac{gh}{c^2}\right) A \cos [k(x-ct)] \tag{5.78}$$

and from Eq. (5.75) for v_x

$$v'_x = \left(1 - \frac{gh}{c^2}\right) \frac{cA}{h'} \cos [k(x-ct)] \tag{5.79}$$

Substituting these values of η, v_x, η', and v'_x into the fourth of the differential equations, Eq. (5.73), we obtain the conditional equation for c^2,

$$\left(1 - \frac{gh}{c^2}\right)\frac{c^2}{h'} = g\frac{\rho}{\rho'} + g\left(1 - \frac{\rho}{\rho'}\right)\left(1 - \frac{gh}{c^2}\right)$$

$$\frac{c^2}{h'} - \frac{gh}{h'} = g - \frac{g^2 h}{c^2}\left(1 - \frac{\rho}{\rho'}\right)$$

$$c^4 - g(h+h')c^2 + \left(1 - \frac{\rho}{\rho'}\right) g^2 hh' = 0$$

whose solution is

$$c^2 = \tfrac{1}{2}\left[g(h+h') \pm \sqrt{g^2(h+h')^2 - 4\left(1 - \frac{\rho}{\rho'}\right)g^2 hh'} \right] \tag{5.80}$$

Since the density contrast between the two fluids, $(\rho'-\rho)/\rho$, is usually small, we may approximate Eq. (5.80) by

$$c^2 = \tfrac{1}{2}\left\{ g(h+h') \pm \left[g(h+h') - 2\left(1 - \frac{\rho}{\rho'}\right)\frac{ghh'}{h+h'} + \cdots \right] \right\} \tag{5.81}$$

from which we obtain the two possible values of c^2, corresponding, respectively, to the plus and minus signs in Eq. (5.81)

$$c_0^2 = g(h+h') \qquad (5.82)$$

and

$$c_i^2 = \left(1 - \frac{\rho}{\rho'}\right)\frac{ghh'}{h+h'} \qquad (5.83)$$

We may then write for the ratios of the vertical displacements and the horizontal particle velocities from Eqs. (5.76), (5.77), (5.78), and (5.79), using Eqs. (5.82) and (5.83),

$$\frac{\eta_0'}{\eta_0} = \frac{c_0^2 - gh}{c_0^2} = \frac{h'}{h+h'} \qquad (5.84)$$

$$\frac{v_{x0}'}{v_{x0}} = \frac{c_0^2 - gh}{gh'} = 1 \qquad (5.85)$$

and

$$\frac{\eta_i'}{\eta_i} = \frac{c_i^2 - gh}{c_i^2} = -\frac{\rho'(h+h')}{(\rho'-\rho)h'} \qquad (5.86)$$

$$\frac{v_{xi}'}{v_{xi}} = \frac{c_i^2 - gh}{gh'} = -\frac{h}{h'} \qquad (5.87)$$

where the latter two expressions are approximations for a small density contrast between the two fluids, so that c_i^2 is small compared with gh.

We see that the first solution, represented by Eqs. (5.82), (5.84), and (5.85), is simply a direct extension to two fluid layers of the solution of Section 5.1 for a single layer. The velocity of wave propagation is given in terms of the total depth $h+h'$ from the free surface to the bottom. The vertical displacement amplitude of the wave on the internal boundary is diminished in direct ratio to the ratio of its distance above the bottom to the total depth, as would be anticipated from the results of Problem 5.1(c), and the horizontal particle velocities of the wave motion at the two surfaces are in phase and of equal magnitude.

The second, or *internal wave*, solution, represented by Eqs. (5.83), (5.86), and (5.87), is a considerably different type of motion. Since the density contrast between the two fluids is small, the velocity of wave propagation will be small. For a given wavelength in the two cases, the period of the internal wave will be much longer than that of surface tidal wave. The vertical displacement amplitude of the wave on the internal boundary will be large compared to the vertical displacement amplitude of the wave on the free surface and will be out of phase with it. Further, the horizontal particle velocities will be out of phase and their ratio inversely proportional to the thicknesses of the two fluid layers.

Physical Oceanography—Waves and Tides ¶ 163

For the first solution, the primary wave motion is at the free surface and that at the internal boundary a continuation of this motion downward. For the second solution, the primary wave motion is at the internal boundary and that at the free surface a continuation of this motion upward. For driven tidal wave motion, the relative excitation of each of these types of motion will depend substantially on the closeness of the velocity of the driving force to the free velocities of the two motions.

Problem 5.5(a) Derive the relation for seiches for this example corresponding to Eq. (5.37).

5.6 Surface Waves

In Section 5.1 we discussed surface wave motion subject to the force of gravity for which the wavelength was very great compared to the water depth. Now we shall direct our attention to shorter wavelength gravity waves, for which now the horizontal motion will not be essentially the same throughout a vertical column. In particular, the vertical accelerations can no longer be neglected. We shall assume as before that the fluid is incompressible. We shall also assume that the motion is irrotational; this is a reasonable assumption in consideration of Kelvin's circulation theorem and the fact that we are only considering an external force, gravity, derivable from a potential. We shall further assume as before that the particle motion is sufficiently small that the time derivative D/Dt may be replaced by $\partial/\partial t$.

As before, we shall treat the specific problem of one-dimensional wave motion for an ocean of constant depth h. We shall here expand the definitions of the horizontal and vertical displacements ξ and η to include motion anywhere in the ocean as well as at the free surface.

As the motion is irrotational, a velocity potential Φ exists, and the particle velocity components may be given by

$$v_x = \frac{D\xi}{Dt} = \frac{\partial \xi}{\partial t} = -\frac{\partial \Phi}{\partial x} \tag{5.88}$$

and

$$v_z = \frac{D\eta}{Dt} = \frac{\partial \eta}{\partial t} = -\frac{\partial \Phi}{\partial z} \tag{5.89}$$

The equation of continuity (3.19), defining the motion, assumes the form

$$\frac{\partial^2 \Phi}{\partial x^2} + \frac{\partial^2 \Phi}{\partial z^2} = 0 \tag{5.90}$$

for the two-dimensional problem under consideration.

The wave motion will be subject to the boundary conditions that at the bottom, the vertical particle velocity component is zero and that at the free surface, the pressure is zero, or a constant atmospheric pressure. Taking coordinates as shown in Fig. 5.1 with the origin at the bottom, the X axis in the direction of the wave propagation, and the Z axis vertically upward, the first boundary condition in terms of the velocity potential is simply

$$v_z = -\frac{\partial \Phi}{\partial z} = 0 \qquad (z = 0) \tag{5.91}$$

For the upper boundary condition, we have from Eq. (3.35), for the pressure in an incompressible medium,

$$\frac{p}{\rho} = \frac{\partial \Phi}{\partial t} - \Omega - \tfrac{1}{2}v^2 + G(t)$$

The additive function $G(t)$ may be included in the time derivative of the velocity potential. As the water particle velocities are small, we can neglect terms in v^2. For our coordinate system, the gravitational potential is simply $g(z+\eta)$ so that we have

$$\frac{p}{\rho} = \frac{\partial \Phi}{\partial t} - g(z+n) \tag{5.92}$$

or

$$\eta = \frac{1}{g}\frac{\partial \Phi}{\partial t} - z - \frac{p}{\rho g} \tag{5.93}$$

Taking the time derivative of Eq. (5.93), we then have for our second boundary condition

$$v_z = \frac{\partial \eta}{\partial t} = \frac{1}{g}\frac{\partial^2 \Phi}{\partial t^2} = -\frac{\partial \Phi}{\partial z} \qquad (z = h) \tag{5.94}$$

where to the first order of small quantities, we have approximated the free surface by the plane $z = h$.

To find a wave motion solution of the equation of continuity satisfying these boundaries, we may make the assumption of a simple harmonic wave propagating in the X direction with an as yet undetermined z dependence. If an adequate solution can be found, it can then be applied to the more general case of arbitrary initial conditions through the use of the Fourier integral theorem. Let us try, then,

$$\Phi = f(z) \cos [k(x-ct)] \tag{5.95}$$

Substituting into Eq. (5.90), we obtain the ordinary differential equation

$$\frac{d^2 f}{dz^2} - k^2 f = 0 \tag{5.96}$$

whose solution is simply

$$f = A \cosh(kz) + B \sinh(kz) \tag{5.97}$$

From the first boundary condition, Eq. (5.91), we see that the coefficient B must vanish so that our solution will be

$$\Phi = A \cosh(kz) \cos[k(x-ct)] \tag{5.98}$$

Substituting Eq. (5.98) into the second boundary condition, Eq. (5.94), we obtain the conditional relation for the wave propagation velocity

$$\frac{k^2 c^2}{g} = k \tanh(kh)$$

or

$$c^2 = \frac{g}{k} \tanh(kh) = \frac{g\lambda}{2\pi} \tanh\left(\frac{2\pi h}{\lambda}\right) \tag{5.99}$$

For the case of the wavelength long compared with the depth, Eq. (5.99) reduces to

$$c^2 = gh \tag{5.100}$$

the same result as obtained in Section 5.1. For the case of the wavelength short compared with the depth, Eq. (5.99) reduces to

$$c^2 = \frac{g}{k} = \frac{g\lambda}{2\pi} \tag{5.101}$$

Substituting Eq. (5.98) into Eqs. (5.88) and (5.89), the particle velocity and displacement components are then simply

$$\begin{aligned} v_x &= kA \cosh(kz) \sin[k(x-ct)] \\ v_z &= -kA \sinh(kz) \cos[k(x-ct)] \end{aligned} \tag{5.102}$$

and

$$\begin{aligned} \xi &= \frac{A}{c} \cosh(kz) \cos[k(x-ct)] \\ \eta &= \frac{A}{c} \sinh(kz) \sin[k(x-ct)] \end{aligned} \tag{5.103}$$

The particle motion is seen to be elliptical in character. As the hyperbolic sine and cosine approach the same limiting value for large values of the argument, the ellipse becomes a circle near the surface of deep water. As the hyperbolic sine approaches zero for small values of the argument, the ellipse reduces to linear horizontal motion at the bottom.

We see from Eq. (5.99) that the velocity of wave propagation is here also a function of the wave number k. From the discussion of Section 7.2, we then know that in the generalized case we shall have a dispersive wave train and that c will now represent the phase velocity. For the case of short wavelengths, Eq. (5.101), the corresponding group velocity will be given by Eq. (7.37) as

$$U = c - \lambda \frac{dc}{d\lambda} = c - \frac{g\lambda}{2\pi} \frac{1}{2c} = \frac{1}{2} c \qquad (5.104)$$

For the more general case of Eq. (5.99), we obtain, after some reduction,

$$U = \tfrac{1}{2} c \left[1 + \frac{2kh}{\sinh (2kh)} \right] \qquad (5.105)$$

Problem 5.6(a) Obtain the equation of the ellipse ξ, η of the fluid particle motion. Sketch and discuss the fluid particle motion as a function of depth and wavelength.

Problem 5.6(b) Carry through the reduction for Eq. (5.105).

5.7 Permanent Waves

We saw in Section 5.1 on tidal waves that when finite amplitude wave motion was considered, the wave velocity became a function of wave height. A simple harmonic wave could not be considered to be propagated without a change in shape. We should like to examine here finite amplitude wave motion for gravity waves and to see if we can determine a *permanent wave* shape, that is, a wave shape for finite amplitude waves which, in the case of steady-state motion, does not change its profile. In effect, such a solution will be the finite amplitude counterpart of sinusoidal wave motion for infinitesimal amplitude.

Changing, for convenience, our origin to the ocean surface, we may rewrite the velocity potential of Eq. (5.98) as

$$\varphi = a \frac{\cosh [k(z+h)]}{\cosh (kh)} \cos [k(x-ct)] \qquad (5.106)$$

For wavelengths short compared with the depth, the normal case for deep ocean waves, Eq. (5.106) reduces to

$$\varphi = ae^{kz} \cos [k(x-ct)] \qquad (5.107)$$

Using the method of Rayleigh from Section 5.1, we see that our potential function will be given by

$$\frac{\varphi}{c} = -x + \beta e^{kz} \sin (kx) \qquad (5.108)$$

representing by the second term a periodic wave motion of the form of Eq.

(5.107) superimposed on a uniform current velocity c, represented by the first term. From the results of Problem 3.2(c), we have that the corresponding stream function is

$$\frac{\psi}{c} = -z + \beta e^{kz} \cos(kx) \qquad (5.109)$$

For the profile of the free surface to represent a permanent wave shape, it must be a streamline. We shall take it to be the line $\psi = 0$. Its form, then, is given by

$$z = \beta e^{kz} \cos(kx) \qquad (5.110)$$

By successive approximations to the expression

$$z = \beta(1 + kz + \tfrac{1}{2}k^2z^2 + \cdots) \cos(kx)$$

we obtain to terms of the order β^3

$$z = \beta \cos(kx) + k\beta^2 \cos^2(kx) + \tfrac{3}{2}k^2\beta^3 \cos^3(kx) \qquad (5.111)$$

Consider next the trochoid traced out by the circular disk, shown in Fig. 5.8, of unit radius rolling on the upper surface and with a tracing arm of

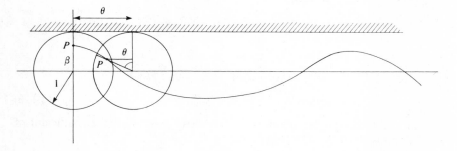

Fig. 5.8

length β. The equation of the trochoid will be given implicitly by the parametric equations

$$z = \beta \cos \theta \qquad x = \theta - \beta \sin \theta \qquad (5.112)$$

Making a Taylor series expansion in terms of β to terms of the order β^3, we obtain for z in terms of x

$$z = -\beta + (\beta - \tfrac{3}{2}\beta^3) \cos x + \beta^2 \cos^2 x + \tfrac{3}{2}\beta^3 \cos^3 x \qquad (5.113)$$

Adjusting the Z origin to eliminate the constant term, making the substitution

$$a = \beta - \tfrac{3}{2}\beta^3$$

so that $a^2 = \beta^2$ and $a^3 = \beta^3$ to terms of the third order, and choosing a circular disk of radius $1/k$ so that the trochoid wavelength becomes λ, we obtain

$$z = a\cos(kx) + ka^2\cos^2(kx) + \tfrac{3}{2}k^2a^3\cos^3(kx) \tag{5.114}$$

the same as expression (5.111). To terms of the third order, the wave shape is that of a *trochoid*, steeper at the crests and flatter in the troughs than a sinusoidal wave. If the expansions of Eqs. (5.111) and (5.114) were carried to fourth-order terms, they would not be the same; the trochoid is an approximate and not an exact solution.

We still have to show, as in the similar discussion of Section 5.1, that the condition of uniform pressure along this streamline can be satisfied by a suitably chosen value of the wave velocity c. From Eq. (5.108) we shall have that the particle velocity v is given by

$$v^2 = v_x^2 + v_y^2 = \left(\frac{\partial \varphi}{\partial x}\right)^2 + \left(\frac{\partial \varphi}{\partial z}\right)^2$$

$$= c^2[1 - 2k\beta e^{kz}\cos(kx) + k^2\beta^2 e^{2kz}] \tag{5.115}$$

Substituting Eqs. (5.115) and (5.110) into Eq. (5.10) and expanding the final term of the resultant expression to terms of the order β^3, we obtain

$$\frac{1}{\rho}p(z) = \text{const} - gz - \tfrac{1}{2}c^2[1 - 2k\beta e^{kz}\cos(kx) + k^2\beta^2 e^{2kz}]$$

$$= \text{const} + (kc^2 - g)z - \tfrac{1}{2}k^2c^2\beta^2 e^{2kz}$$

$$= \text{const} + (kc^2 - g)z - \tfrac{1}{2}k^2c^2\beta^2(1 + 2kz + \cdots)$$

$$= \text{const} + (kc^2 - g - k^3c^2\beta^2)z \tag{5.116}$$

Hence, the condition for a stationary free surface is met if the coefficient of the second term of Eq. (5.116) is zero, or

$$kc^2 - g - k^3c^2\beta^2 = 0$$

which gives for c, to the same order of approximation in β,

$$c^2 = \frac{g}{k}(1 + k^2\beta^2) \tag{5.117}$$

This determines the velocity of progressive waves of permanent type. It shows that it is a constant for a given trochoidal wave and that it does increase somewhat with the amplitude coefficient β of the wave.

Problem 5.7(a)   Carry through the successive approximations to obtain expression (5.111).

Problem 5.7(b)   Carry through the Taylor series expansion of Eqs. (5.112) to obtain Eq. (5.113).

Problem 5.7(c) Show that to the same order of approximation, expression (5.113) may be written as

$$z = b\cos(kx) + \tfrac{1}{2}kb^2\cos(2kx) + \tfrac{3}{8}k^2b^3\cos(3kx)$$

5.8 Waves Due to a Local Disturbance

Up to this point we have considered only steady-state motion. We should now like to consider the wave motion that will result from and be propagated out from a local initial disturbance. This will necessitate the use of some of the concepts of Chapter 7, namely, phase and group velocities, dispersion, and stationary phase. Some of the features of the resultant wave motion will be similar to those obtained for steady-state motion; many others, understandably, will not be similar. A physical understanding of these differences is essential to the consideration of more complex wave propagation problems.

We shall wish to consider gravity wave propagation in deep water produced by a local disturbance. This problem is sometimes referred to as the *Cauchy-Poisson problem*. We shall take our origin in the ocean surface with the approximation of infinitesimal amplitudes. Further, we shall take only one-dimensional propagation; our source then is a line source along the Y axis. The velocity potential then will be of the form of Eq. (5.107). For convenience here, we shall take a potential given by one term of the $\sin[k(x-ct)]$ expansion, or

$$\varphi = ae^{kz}\sin(\omega t)\cos(kx) \tag{5.118}$$

where from Eq. (5.101) we have for the wave propagation velocity c, or phase velocity, in terms of any two of the quantities g, k, ω,

$$c = \frac{\omega}{k} = \left(\frac{g}{k}\right)^{1/2} = \frac{g}{\omega} \tag{5.119}$$

We shall wish to consider only the displacement of the ocean surface η_0, which from Eq. (5.93) is given in terms of φ as

$$\eta_0 = \frac{1}{g}\left(\frac{\partial\varphi}{\partial t}\right)_0 = \frac{1}{g}\frac{\partial\varphi_0}{\partial t} \tag{5.120}$$

Further, we shall want to have a unit amplitude displacement for η_0 so that $a = g/\omega$. From Eqs. (5.120) and (5.118) we then have

$$\eta_0 = \cos(\omega t)\cos(kx) \tag{5.121}$$

and

$$\varphi_0 = g\frac{\sin(\omega t)}{\omega}\cos(kx) \tag{5.122}$$

Let us take as initial conditions an initial elevation of the sea surface without initial velocity. The initial conditions may then be written as

$$\eta_0 = f(x) \qquad (t = 0) \qquad (5.123)$$
$$\varphi_0 = 0$$

With reference to the discussions of Sections 1.9 and 7.2, let us generalize η_0 and φ_0 of Eqs. (5.121) and (5.122) in order to meet these initial conditions by the following integrations with respect to the dummy variables k and ξ, where we have also made the substitution $x - \xi$ for x,

$$\eta_0 = \frac{1}{\pi} \int_0^\infty \cos(\omega t)\, dk \int_{-\infty}^\infty f(\xi) \cos[k(x-\xi)]\, d\xi \qquad (5.124)$$

and

$$\varphi_0 = \frac{g}{\pi} \int_0^\infty \frac{\sin(\omega t)}{\omega}\, dk \int_{-\infty}^\infty f(\xi) \cos[k(x-\xi)]\, d\xi \qquad (5.125)$$

The result of an integration with respect to the variables k and ξ, which do not enter into the original differential equation, will also, of course, be a solution of that equation. We see that at $t = 0$ Eq. (5.125) reduces to zero and from the Fourier integral theorem, Eq. (1.161), that Eq. (5.124) reduces to $f(x)$, the initial conditions. Taking an initial elevation confined to the immediate neighborhood of the origin such that $f(x)$ vanishes for all but infinitesimal values of x in such a way that

$$\int_{-\infty}^\infty f(\xi)\, d\xi = 1 \qquad (5.126)$$

expression (5.125) for φ_0 further reduces to

$$\varphi_0 = \frac{g}{\pi} \int_0^\infty \frac{\sin(\omega t)}{\omega} \cos(kx)\, dk \qquad (5.127)$$

We may now proceed to evaluate this integral. Making a change in the variable of integration from k to ω, we have from Eq. (5.119)

$$\varphi_0 = \frac{2}{\pi} \int_0^\infty \sin(\omega t) \cos\left(\frac{\omega^2 x}{g}\right) d\omega$$
$$= \frac{1}{\pi} \left[\int_0^\infty \sin\left(\frac{\omega^2 x}{g} + \omega t\right) d\omega - \int_0^\infty \sin\left(\frac{\omega^2 x}{g} - \omega t\right) d\omega \right] (5.128)$$

Making a further change of variable from ω to σ, where σ is defined by

$$\sigma = \frac{x^{1/2}}{g^{1/2}} \left(\omega \pm \frac{gt}{2x} \right)$$

or

$$\sigma^2 = \frac{\omega^2 x}{g} \pm \omega t + \frac{gt^2}{4x}$$

(5.129)

we shall have, using the positive sign for the first integral of Eq. (5.128) and the minus sign for the second,

$$\varphi_0 = \frac{g^{1/2}}{\pi x^{1/2}} \left[\int_\tau^\infty \sin(\sigma^2 - \tau^2)\, d\sigma - \int_{-\tau}^\infty \sin(\sigma^2 - \tau^2)\, d\sigma \right]$$

$$= -\frac{2g^{1/2}}{\pi x^{1/2}} \int_0^\tau \sin(\sigma^2 - \tau^2)\, d\sigma \qquad (5.130)$$

where τ is defined by

$$\tau^2 = \frac{gt^2}{4x} \qquad (5.131)$$

From Eq. (5.120) we shall then have for η_0

$$\eta_0 = \frac{g^{1/2} t}{\pi x^{3/2}} \int_0^\tau \cos(\sigma^2 - \tau^2)\, d\sigma$$

$$= \frac{g^{1/2} t}{\pi x^{3/2}} \left[\cos \tau^2 \int_0^\tau \cos \sigma^2\, d\sigma + \sin \tau^2 \int_0^\tau \sin \sigma^2\, d\sigma \right] \qquad (5.132)$$

The two integrals of Eq. (5.132) are a tabulated form; they differ from the Fresnel integrals by a constant multiplier. When τ is large, the upper limit of each integral may be approximated by infinity, the value of each integral then being $\pi^{1/2}/2^{3/2}$. For τ large, then, Eq. (5.132) reduces to

$$\eta_0 = \frac{g^{1/2} t}{2^{3/2} \pi^{1/2} x^{3/2}} \left[\cos \frac{gt^2}{4x} + \sin \frac{gt^2}{4x} \right]$$

$$= \frac{g^{1/2} t}{2\pi^{1/2} x^{3/2}} \cos\left(\frac{gt^2}{4x} - \frac{\pi}{4} \right) \qquad (5.133)$$

The expressions (5.132) and (5.133) define a wave motion that is quite different from what we are familiar with from steady-state conditions. For small values of τ, expression (5.132) is not subject to any direct, simple interpretation. For large values of τ, we can get some insight into the characteristics of the motion from an examination of Eq. (5.133). We see that we have a series of oscillations of slowly varying amplitude and wavelength. The motion is essentially repeated for an increase of the argument by 2π, or

$$\Delta\left(\frac{gt^2}{4x}\right) = 2\pi \qquad (5.134)$$

If we were to hold x constant, we would find the period of the oscillation P from Eq. (5.134) as

$$\frac{2gt}{4x} \Delta t = 2\pi$$

or

$$P = \Delta t = \frac{4\pi x}{gt} \quad (5.135)$$

Similarly, if we were to hold t constant, we would find the wavelength of the oscillation λ from Eq. (5.134) as

$$-\frac{gt^2}{4x^2}\Delta x = 2\pi$$

or

$$\lambda = -\Delta x = \frac{8\pi x^2}{gt^2} \quad (5.136)$$

Further, if were to follow a particular phase of the motion, we would find the wave, or phase, velocity c from Eq. (5.133) as

$$-\frac{gt^2}{4x^2}\Delta x + \frac{2gt}{4x}\Delta t = 0$$

or

$$c = \frac{\Delta x}{\Delta t} = \frac{2x}{t} = \left(\frac{g\lambda}{2\pi}\right)^{1/2} \quad (5.137)$$

where the last term is obtained by substituting from Eq. (5.136). We see, as we should have expected, that the phase velocity here is the same as the wave propagation velocity [Eq. (5.101)] for the steady-state case. If we should next concentrate on a group of waves having approximately the same wavelength λ, we would find for their velocity of propagation U from Eq. (5.137)

$$U = \frac{x}{t} = \frac{1}{2}\left(\frac{g\lambda}{2\pi}\right)^{1/2} = \frac{1}{2}c \quad (5.138)$$

the same as obtained for the group velocity in Eq. (5.104).

We could also have approached this problem by the approximation of the method of stationary phase, discussed in Section 7.2. From Eq. (7.33) we would have following the formulation of Eq. (7.28)

$$\eta_0 = \frac{2^{1/2}\pi^{1/2}}{t^{1/2}}\frac{g(k)}{|d^2\omega/d^2k|^{1/2}}e^{i(kx-\omega t \mp \pi/4)} \quad (5.139)$$

From the results of Problems 1.9(c) and 1.9(d) and following the formulation of Eq. (7.26), $g(k)$ will be given by

$$g(k) = \frac{1}{2\pi} \quad (5.140)$$

The stationary phase condition corresponding to Eq. (7.34) will be

$$x - t\frac{d\omega}{dk} = 0 \quad (5.141)$$

From Eqs. (5.141) and (5.119) we then obtain

$$x - t\frac{g}{2\omega} = 0$$

$$\omega = \frac{gt}{2x} \tag{5.142}$$

and

$$k = \frac{\omega^2}{g} = \frac{gt^2}{4x^2} \tag{5.143}$$

and

$$\frac{d\omega}{dk} = \frac{g}{2\omega} = \frac{g^{1/2}}{2k^{1/2}}$$

$$\frac{d^2\omega}{dk^2} = -\frac{g^{1/2}}{4k^{3/2}} = -\frac{2x^3}{gt^3} \tag{5.144}$$

Substituting Eqs. (5.140), (5.142), (5.143), and (5.144) into Eq. (5.139) and taking the real value, we obtain

$$\eta_0 = \frac{g^{1/2}t}{2\pi^{1/2}x^{3/2}} \cos\left(\frac{gt^2}{4x} - \frac{\pi}{4}\right) \tag{5.145}$$

the same as the approximation (5.133).

Problem 5.8(a) Carry through the derivation for an initial impulse without initial elevation.

5.9 Equilibrium Theory of the Tides

In this preliminary investigation of tidal action, we shall neglect the effect of the rotation of the earth on its axis. We shall, however, want to make use of the concept of gravitational potential, and the reader may wish to refer to Section 8.1 to review these general relations between force per unit mass and potential. Let O be the center of the earth, C the center of the moon, P the position of a particle on the surface of a spherical earth, θ the polar angle measured from the line OC as axis, a the radius of the earth, R the distance between O and C, and r the distance between P and C, all as shown in Fig. 5.9. The acceleration, or gravitational force per unit mass, $\mathbf{f}_0$ of a particle P on the surface of the earth relative to axes fixed in an inertial system is equal to the sum of the acceleration $\ddot{\boldsymbol{\rho}}$ of the center of the earth and the acceleration $\mathbf{f}$ of P relative to the center of the earth, that is,

$$\mathbf{f}_0 = \ddot{\boldsymbol{\rho}} + \mathbf{f} \tag{5.146}$$

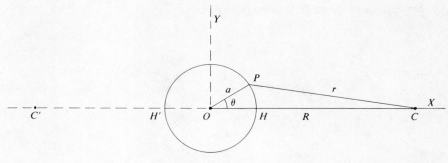

Fig. 5.9

as was shown in Section 1.13. If $\mathbf{K}_0$ is the force that the moon exerts on the earth as a whole, we shall have

$$\mathbf{K}_0 = \mathbf{i}\gamma \frac{ME}{R^2} = E\ddot{\boldsymbol{\rho}}$$

where M is the mass of the moon and E that of the earth and where the X axis has been taken along the line OC so that the gravitational attraction per unit mass $\ddot{\boldsymbol{\rho}}$ is then given in terms of its gravitational potential Ω_0 by

$$\ddot{\boldsymbol{\rho}} = \mathbf{i}\frac{\gamma M}{R^2} = -\mathbf{i}\frac{\partial}{\partial x}\left(-\frac{\gamma M}{R^2}x\right) = -\nabla\Omega_0 \tag{5.147}$$

We then have for Ω_0

$$\Omega_0 = -\frac{\gamma M}{R^2}x = -\frac{\gamma M}{R^2}a\cos\theta \tag{5.148}$$

Now the gravitational potential at P due to the moon's attraction is

$$\Omega = -\frac{\gamma M}{r} = -\frac{\gamma M}{R\sqrt{1-2\frac{a}{R}\cos\theta + \frac{a^2}{R^2}}}$$

$$= -\frac{\gamma M}{R}\left[1 - \tfrac{1}{2}\left(-2\frac{a}{R}\cos\theta + \frac{a^2}{R^2}\right) + \tfrac{3}{8}\left(4\frac{a^2}{R^2}\cos^2\theta + \cdots\right) + \cdots\right]$$

$$= -\frac{\gamma M}{R}\left[1 + \frac{a}{R}\cos\theta - \tfrac{1}{2}\frac{a^2}{R^2}(1 - 3\cos^2\theta)\right] \tag{5.149}$$

carrying out the expansion to terms of the order a^2/R^2.

A particle at P is subject to the force of earth gravity as well as the gravitational attraction of the moon. If the potential due to the earth's gravitational field is denoted by Ψ, the total force due to both causes is $-\nabla(\Omega + \Psi)$. Therefore, the equation of motion of a particle at P not subject to constraints and with reference to a fixed inertial system is

$$\mathbf{f}_0 = -\nabla \Omega - \nabla \Psi$$

and the acceleration $\mathbf{f}$ of P relative to a nonrotating earth is

$$\mathbf{f} = \mathbf{f}_0 - \ddot{\boldsymbol{\rho}} = -\nabla(\Omega - \Omega_0 + \Psi) \tag{5.150}$$

Consequently, a particle at P has the same motion relative of the earth as if the earth were fixed and the particle acted upon by a field of potential

$$\Omega' = \Omega - \Omega_0 + \Psi$$

$$= -\frac{\gamma M}{R}\left[1 - \tfrac{1}{2}\frac{a^2}{R^2}(1 - 3\cos^2\theta)\right] + \Psi \tag{5.151}$$

If the earth is covered with water, the surface of the liquid must be an equipotential surface for equilibrium to exist; for if it were not, there would be a tangential force on the surface due to the effective field. The potential due to gravity is $\Psi = g\zeta$, where ζ is the elevation of the water above mean sea level. Therefore, the equation of the equilibrium surface is

$$\frac{\gamma M a^2}{2R^3}(1 - 3\cos^2\theta) + g\zeta = C$$

where C is a constant since the first term in the brackets of Eq. (5.151) is a constant; or the elevation above sea level is

$$\zeta = \frac{\gamma M a^2}{2gR^3}(3\cos^2\theta - 1) + \frac{C}{g}$$

As the mass E of a spherical earth is $a^2 g/\gamma$, this may be written as

$$\zeta = \tfrac{1}{2}\frac{M}{E}\left(\frac{a}{R}\right)^3 a(3\cos^2\theta - 1) + \frac{C}{g} \tag{5.152}$$

Since ζ is the elevation above mean sea level, the average value of ζ over the surface of the earth must vanish. Therefore,

$$\int_0^\pi \zeta 2\pi a^2 \sin\theta\, d\theta = 0$$

Substituting the value of ζ above and integrating it, it is seen that C is zero. Therefore, Eq. (5.152) becomes

$$\zeta = \tfrac{1}{2}\frac{M}{E}\left(\frac{a}{R}\right)^3 a(3\cos^2\theta - 1) \tag{5.153}$$

The tide should be high, then, at H and H' and low at points on the great circle perpendicular to HH'. Actually, however, the tide is nearly low at H and H'. While the equilibrium theory accounts for two tides daily as actually observed, it gives a wrong phase to the tides. A closer approximation is given by the dynamical theory to be developed in the next section.

Problem 5.9(a) Show that the potential [Eq. (5.151)] is the same as if the earth were at rest in an inertial system and two equidistant moons of mass $\frac{1}{2}M$ each were located at C and C' in Fig. 5.9.

Problem 5.9(b) On the equilibrium theory of the tides, find the elevation at high tide and depression at low tide from mean sea level. The ratio of the masses of the earth to moon is 80, and the ratio of the distance of the moon from the center of the earth to the radius of the earth is 60.
Ans. 1.22 ft, 0.61 ft

Problem 5.9(c) What is the ratio of the height of the solar tide to that of the lunar tide? The ratio of the masses of sun to earth is 332,000, and the ratio of the distance of the sun from the earth to the radius of the earth is 23,200.
Ans. 0.46

5.10 Dynamical Theory of the Tides

On the dynamical theory, the tides are attributed to a wave motion produced by the attraction of the moon. We shall not consider an earth covered with water, but shall limit ourselves to the simpler case of a canal with perfectly smooth walls encircling the earth at the equator. As the centrifugal force due to the rotation of the earth is taken account of in the measured value of g and as the Coriolis force is negligible for the small relative velocities involved, we may consider the earth to be without rotation and consider the moon revolving around it in a little more than a day. The field acting on the water particles is then derivable from the potential function of Eq. (5.151). Let O and C be the centers of the earth and moon respectively, φ the longitude of a point P measured in the equatorial plane of the moon's orbit, and ω the westward angular velocity of the moon relative to a meridian fixed on the earth, all as shown in Fig. 5.10. As so defined, ω is the excess of the eastward angular velocity of the earth relative to an inertial system over that of the moon. As

$$\theta = \omega t + \varphi$$

the tangential force on a unit mass of water at P will be

$$-\frac{\partial \Omega'}{a \partial \varphi} = -\frac{\partial \Omega'}{a \partial \theta} = -\frac{3}{2}\frac{\gamma M a}{R^3}\sin 2\theta$$

$$= -\frac{3}{2}\frac{\gamma M a}{R^3}\sin [2(\omega t + \varphi)] \qquad (5.154)$$

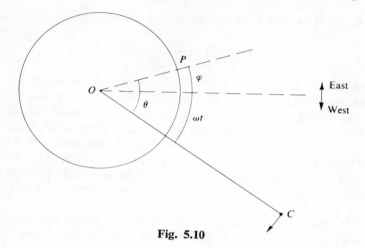

Fig. 5.10

from Eq. (5.151), as the term Ψ' gives rise to a radial force only. This force has to be substituted for the external driving force on the right-hand side of Eq. (5.29), giving for the equation of motion

$$\frac{\partial^2 \xi}{\partial t^2} = c^2 \frac{\partial^2 \xi}{a^2 \partial \varphi^2} - f \sin[2(\omega t + \varphi)] \qquad (5.155)$$

where $a d\varphi$ has been put for the horizontal element of distance dx and where f is given by

$$f = \frac{3}{2} \frac{\gamma M a}{R^3} \qquad (5.156)$$

The forced wave due to the moon motion is given by a particular solution of this differential equation having the same frequency as the driving force due to the moon's attraction. Substituting a solution of the form

$$\xi = C \sin[2(\omega t + \varphi)]$$

into Eq. (5.155), we see that the differential equation is satisfied provided

$$-4\omega^2 C = -4 \frac{c^2}{a^2} C - f$$

or

$$C = -\frac{1}{4} \frac{f a^2}{c^2 - \omega^2 a^2}$$

giving then for the horizontal displacement ξ

$$\xi = -\frac{1}{4} \frac{f a^2}{c^2 - \omega^2 a^2} \sin[2(\omega t + \varphi)] \qquad (5.157)$$

178 ¶ Thermodynamics and Hydrodynamics

From Eq. (5.3) we then have for the vertical displacement η

$$\eta = -h\frac{\partial \xi}{a\partial \varphi} = \tfrac{1}{2}\frac{fah}{c^2-\omega^2 a^2}\cos\,[2(\omega t+\varphi)] \tag{5.158}$$

These equations show that the frequency of the tides is double that of the apparent revolution of the moon. Therefore, there are two tides a day, in agreement with the equilibrium theory. As we have discussed before, if c^2 is greater than $\omega^2 a^2$, the tides are direct but if c^2 is less than $\omega^2 a^2$, the tides are inverted. For the actual ocean depths, c^2 is small compared with $\omega^2 a^2$; the normal tide is inverted. Also as discussed previously, added on to the particular solution of the equation of motion, which represents the forced waves due to the attraction of the moon, we may have solutions representing free waves.

Problem 5.10(a) Find the amplitudes of the horizontal and vertical displacements due to lunar tides in an equatorial canal of 2-mi depth.

Ans. 165 ft, 0.16 ft

Problem 5.10(b) Find the depth at which the inverted tide would change to a direct tide.

Ans. 13 mi

References

Darwin, G. H. 1911. *The Tides and Kindred Phenomena in the Solar System.* New York: Houghton-Mifflin.
Defant, A. 1960. *Physical Oceanography*, vol. 2. New York: Pergamon.
Hill, M. N., ed. 1962. *The Sea*, vol. 1, *Physical Oceanography.* New York: Wiley
Lamb, H. 1932. *Hydrodynamics.* New York: Dover.
Page, L. 1935. *Introduction to Theoretical Physics.* New York: Van Nostrand.
Proudman, J. 1953. *Dynamical Oceanography.* New York: Wiley.
Stokes, G. G. 1847. "On the theory of oscillatory waves," *Cambridge Trans.*, vol. 8, (collected papers, I).
Sverdrup, H. U., M. W. Johnson, and R. H. Fleming. 1942. *The Oceans.* Englewood Cliffs, N.J.: Prentice-Hall.

Part Three

Seismology, Gravity, and Magnetism

CHAPTER 6

SEISMOLOGY—RAY THEORY

6.1 Dynamics

We shall restate here for review purposes some of the basic relations in the dynamics of particles, rigid bodies, and elastic bodies that will be of use to us.

In Cartesian coordinates, the unit vectors defining the position of a particle are fixed so that the velocity $\mathbf{v}$ and acceleration $\mathbf{f}$ are given simply by

$$\mathbf{v} = \mathbf{i}\frac{dx}{dt} + \mathbf{j}\frac{dy}{dt} + \mathbf{k}\frac{dz}{dt} \tag{6.1}$$

and

$$\mathbf{f} = \mathbf{i}\frac{d^2x}{dt^2} + \mathbf{j}\frac{d^2y}{dt^2} + \mathbf{k}\frac{d^2z}{dt^2} \tag{6.2}$$

whose position vector is given by $\mathbf{r} = \mathbf{i}x + \mathbf{j}y + \mathbf{k}z$. In spherical coordinates, however, the position vectors $\mathbf{r}_1$, $\mathbf{\theta}_1$, $\mathbf{\varphi}_1$, change in direction, and $\mathbf{v}$ and $\mathbf{f}$ will now be given by

$$\mathbf{v} = \mathbf{r}_1 \frac{dr}{dt} + \mathbf{\theta}_1 r \frac{d\theta}{dt} + \mathbf{\varphi}_1 r \sin\theta \frac{d\varphi}{dt} \tag{6.3}$$

and

$$\mathbf{f} = \mathbf{r}_1 \left[\frac{d^2r}{dt^2} - r\left(\frac{d\theta}{dt}\right)^2 - r\sin^2\theta \left(\frac{d\varphi}{dt}\right)^2 \right]$$
$$+ \mathbf{\theta}_1 \left[2\frac{dr}{dt}\frac{d\theta}{dt} + r\frac{d^2\theta}{dt^2} - r\sin\theta\cos\theta\left(\frac{d\varphi}{dt}\right)^2 \right]$$
$$+ \mathbf{\varphi}_1 \left[2\sin\theta\frac{dr}{dt}\frac{d\varphi}{dt} + 2r\cos\theta\frac{d\theta}{dt}\frac{d\varphi}{dt} + r\sin\theta\frac{d^2\varphi}{dt^2} \right] \tag{6.4}$$

where the position vector is now given by $\mathbf{r} = \mathbf{r}_1 r$. Resolving components tangent and normal to the path of the particle, we obtain

$$\mathbf{v} = \mathbf{t}_1 v \tag{6.5}$$

and

$$\mathbf{f} = \mathbf{t}_1 \frac{dv}{dt} + \mathbf{n}_1 \frac{v^2}{\rho} \tag{6.6}$$

where $\mathbf{t}_1$ and $\mathbf{n}_1$ are the unit vectors tangential and normal to the path and where ρ is the radius of curvature of the path.

For the motion of a particle of mass m under the action of a force $\mathbf{F}$, we have the familiar relation that

$$\tfrac{1}{2}mv^2 - \tfrac{1}{2}mv_0^2 = \int_{t_0}^{t}(F_x dx + F_y dy + F_z dz) \tag{6.7}$$

where the left-hand side is the change in *kinetic energy*, and the right-hand side is the work done on the particle. If the components of the force are functions of the coordinates only, such that $F_x dx + F_y dy + F_z dz$ is an exact differential $-dV$, the force field is said to be *conservative*. In this case, the right-hand side, $-(V - V_0)$, represents the change in *potential energy*, and Eq. (6.7) then states that the sum of the kinetic and potential energies remain unchanged during the motion.

For a group of particles, our dynamic equation of motion states that the motion of the *center of mass* $\bar{\mathbf{r}}$ defined by

$$\bar{\mathbf{r}} = \frac{1}{m} \sum m_i \mathbf{r}_i \tag{6.8}$$

is the same as if all the mass were concentrated at that point and all the external forces applied there, or

$$m \frac{d^2 \bar{\mathbf{r}}}{dt^2} = \sum \mathbf{F}_i = \mathbf{F} \tag{6.9}$$

We also have that the resultant of the external forces acting on the group of particles is equal to the time rate of increase of linear momentum, or

$$\mathbf{F} = \frac{d\mathbf{G}}{dt} \tag{6.10}$$

where the *linear momentum* $\mathbf{G}$ of the group of particles is defined by

$$\mathbf{G} = \sum m_i \mathbf{v}_i \tag{6.11}$$

And we also have that the resultant of the external torques is equal to the time rate of increase of angular momentum, or

$$\mathbf{L} = \frac{d\mathbf{H}}{dt} \tag{6.12}$$

where the *angular momentum* $\mathbf{H}$ of the group of particles is defined by

$$\mathbf{H} = \sum m_i (\mathbf{r}_i \times \mathbf{v}_i) \tag{6.13}$$

Equation (6.12) applies equally well whether computed about a fixed point or about the moving center of mass.

For a rigid body, the translation of the center of mass is given by expression (6.9) and the rotation of the rigid body by Eq. (6.12). From the definition of angular momentum, we can easily obtain a vector relation between **H** and the angular velocity of rotation of the rigid body ω, which is

$$\mathbf{H} = \Phi \cdot \boldsymbol{\omega} \tag{6.14}$$

where the *momental dyadic* Φ is given by

$$\begin{aligned}\Phi &= \int_\tau (r^2 I - \mathbf{rr})\rho \, d\tau \\ &= \mathbf{ii}\int_\tau (y^2+z^2)\rho \, d\tau - \mathbf{ij}\int_\tau xy\rho \, d\tau - \mathbf{ik}\int_\tau xz\rho \, d\tau \\ &\quad - \mathbf{ji}\int_\tau yx\rho \, d\tau + \mathbf{jj}\int_\tau (z^2+x^2)\rho \, d\tau - \mathbf{jk}\int_\tau yz\rho \, d\tau \\ &\quad - \mathbf{ki}\int_\tau zx\rho \, d\tau - \mathbf{kj}\int_\tau zy\rho \, d\tau + \mathbf{kk}\int_\tau (x^2+y^2)\rho \, d\tau \end{aligned} \tag{6.15}$$

The elements

$$\begin{aligned} A &= \int_\tau (y^2+z^2)\rho \, d\tau \\ B &= \int_\tau (z^2+x^2)\rho \, d\tau \\ C &= \int_\tau (x^2+y^2)\rho \, d\tau \end{aligned} \tag{6.16}$$

are known as the *moments of inertia* of the body about the axes X, Y, Z, respectively, and the elements

$$\begin{aligned} D &= \int_\tau yz\rho \, d\tau \\ E &= \int_\tau zx\rho \, d\tau \\ F &= \int_\tau xy\rho \, d\tau \end{aligned} \tag{6.17}$$

as the *products of inertia*. As the dyadic Φ is symmetric, it is always possible to orient the axes XYZ attached to the rigid body, so that products of inertia vanish, and the dyadic is reduced to its principal axis form, the quantities A, B, C then being referred to as the principal *moments of inertia*. Using Eq. (1.241), Eq. (6.12) then becomes, for a set of axes rotating with the body with origin at a fixed point in the body or at the center of mass,

$$\mathbf{L} = \frac{d\mathbf{H}}{d\tau} + \boldsymbol{\omega} \times \mathbf{H} \tag{6.18}$$

If the fixed axes are also the principal axes, Eq. (6.18) reduces down to
$$\mathbf{L} = \mathbf{i}[A\dot{\omega}_x+(C-B)\omega_y\omega_z]+\mathbf{j}[B\dot{\omega}_y+(A-C)\omega_z\omega_x]$$
$$+\mathbf{k}[C\dot{\omega}_z+(B-A)\omega_x\omega_y] \quad (6.19)$$
which is known as *Euler's equation*.

For elastic bodies, we must be concerned with the concepts of stress and strain. Stress is defined as force per unit area. Usually, the stress is resolved into components normal and tangential to a reference plane in the elastic medium. The former is designated as a *tension* or *tensional stress* if positive, or a *pressure* or *compressive stress* if negative; the latter is known as a *shearing stress*. Since stress refers a vector force to a vector surface, we must consider it as a dyadic in the form

$$\begin{aligned}\Psi = &\ \mathbf{ii}\sigma_{xx}+\mathbf{ij}\sigma_{yx}+\mathbf{ik}\sigma_{zx}\\ &+\mathbf{ji}\sigma_{xy}+\mathbf{jj}\sigma_{yy}+\mathbf{jk}\sigma_{zy}\\ &+\mathbf{ki}\sigma_{xz}+\mathbf{kj}\sigma_{yz}+\mathbf{kk}\sigma_{zz}\end{aligned} \quad (6.20)$$

which is known as the *stress dyadic*. In each case, the first subscript of the component terms indicates the direction of the stress, and the second, the direction of the positive normal to the surface on which it acts. It can be shown that the stress dyadic is symmetric, which leads to the fact that it can be reduced to a principal axis form. When the shearing stresses are all zero and the tensional stresses are all equal and negative, the stress distribution is known as a *hydrostatic pressure*.

For an elastic medium that has undergone deformation, a point P whose position is given by $P = \mathbf{i}x+\mathbf{j}y+\mathbf{k}z$ will undergo a displacement, which we may designate by $\mathbf{s} = \mathbf{i}u+\mathbf{j}v+\mathbf{k}w$, so that the differential displacement of neighboring points with respect to P will be simply to first-order terms

$$d\mathbf{s} = \frac{\partial \mathbf{s}}{\partial x}dx+\frac{\partial \mathbf{s}}{\partial y}dy+\frac{d\mathbf{s}}{\partial z}dz = d\mathbf{r}\cdot\nabla\mathbf{s} \quad (6.21)$$

where $\nabla\mathbf{s}$ is given by

$$\begin{aligned}\nabla\mathbf{s} = &\ \mathbf{ii}\frac{\partial u}{\partial x}+\mathbf{ij}\frac{\partial v}{\partial x}+\mathbf{ik}\frac{\partial w}{\partial x}\\ &+\mathbf{ji}\frac{\partial u}{\partial y}+\mathbf{jj}\frac{\partial v}{\partial y}+\mathbf{jk}\frac{\partial w}{\partial y}\\ &+\mathbf{ki}\frac{\partial u}{\partial z}+\mathbf{kj}\frac{\partial v}{\partial z}+\mathbf{kk}\frac{\partial w}{\partial z}\end{aligned} \quad (6.22)$$

and is known as the *strain dyadic*. The theory developed using only the first-order terms above is known as *infinitesimal strain theory*. Each of the strain components represents a deformation per unit length. The diagonal components represent elongations per unit length and are known as *extensions* if positive, and *contractions* if negative. The other components represent sidewise distortions and are known as *shears*. The dyadic $\nabla \mathbf{s}$ may be written as the sum of a symmetric dyadic

$$\Phi = \tfrac{1}{2}[\nabla \mathbf{s} + (\nabla \mathbf{s})_c] \tag{6.23}$$

and a skew-symmetric dyadic

$$\Theta = \tfrac{1}{2}[\nabla \mathbf{s} - (\nabla \mathbf{s})_c] \tag{6.24}$$

The dyadic Θ represents a rotation only without distortion or change in volume and is known as the *rotation dyadic*. The dyadic Φ is known as the *pure strain dyadic*. A useful strain relation is that of *dilatation* Δ, defined as the increase in volume per unit volume; it can be shown to be given by

$$\Delta = \nabla \cdot \mathbf{s} = \frac{\partial u}{\partial x} + \frac{\partial v}{\partial y} + \frac{\partial w}{\partial z} \tag{6.25}$$

In practice, it is observed that within certain limits, the strains produced in an elastic body are directly proportional to the applied stresses. This is known as *Hooke's law*. Its generalized form states that each of the six independent components of stress at any point in a body will be a linear function of the six independent components of pure strain. For an isotropic body, that is, one whose properties are the same in all directions, these 36 coefficients relating the stress components to the strain components, known as the *elastic constants*, reduce down to two independent constants. The stress-strain relations, then, reduce to in vector form

$$\Psi = \lambda \Delta I + 2\mu \Phi = \lambda \nabla \cdot \mathbf{s} I + 2\mu \Phi \tag{6.26}$$

where I is the idemfactor. The coefficients λ and μ are known as the *Lamé constants*.

Other elastic constants that are often used for an isotropic solid are Young's modulus, Poisson's ratio, and the bulk modulus. *Young's modulus* E is defined as the ratio of the longitudinal stress on a cylinder to the longitudinal extension produced, all other stresses being zero, and *Poisson's ratio* σ as the ratio of the resultant lateral contraction to the longitudinal extension. Both are given in terms of the Lamé constants by

$$E = \frac{\sigma_{xx}}{\partial u/\partial x} = \frac{\mu(3\lambda + 2\mu)}{\lambda + \mu} \tag{6.27}$$

and

$$\sigma = -\frac{\partial w/\partial z}{\partial u/\partial x} = -\frac{\partial v/\partial y}{\partial u/\partial x} = \frac{\lambda}{2(\lambda + \mu)} \tag{6.28}$$

The *bulk modulus* or *incompressibility* k is defined as the ratio of the hydro-

186 ¶ *Seismology, Gravity, and Magnetism*

static pressure on a body to the resultant fractional change in volume and is given by

$$k = \frac{p}{-\Delta} = \frac{3\lambda + 2\mu}{3} \tag{6.29}$$

The *shear modulus* or *rigidity* is defined as the ratio between an applied shear stress and the resultant shear strain and is seen from expression (6.26) to be simply the Lamé constant μ. In some few problems, it is convenient to assume that $\lambda = \mu$. This relation, known variously as *Poisson's* or *Cauchy's relation*, produces the following simple relations among the several constants

$$\lambda = \mu \quad \sigma = \tfrac{1}{4} \quad k = \tfrac{5}{3}\mu \quad E = \tfrac{5}{2}\mu \tag{6.30}$$

Problem 6.1(a)   Derive the expressions (6.3) and (6.4).

Problem 6.1(b)   Derive the expression (6.6).

Problem 6.1(c)   Derive the relation (6.25).

Problem 6.1(d)   Derive the relations for Young's modulus, Poisson's ratio, and the bulk modulus in terms of the Lamé constants of Eq. (6.26).

6.2 Bodily Elastic Waves

We are interested in deriving and solving the equations of motion for a homogeneous, isotropic solid. The equations of motion are obtained as usual from Newton's second law of motion, equating the products of density and acceleration in a given direction to the net forces acting per unit volume in that direction. Referring to Fig. 6.1, it is seen that there are stresses acting in the x direction on each of the six faces of the rectangular parallelopiped, the stress on a back face being increased by the appropriate amount $(\partial/\partial r)\delta r$ to the front face. The total force in the x direction is then

$$\left(\sigma_{xx} + \frac{\partial \sigma_{xx}}{\partial x}\delta x\right)\delta y \delta z - \sigma_{xx}\delta y \delta z + \left(\sigma_{xy} + \frac{\partial \sigma_{xy}}{\partial y}\delta y\right)\delta x \delta z - \sigma_{xy}\delta x \delta z$$
$$+ \left(\sigma_{xz} + \frac{\partial \sigma_{xz}}{\partial z}\delta z\right)\delta x \delta y - \sigma_{xy}\delta x \delta y = \left(\frac{\partial \sigma_{xx}}{\partial x} + \frac{\partial \sigma_{xy}}{\partial y} + \frac{\partial \sigma_{xy}}{\partial z}\right)\delta x \delta y \delta z$$

Newton's second law then gives

$$\rho \frac{d^2 u}{dt^2} = \frac{\partial \sigma_{xx}}{\partial x} + \frac{\partial \sigma_{xy}}{\partial y} + \frac{\partial \sigma_{xz}}{\partial z} + \rho X$$

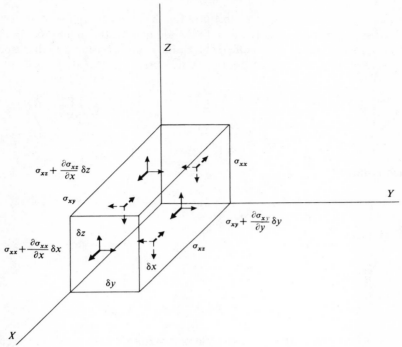

Fig. 6.1

and similarly

$$\rho \frac{d^2 v}{dt^2} = \frac{\partial \sigma_{yx}}{\partial x} + \frac{\partial \sigma_{yy}}{\partial y} + \frac{\partial \sigma_{yz}}{\partial z} + \rho Y \qquad (6.31)$$

$$\rho \frac{d^2 w}{dt^2} = \frac{\partial \sigma_{zx}}{\partial x} + \frac{\partial \sigma_{zy}}{\partial y} + \frac{\partial \sigma_{zz}}{\partial z} + \rho Z$$

for the y and z directions, where X, Y, Z are the body forces per unit mass. In elastic wave propagation, the body forces, such as gravity, can usually be neglected. In addition, for small displacements and velocities, which is usually the case in elasticity but not, for example, in hydrodynamics, the differential d/dt can be replaced by $\partial/\partial t$. Equations (6.31) then reduce to

$$\rho \frac{\partial^2 u}{\partial t^2} = \frac{\partial \sigma_{xx}}{\partial x} + \frac{\partial \sigma_{xy}}{\partial y} + \frac{\partial \sigma_{xz}}{\partial z}$$

$$\rho \frac{\partial^2 v}{\partial t^2} = \frac{\partial \sigma_{yx}}{\partial x} + \frac{\partial \sigma_{yy}}{\partial y} + \frac{\partial \sigma_{yz}}{\partial z} \qquad (6.32)$$

$$\rho \frac{\partial^2 w}{\partial t^2} = \frac{\partial \sigma_{zx}}{\partial x} + \frac{\partial \sigma_{zy}}{\partial y} + \frac{\partial \sigma_{zz}}{\partial z}$$

For an isotropic solid, we can substitute for the stress components in terms of the displacements by Eq. (6.26) with the aid of Eqs. (6.20) and (6.23). For a homogeneous solid, the coefficients λ and μ are constants with respect to x, y, and z. For the first of Eq. (6.32), we obtain

$$\rho \frac{\partial^2 u}{\partial t^2} = \frac{\partial}{\partial x}\left(\lambda \Delta + 2\mu \frac{\partial u}{\partial x}\right) + \frac{\partial}{\partial y}\left[\mu\left(\frac{\partial v}{\partial x} + \frac{\partial u}{\partial y}\right)\right]$$
$$+ \frac{\partial}{\partial z}\left[\mu\left(\frac{\partial w}{\partial x} + \frac{\partial u}{\partial z}\right)\right] \tag{6.33}$$

which reduces to

$$\rho \frac{\partial^2 u}{\partial t^2} = (\lambda+\mu)\frac{\partial \Delta}{\partial x} + \mu \nabla^2 u$$

and similarly

$$\rho \frac{\partial^2 v}{\partial t^2} = (\lambda+\mu)\frac{\partial \Delta}{\partial y} + \mu \nabla^2 v \tag{6.34}$$

$$\rho \frac{\partial^2 w}{\partial t^2} = (\lambda+\mu)\frac{\partial \Delta}{\partial z} + \mu \nabla^2 w$$

for the second and third equations. We may write these equations in vector form by multiplying the first by **i**, the second by **j**, and the third by **k** and adding, obtaining

$$\rho \frac{\partial^2 \mathbf{s}}{\partial t^2} = (\lambda+\mu)\nabla \Delta + \mu \nabla^2 \mathbf{s}$$
$$= (\lambda+\mu)\nabla(\nabla \cdot \mathbf{s}) + \mu \nabla^2 \mathbf{s}$$
$$= (\lambda+2\mu)\nabla(\nabla \cdot \mathbf{s}) - \mu \nabla \times \nabla \times \mathbf{s} \tag{6.35}$$

with the aid of Eqs. (6.25) and (1.37).

Let us assume for the moment that the displacement **s** is irrotational. Then, by definition, $\nabla \times \mathbf{s}$ vanishes, and we shall have from Eq. (1.37)

$$\nabla \times \nabla \times \mathbf{s} = \nabla(\nabla \cdot \mathbf{s}) - \nabla \cdot \nabla \mathbf{s} = 0$$

or

$$\nabla(\nabla \cdot \mathbf{s}) = \nabla \cdot \nabla \mathbf{s}$$

Substituting into Eq. (6.35), we obtain

$$\rho \frac{\partial^2 \mathbf{s}}{\partial t^2} = (\lambda+2\mu)\nabla^2 \mathbf{s} \tag{6.36}$$

From the discussion of Section 1.10, we see that this is simply the wave equation and that it defines a displacement propagating through the medium with a velocity α given by

$$\alpha = \left(\frac{\lambda+2\mu}{\rho}\right)^{1/2} \tag{6.37}$$

Let us next assume that the displacement **s** is solenoidal. Then, by definition, $\nabla \cdot \mathbf{s}$ vanishes and we shall have for Eq. (6.35)

$$\rho \frac{\partial^2 \mathbf{s}}{\partial t^2} = \mu \nabla^2 \mathbf{s} \tag{6.38}$$

Again, this is simply the wave equation defining a displacement propagating through the medium with a velocity β given by

$$\beta = \left(\frac{\mu}{\rho}\right)^{1/2} \tag{6.39}$$

This reduction gives a complete description of wave propagation in a homogeneous, isotropic solid, for we have stated in Section 1.1 that any vector displacement may be represented by an irrotational and a solenoidal component. The first type of wave, propagated with the higher velocity α, is variously known as the *dilatational, irrotational*, or *P* (primary) wave; the second type of wave traveling with the slower velocity β is variously known as the *distortional, equivoluminal, rotational*, or *S* (secondary) wave.

The two different types of propagation can be illustrated in a simple manner by consideration of plane wave propagation. Let us assume a plane wave propagated in the x direction. Then the displacement components u, v, w will be functions of x and t only. The equations of motion (6.34) reduce to

$$\rho \frac{\partial^2 u}{\partial t^2} = (\lambda+\mu)\frac{\partial^2 u}{\partial x^2} + \mu \frac{\partial^2 u}{\partial x^2} = (\lambda+2\mu)\frac{\partial^2 u}{\partial x^2}$$

$$\rho \frac{\partial^2 v}{\partial t^2} = \mu \frac{\partial^2 v}{\partial x^2} \tag{6.40}$$

$$\rho \frac{\partial^2 w}{\partial t^2} = \mu \frac{\partial^2 w}{\partial x^2}$$

The displacement in the direction of propagation or *longitudinal* vibration is propagated with the *P* velocity α; the displacements at right angles to the direction of propagation or *transverse* vibration are propagated with the *S* velocity β.

For a perfect liquid, $\mu = 0$, the equations of motion reduce to

$$\rho \frac{\partial^2 \Delta}{\partial t^2} = k \nabla^2 \Delta \tag{6.41}$$

This again is the wave equation, the dilatation being propagated with the velocity

$$c = \left(\frac{k}{\rho}\right)^{1/2} \tag{6.42}$$

Problem 6.2(a)   Derive the equation of motion (6.35) directly from the vector expressions of stress and strain.

Problem 6.2(b)   Obtain the dilatational and rotational wave equations from the scalar equations of motion (6.34).

Problem 6.2(c)   Derive Eq. (6.41).

Problem 6.2(d)   Show from the scalar equations of motion (6.34) that the inclusion of a body force term derivable from a potential, such as gravity, will affect only the dilatational waves and not the rotational waves.

6.3 Reflection and Refraction of Elastic Waves

Let us next investigate the effect on elastic wave propagation of an interface between two media. Consider two homogeneous, isotropic media M and M_1 in welded contact separated by a plane boundary, as shown in Fig. 6.2. If we were to have a P wave incident on the boundary from the lower medium, we might expect, in the most general case, that both a P and an S wave would be returned back to the lower medium from the boundary and that both a P and an S wave would be transmitted to the upper medium. The waves returned to the lower medium are known as *reflections*, and the waves transmitted to the upper medium are known as *refractions*. For an incident S wave, we would expect again both the two reflected and the two refracted P and S waves.

Let us take an incident P wave. We shall assume, for simplicity, that it is a simple harmonic plane wave. The normal to the incident plane wave is shown by the incident arrow line in Fig. 6.2 and the normals to the four resultant waves as indicated. We shall also assume, for convenience, that the incident wave front is normal to the XZ plane, so that there is no propagation dependence on the y coordinate, and we shall assume that the boundary between the two media is coincident with the XY plane. From Eq. (1.197) the incident P wave may be represented by

$$\varphi = A_0 e^{i(k_x x - k_z z - \omega t)}$$
$$= A_0 e^{i\kappa(x - z \tan e - ct)} \tag{6.43}$$

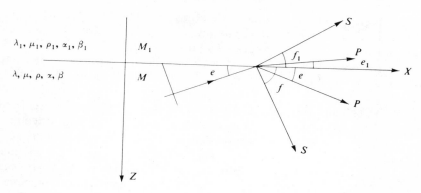

Fig. 6.2

where κ has been used for k_x for convenience and, from Eqs. (1.199), (1.189), and (6.37), is given by

$$\kappa = k_x = \frac{2\pi \cos e}{\lambda} = k \cos e = \omega \frac{\cos e}{\alpha} \qquad (6.44)$$

where from Eq. (1.199), k_z is given by

$$k_z = \frac{2\pi \sin e}{\lambda} = k_x \frac{\sin e}{\cos e} = \kappa \tan e \qquad (6.45)$$

e being the angle between the normal to the incident wave and the boundary and λ in Eqs. (6.44) and (6.45) being the wavelength, and where from Eqs. (1.189), (1.186), and (6.37) c is given by

$$c = \frac{\omega}{\kappa} = \frac{\alpha k}{\kappa} = \frac{\alpha k_x}{\kappa \cos e} = \frac{\alpha}{\cos e} \qquad (6.46)$$

We see from Eqs. (6.43) and (6.46) that c is the velocity of propagation of the incident wave front along the horizontal boundary between the two media. Along this boundary $z = 0$, and the argument of the exponential in Eq. (6.43) will reduce to $\kappa(x-ct)$. It is apparent that we could write down expressions similar to Eq. (6.43) for the two reflected and the two refracted waves involving terms in $e, f, e_1, f_1, \alpha, \beta, \alpha_1$, and β_1. In all of these expressions, we can obtain an argument of the exponential of the form $\kappa(x-ct)$ for $z = 0$. If our solutions involving these five waves are to satisfy the boundary conditions for all x and t, the argument for all five terms must be the same along the plane $z = 0$. From this condition and Eqs. (6.46) and (6.44), we have

$$\frac{\cos e}{\alpha} = \frac{\cos f}{\beta} = \frac{\cos e_1}{\alpha_1} = \frac{\cos f_1}{\beta_1} = \frac{1}{c} \qquad (6.47)$$

This expression tells us that the angle of the reflected P wave is the same as the angle of the incident P wave and that the angles of reflection and refraction of the other waves are as given by Eq. (6.47) in terms of the angle of the incident P wave. If we had considered an incident S wave, we would have obtained the same expressions in terms of an angle of incidence f of the S wave.

We shall want to obtain solutions for the characteristics of the reflected P and S waves, or for quantities related to them, rather than attempt to obtain solutions for the vector displacements $\mathbf{s}$ or it components u, v, w, all of which will, in general, be a combination of both P and S type motions. We may do this through the introduction of the concept of *displacement potential*. The displacement potentials, similar to gravitational potentials, give the desired displacement components in terms of combinations of their space derivatives and are, in themselves, solutions of the wave equations. For a given problem, we are at some liberty in the definition of these potentials. For our present problem of a plane, horizontal boundary and for wave motion independent of y, it is convenient to define two potentials, φ and ψ, by

$$u = \frac{\partial \varphi}{\partial x} + \frac{\partial \psi}{\partial z} \quad v = v \quad w = \frac{\partial \varphi}{\partial z} - \frac{\partial \psi}{\partial x} \qquad (6.48)$$

Substituting Eqs. (6.48) into the equations of motion (6.34), we obtain for the first and third equations

$$\rho \frac{\partial^2}{\partial t^2}\left(\frac{\partial \varphi}{\partial x} + \frac{\partial \psi}{\partial z}\right) = (\lambda+\mu)\frac{\partial}{\partial x}\nabla^2\varphi + \mu\nabla^2\left(\frac{\partial \varphi}{\partial x} + \frac{\partial \psi}{\partial z}\right)$$

$$\rho \frac{\partial^2}{\partial t^2}\left(\frac{\partial \varphi}{\partial z} - \frac{\partial \psi}{\partial x}\right) = (\lambda+\mu)\frac{\partial}{\partial z}\nabla^2\varphi + \mu\nabla^2\left(\frac{\partial \varphi}{\partial z} - \frac{\partial \psi}{\partial x}\right)$$

remembering that all derivatives with respect to y are zero. These equations will be satisfied if φ and ψ satisfy the wave equations

$$\rho \frac{\partial^2 \varphi}{\partial t^2} = (\lambda + \partial\mu)\nabla^2\varphi \qquad (6.49)$$

and

$$\rho \frac{\partial^2 \psi}{\partial t^2} = \mu\nabla^2\psi \qquad (6.50)$$

as can be seen by substituting Eqs. (6.49) and (6.50) back into these two equations. The second of Eqs. (6.34) reduces simply to

$$\rho \frac{\partial^2 v}{\partial t^2} = \mu\nabla^2 v \qquad (6.51)$$

We see that the displacement potential φ defines the P wave and that the displacement potential ψ defines an S wave. We also see that the displacement component v is an S wave only. It is useful to distinguish between these two types of S motion. For S motion that is parallel, or horizontal, to the boundary, that is, the v displacement component, we shall designate the motion by SH. For S motion that is normal, or vertical, to the boundary, that is, the ψ displacement potential, we shall designate the motion by SV.

We shall consider, as our first example, a reflection at a free surface. In such case, there is no upper medium M_1, and the boundary conditions become simply that the stress components must vanish over the surface $z = 0$. If the stress components did not vanish, there would be an infinite acceleration in the negative z direction across the boundary. From Eqs. (6.26), (6.20), and (6.23), these boundary conditions are

$$\sigma_{zz} = \lambda\Delta + 2\mu\frac{\partial w}{\partial z} = 0$$

$$\sigma_{xz} = \mu\left(\frac{\partial w}{\partial x} + \frac{\partial u}{\partial z}\right) = 0 \quad (6.52)$$

$$\sigma_{yz} = \mu\left(\frac{\partial w}{\partial y} + \frac{\partial v}{\partial z}\right) = 0$$

Substituting from Eqs. (6.48), we obtain

$$\lambda\nabla^2\varphi + 2\mu\left(\frac{\partial^2\varphi}{\partial z^2} - \frac{\partial^2\psi}{\partial x\partial z}\right) = 0 \quad (6.53)$$

$$\mu\left(2\frac{\partial^2\varphi}{\partial x\partial z} - \frac{\partial^2\psi}{\partial x^2} + \frac{\partial^2\psi}{\partial z^2}\right) = 0 \quad (6.54)$$

$$\mu\frac{\partial v}{\partial z} = 0 \quad (6.55)$$

We see that φ and ψ are involved only in the first two equations and v only in the third. We may conclude then that an incident SH wave will only provide a reflected SH wave and that an incident P or SV wave will only produce reflected P and SV waves. As we shall see later, this relation also applies to the refracted waves when we have an upper medium M_1 across the boundary.

For an incident P wave against a free surface, we may then write, for φ in terms of the incident and reflected P waves and for ψ in terms of the reflected SV wave,

$$\varphi = A_0 e^{i\kappa(x - z\tan e - ct)} + A e^{i\kappa(x + z\tan e - ct)} \quad (6.56)$$

$$\psi = B e^{i\kappa(x + z\tan f - ct)} \quad (6.57)$$

where A_0, A, and B are the amplitudes of the incident P, reflected P, and reflected SV waves, respectively. Substituting Eqs. (6.56) and (6.57) into Eqs. (6.53) and (6.54) we obtain

$$\lambda(A_0+A)(1+\tan^2 e)+2\mu(A_0+A)\tan^2 e - 2\mu B \tan f = 0 \quad (6.58)$$

$$-2(A_0-A)\tan e + B(\tan^2 f - 1) = 0 \quad (6.59)$$

The equations can be solved numerically for the ratios of the amplitudes of the reflected waves A and B to the amplitude of the incident wave A_0 as a function of e. Such ratios are referred to as *reflection coefficients* or, in the case of a refracted wave, as *refraction coefficients*.

From Eq. (6.47) we can determine some of the general characteristics of the reflected wave directions. Since $\beta < \alpha$, $\cos e < \cos f$; there will always be a reflected SV direction for all incident P directions. For $e = \pi/2, f = \pi/2$; for $e = 0, f = \cos^{-1}(\beta/\alpha)$. The reflected SV wave directions are confined between these two angles as the incidence P wave direction varies from normal incidence, $e = \pi/2$, to grazing incidence, $e = 0$.

Using Eqs. (6.37), (6.39), and (6.47), we may rewrite Eqs. (6.58) and (6.59) as

$$A_0 + A = \frac{2 \tan f}{\tan^2 f - 1} B \quad (6.60)$$

and

$$A_0 - A = \frac{\tan^2 f - 1}{2 \tan e} B \quad (6.61)$$

or combining the two,

$$\frac{A_0 - A}{A_0 + A} = \frac{(\tan^2 f - 1)^2}{4 \tan e \tan f} \quad (6.62)$$

We see that $A_0 = A$ only if $\tan f = 1$. Such values of f are usually not possible for earth, elastic materials. We see that $A_0 = -A$ for $\tan e = 0$, $\tan f = 0$, or $\tan f = \infty$. The first and third conditions correspond to grazing and normal incidence, respectively. The second solution is an impossibility. We see then that there will be a reflected SV wave except at grazing and normal incidence. Using specific elastic medium constants, one generally finds that the amplitude of the reflected SV is appreciable over all angles of incidence except those near grazing and normal incidence.

From Eqs. (6.48), (6.56), and (6.57) the surface movement along the plane $z = 0$ is given by, where the $i\kappa$ multiplier is included in the constants,

$$u = (A_0 + A + B \tan f)e^{i\kappa(x-ct)} \quad (6.63)$$

and

$$w = -[(A_0 - A)\tan e + B]e^{i\kappa(x-ct)} \quad (6.64)$$

We can speak of an *apparent* angle of emergence, or incidence, $\bar{e}$, given in terms of the ratios of the displacement components, as

$$\tan \bar{e} = -\frac{w}{u} \tag{6.65}$$

remembering that the direction of Z axis is down in Fig. 6.2. Substituting from Eqs. (6.60) and (6.61), we obtain

$$\tan \bar{e} = \frac{(A_0 - A)\tan e + B}{A_0 + A + B \tan f} = \frac{\dfrac{\tan^2 f - 1}{2 \tan e}\tan e + 1}{\dfrac{2 \tan f}{\tan^2 f - 1} + \tan f}$$

$$= \frac{\tan^2 f - 1}{2 \tan f} = -\cot 2f$$

or

$$\bar{e} = 2f - \frac{\pi}{2} \tag{6.66}$$

Equation (6.66) gives us a relation between the measured quantity $\bar{e}$ and the propagation angle f, or e.

For an incident SV wave against a free surface, the expressions for φ and ψ will be

$$\psi = B_0 e^{i\kappa(x - z \tan f - ct)} + B e^{i\kappa(x + z \tan f - ct)} \tag{6.67}$$

$$\varphi = A e^{i\kappa(x + z \tan e - ct)} \tag{6.68}$$

where B_0, B, and A are the amplitudes of the incident SV, reflected SV, and reflected P waves, respectively. Following the same procedure as before, we can obtain the expressions

$$B_0 + B = -\frac{2 \tan e}{\tan^2 f - 1} A \tag{6.69}$$

and

$$B_0 - B = -\frac{\tan^2 f - 1}{2 \tan f} A \tag{6.70}$$

or

$$\frac{B_0 - B}{B_0 + B} = \frac{(\tan^2 f - 1)^2}{4 \tan e \tan f} \tag{6.71}$$

Expression (6.71) is the same as Eq. (6.62). We see that there is complete reflection of the SV wave at grazing and normal incidence.

We have a new complication with regard to the P wave. As the angle of incidence f is decreased from normal incidence, $f = \pi/2$, where from Eq.

(6.47) we also have $e = \pi/2$, we reach an angle $f = \cos^{-1}(\beta/\alpha)$ for which $e = 0$. For more grazing angles of incidence, e will be imaginary; there will be no reflected P wave. From Eq. (6.68) we see that for such angles, the P motion is not a simple harmonic wave, but a disturbance confined near the plane $z = 0$, propagating in the x direction and decreasing exponentially away from the boundary in the z direction. Remembering that the sum of two complex numbers, which are conjugate, is a real number and that their difference is an imaginary number, we see that since $\tan e$ is imaginary that Eq. (6.69) is imaginary and Eq. (6.70) real or that iB_0/A is the complex conjugate of iB/A. The amplitude of the reflected SV wave is equal to the amplitude of the incident SV. This condition is called *total reflection*. We see that for angles of incidence more grazing than $f = \cos^{-1}(\beta/\alpha)$, which is referred to as the *critical angle*, there is total reflection.

For an incident SH wave against a free surface, the expression for v will be

$$v = C_0 e^{i\kappa(x - z\tan f - ct)} + C e^{i\kappa(x + z\tan f - ct)} \tag{6.72}$$

where C_0 and C are the amplitudes of the incident and reflected SH waves. Substituting Eq. (6.72) into the boundary condition (6.55), we obtain simply that

$$C_0 = C \tag{6.73}$$

The amplitude of the reflected wave is equal to the amplitude of the incident wave.

Returning to the more general case of two elastic media M and M_1 separated by a plane boundary, we might consider an incident, simple harmonic wave of the combined P and SV type with amplitude coefficients A_0 and B_0. There would result, in general, a reflected and refracted P wave with amplitude coefficients A and A_1 and a reflected and refracted SV wave with amplitude coefficients B and B_1. The expressions for these four waves would be similar to those given in the above derivations. The four boundary conditions that lead to the four conditional equations for A, A_1, B, and B_1 in terms of A_0 and B_0 are continuity across the boundary of the displacement components u and w and the stress components σ_{xz} and σ_{zz}. These conditional equations do not lead to any simple, direct interpretation and must be solved numerically for the elastic media constants involved. The existence of any one reflected or refracted wave is governed by the relations of Eq. (6.47).

We shall examine here the case of an incident SH wave to a plane boundary separating two elastic media M and M_1. We shall have, in general, a reflected and refracted SH, the SH displacements in media M and M_1 being given, respectively, by

$$v = C_0 e^{i\kappa(x - z\tan f - ct)} + C e^{i\kappa(x + z\tan f - ct)} \tag{6.74}$$

and

$$v_1 = C_1 e^{i\kappa(x - z\tan f_1 - ct)} \tag{6.75}$$

The boundary conditions are continuity of v and σ_{yz}, or

$$v = v_1 \tag{6.76}$$

and

$$\mu \frac{\partial v}{\partial z} = \mu_1 \frac{\partial v_1}{\partial z} \tag{6.77}$$

Substituting Eqs. (6.74) and (6.75) into Eqs. (6.76) and (6.77), we obtain

$$C_0 + C = C_1 \tag{6.78}$$

and

$$\mu(C_0 - C)\tan f = \mu_1 C_1 \tan f_1 \tag{6.79}$$

Multiplying Eq. (6.78) by $\mu_1 \tan f_1$ and subtracting from Eq. (6.79) and multiplying by $\mu \tan f$ and adding, we obtain for the reflection and refraction coefficients, respectively,

$$\frac{C}{C_0} = \frac{\mu \tan f - \mu_1 \tan f_1}{\mu \tan f + \mu_1 \tan f_1} \tag{6.80}$$

and

$$\frac{C_1}{C_0} = \frac{2\mu \tan f}{\mu \tan f + \mu_1 \tan f_1} \tag{6.81}$$

These coefficients are similar in form to those to be derived in the following paragraphs for two fluid media. Their characteristics may be determined by analogy from that discussion.

Let us consider two fluid media in contact. Since $\mu = 0$, there will be no ψ wave, and our displacement potentials of Eqs. (6.48) will reduce to

$$u = \frac{\partial \varphi}{\partial x} \qquad w = \frac{\partial \varphi}{\partial z} \tag{6.82}$$

Since in this case the two media are not in welded contact, our boundary conditions refer only to continuity of the normal components of displacement and stress. From Eqs. (6.82) and (6.52) these will be simply

$$\frac{\partial \varphi}{\partial z} = \frac{\partial \varphi_1}{\partial z} \tag{6.83}$$

and

$$\lambda \nabla^2 \varphi = \lambda_1 \nabla^2 \varphi_1 \tag{6.84}$$

In general, we shall have both a reflected and a refracted wave so that our displacement potentials in the two media may be expressed by

$$\varphi = A_0 e^{i\kappa(x - z \tan e - ct)} + A e^{i\kappa(x + z \tan e - ct)} \tag{6.85}$$

and

$$\varphi_1 = A_1 e^{i\kappa(x - z \tan e_1 - ct)} \tag{6.86}$$

Substituting Eqs. (6.85) and (6.86) into Eqs. (6.83) and (6.84), we obtain

$$(A_0 - A) \tan e = A_1 \tan e_1 \tag{6.87}$$

and

$$\lambda(A_0 + A)(1 + \tan^2 e) = \lambda_1 A_1 (1 + \tan^2 e_1)$$

$$(A_0 + A) \frac{\lambda}{\cos^2 e} = A_1 \frac{\lambda_1}{\cos^2 e_1}$$

$$\rho(A_0 + A) = \rho_1 A_1 \tag{6.88}$$

where we have substituted from Eq. (6.47) and where in this case $\alpha^2 = \lambda/\rho$ and $\alpha_1^2 = \lambda_1/\rho_1$. Multiplying Eq. (6.87) by ρ_1 and Eq. (6.88) by $\tan e_1$, and subtracting and multiplying Eq. (6.87) by ρ and Eq. (6.88) by $\tan e$ and adding, we obtain for the reflection and refraction coefficients, respectively,

$$\frac{A}{A_0} = \frac{\rho_1 \tan e - \rho \tan e_1}{\rho_1 \tan e + \rho \tan e_1} \tag{6.89}$$

and

$$\frac{A_1}{A_0} = \frac{2\rho \tan e}{\rho_1 \tan e + \rho \tan e_1} \tag{6.90}$$

At this point in our discussion of reflection and refraction between two fluid media, it is convenient to change our notation to that more commonly associated with acoustics. As shown in Fig. 6.3, we shall let $\theta_1 = \pi/2 - e$, $\theta_2 = \pi/2 - e_1$, $c_1 = \alpha$, $c_2 = \alpha_1$, $\rho_1 = \rho$, $\rho_2 = \rho_1$, $A_1 = A$, and $A_2 = A_1$. Then Eqs. (6.89) and (6.90) become

$$\frac{A_1}{A_0} = \frac{\rho_2 \cot \theta_1 - \rho_1 \cot \theta_2}{\rho_2 \cot \theta_1 + \rho_1 \cot \theta_2} \tag{6.91}$$

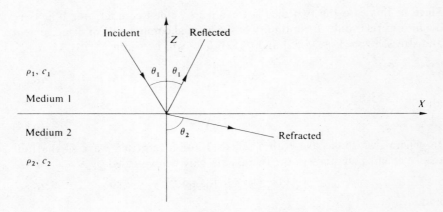

Fig. 6.3

and

$$\frac{A_2}{A_0} = \frac{2\rho_1 \cot \theta_1}{\rho_2 \cot \theta_1 + \rho_1 \cot \theta_2} \tag{6.92}$$

and where Eq. (6.47) becomes

$$\frac{\sin \theta_1}{c_1} = \frac{\sin \theta_2}{c_2} \tag{6.93}$$

Expressions (6.91) and (6.92) may be written in a somewhat more useful form in terms of θ_1 using Eq. (6.93)

$$\frac{A_1}{A_0} = \frac{\dfrac{\rho_2}{\rho_1} - \dfrac{\cot \theta_2}{\cot \theta_1}}{\dfrac{\rho_2}{\rho_1} + \dfrac{\cot \theta_2}{\cot \theta_1}}$$

$$= \frac{\dfrac{\rho_2}{\rho_1} - \dfrac{c_1 \cos \theta_2}{c_2 \cos \theta_1}}{\dfrac{\rho_2}{\rho_1} + \dfrac{c_1 \cos \theta_2}{c_2 \cos \theta_1}}$$

$$= \frac{\dfrac{\rho_2}{\rho_1} - \dfrac{\sqrt{(c_1^2/c_2^2) - \sin^2 \theta_1}}{\sqrt{1 - \sin^2 \theta_1}}}{\dfrac{\rho_2}{\rho_1} + \dfrac{\sqrt{(c_1^2/c_2^2) - \sin^2 \theta_1}}{\sqrt{1 - \sin^2 \theta_1}}} \tag{6.94}$$

and

$$\frac{A_2}{A_0} = \frac{2}{\dfrac{\rho_2}{\rho_1} + \dfrac{\sqrt{(c_1^2/c_2^2) - \sin^2 \theta_1}}{\sqrt{1 - \sin^2 \theta_1}}} \tag{6.95}$$

Expression (6.94) is referred to as the *Rayleigh reflection coefficient*. For normal incidence $\theta_1 = 0$, Eqs. (6.94) and (6.95) reduce to

$$\frac{A_1}{A_0} = \frac{\rho_2 c_2 - \rho_1 c_1}{\rho_2 c_2 + \rho_1 c_1} \tag{6.96}$$

and

$$\frac{A_2}{A_0} = \frac{2\rho_1 c_2}{\rho_2 c_2 + \rho_1 c_1} \tag{6.97}$$

From Eq. (6.94) it is apparent that if $c_2 < c_1$ and $\rho_2 c_2 > \rho_1 c_1$, the reflection coefficient A_1/A_0 will start out as a positive value and will decrease as the angle increases from normal incidence since the radical in the numerator decreases less rapidly than the radical in the denominator. The reflection coefficient will decrease to zero at an angle of incidence such that

$$\frac{\rho_2}{\rho_1} = \frac{\sqrt{(c_1^2/c_2^2) - \sin^2\theta_1}}{\sqrt{1 - \sin^2\theta_1}} \tag{6.98}$$

This angle is often referred to as an *intramission angle*. Beyond the intramission angle, the reflection coefficient will increase to a value of unity at grazing incidence, $\theta_1 = \pi/2$. The sign of the reflection coefficient in this region will be negative, indicating that the reflected wave has been reflected from the boundary with a phase change of π with respect to the incident wave. For the more usual case of $\rho_2 c_2 < \rho_1 c_1$ when $c_2 < c_1$, the reflection coefficient will start out with a negative value at normal incidence and increase to negative unity at grazing incidence

For the case when $c_2 > c_1$, the reflection coefficient will increase from normal incidence to a value of unity at an intermediate angle such that

$$\sin\theta_1 = \frac{c_1}{c_2} \tag{6.99}$$

As before, this angle is referred to as the critical angle. Beyond this angle of incidence, the radical in the numerator of the second term of Eq. (6.94) becomes imaginary. The imaginary radical can be expressed by $\pm i\sqrt{\sin^2\theta_1 - (c_1^2/c_2^2)}$, where $\sqrt{\sin_2\theta_1 - (c_1^2/c_2^2)}$ is a real quantity. The quantity $i\sqrt{\sin^2\theta_1 - (c_1^2/c_2^2)}$ is the correct representation for the physical problem, giving for the refracted wave an amplitude that decreases exponentially away from the boundary into the lower medium and not the physically inconsistent picture of an amplitude that increases continuously toward infinity. Thus, Eq. (6.94) becomes, for incidence more grazing than the critical angle,

$$\frac{A_1}{A_0} = \frac{\dfrac{\rho_2}{\rho_1} - i\dfrac{\sqrt{\sin^2\theta_1 - (c_1^2/c_2^2)}}{\sqrt{1 - \sin^2\theta_1}}}{\dfrac{\rho_2}{\rho_1} + i\dfrac{\sqrt{\sin^2\theta_1 - (c_1^2/c_2^2)}}{\sqrt{1 - \sin^2\theta_1}}} \tag{6.100}$$

which can be written in the following form, after a small amount of algebra,

$$\frac{A_1}{A_0} = e^{-i\epsilon} \tag{6.101}$$

where

$$\frac{\epsilon}{2} = \tan^{-1}\frac{\rho_1\sqrt{\sin^2\theta_1 - (c_1^2/c_2^2)}}{\rho_2\sqrt{1 - \sin^2\theta_1}} \tag{6.102}$$

Thus, at angles more grazing than the critical angle, the amplitude of the reflected wave remains unity, but there is a phase change ϵ between the reflected and incident waves. This phase change is 0 at the critical angle and increases to π at grazing incidence. Similarly, the expression for the refraction coefficient [Eq. (6.95)] becomes

$$\frac{A_2}{A_0} = \frac{2}{\left[\frac{\rho_2^2}{\rho_1^2} + \frac{\sin^2\theta_1 - (c_1^2/c_2^2)}{1 - \sin^2\theta_1}\right]^{1/2}} e^{-i(\epsilon/2)} \qquad (6.103)$$

For the refracted wave, the coefficient of the z term in the exponential of Eq. (6.86) becomes imaginary and is replaced by, in our new notation from Eqs. (6.44) and (6.93),

$$-i\kappa \tan e_1 = -i\frac{\omega}{c_2}\sin\theta_2 \cot\theta_2 = -ik_1 \frac{c_1}{c_2}\sqrt{1 - \sin^2\theta_2}$$

$$= k_1 \sqrt{\sin^2\theta_1 - (c_1^2/c_2^2)}$$

Substituting this expression and Eq. (6.103) into Eq. (6.86), we get for the refracted wave

$$\varphi_1 = \frac{2A_0 e^{-i(\epsilon/2)k_1\sqrt{\sin^2\theta_1 - (c_1^2c_2^2)}\, z}}{\left[\frac{\rho_2}{\rho_1} + \frac{\sin^2\theta_1 - (c_1^2/c_2^2)}{1 - \sin^2\theta_1}\right]^{1/2}} e^{i\kappa(x - ct)} \qquad (6.104)$$

We have then, finally, that the amplitude of the reflected wave is equal to that of the incident wave with a phase change of ϵ. There is total reflection and no transmission of energy into the lower medium. Equation (6.104) represents a somewhat different wave. It is *bound* to the discontinuity surface and propagates parallel to this surface with a velocity less than that of the lower medium, except at the critical angle for which the bound wave velocity is equal to that of the lower medium. There is no transmission of energy into the lower medium, and the amplitude decreases exponentially away from the surface into the lower medium. This wave is in no sense a *free* wave; it can exist only as a consequence of reflection at angles of incidence more grazing than critical and is constrained to move along with the trace velocity of the reflection along the boundary. It is a diffraction effect across the interface associated with and bound to the reflected wave.

Problem 6.3(a) Reduce expressions (6.58) and (6.59) to Eqs. (6.60) and (6.61).

Problem 6.3(b) Derive the conditional equations for the general case of a P and SV incident against a boundary between two elastic media.

Problem 6.3(c) Follow through the reductions to obtain expressions (6.101) and (6.103).

6.4 Development of Solution in Terms of Rays

Most of the developments in this and the following chapter are based in one way or another on solutions of the wave equation. As we have discussed previously, there are, in general, a variety of different types of solutions to the wave equation. For a particular problem, one or more of these may be formalized in various ways to meet the environment, boundary, and initial conditions. For some of the more elementary problems, an explicit solution satisfying the above conditions can be given; more often this is not the case. For these latter cases, there are two types of solution to the wave equation that are of particular importance in seismology. One is a transformation of the wave equation to the eikonal equation and a solution in terms of wave surfaces and rays. The other is a development through specific boundary conditions into a solution in terms of normal modes. In some instances, the physical conditions of the problem lead to a simpler solution in terms of rays; in others, a solution in terms of normal modes is more satisfactory.

The *wave surfaces* are the loci of points that are undergoing the same motion in a one-to-one correspondence at a given instant of time. The *rays* are the normals to the wave surfaces and give the direction of propagation of energy through the medium. A *normal mode* defines a preferred frequency of vibration for the system, and a solution in terms of normal modes is a summation of the contributions from the various preferred frequencies of vibration of the system.

In general, both these solutions are complicated in application to specific seismological problems through the introduction of variation in seismic velocity and specific boundary conditions into the wave equation. The general nature of these complications and solutions in terms of rays or waves are given in this and the following chapter.

For the three-dimensional wave equation of Eq. (1.173), we have in rectangular coordinates

$$c^2 \left(\frac{\partial^2 \psi}{\partial x^2} + \frac{\partial^2 \psi}{\partial y^2} + \frac{\partial^2 \psi}{\partial z^2} \right) = \frac{\partial^2 \psi}{\partial t^2} \tag{6.105}$$

Extending the general solution [Eq. (1.175)] to three dimensions, and using for convenience only the first of the two solutions [Eq. (1.175)], we have

$$\psi = f(lx + my + nz - ct) \tag{6.106}$$

as can be seen by substitution back into Eq. (6.105) and where l, m, n are the direction cosines defining the wave propagation direction. We see that ψ in the form of Eq. (6.106) will also be a solution of the equation

$$\left(\frac{\partial \psi}{\partial x} \right)^2 + \left(\frac{\partial \psi}{\partial y} \right)^2 + \left(\frac{\partial \psi}{\partial z} \right)^2 = \frac{1}{c^2} \left(\frac{\partial \psi}{\partial t} \right)^2 \tag{6.107}$$

Equation (6.107) is referred to as the equation of characteristics of Eq.

(6.105). For later applications where c is not a constant, and as we shall see in the discussion that follows the wave propagation is not defined by a straight line, we shall want to replace the argument of Eq. (6.106) by a more general expression such that

$$\psi = f[W(x, y, z) - c_0 t] \tag{6.108}$$

where c_0 is a constant, reference velocity. Substituting Eq. (6.108) into Eq. (6.107), we obtain the time-independent conditional equation

$$\left(\frac{\partial W}{\partial x}\right)^2 + \left(\frac{\partial W}{\partial y}\right)^2 + \left(\frac{\partial W}{\partial z}\right)^2 = \frac{c_0^2}{c^2} = n^2 \tag{6.109}$$

where n is the *index of refraction* and is defined as

$$n = \frac{c_0}{c} \tag{6.110}$$

Equation (6.109) is the *eikonal equation*. This equation is of fundamental importance, for it leads directly to the concept of rays. It is useful in the solution of problems where c is not a constant but a function of the space coordinates.

In those cases where c is not a constant, $c = c(x, y, z)$, such as the earth where the velocity is a function of the elastic properties and the density of the different materials encountered or the ocean where the velocity is a function of temperature, salinity, and depth, a solution of the eikonal equation will not, in general, be a solution of the wave equation. It should also be mentioned that the wave equation itself will be an approximation to the equations of motion, for under these conditions that the elastic parameters are functions of the space coordinates, we should not be able to make the reduction from Eq. (6.33) to Eq. (6.34). It is valid only under the assumption that the space rate of change of these parameters is small with respect to the parameters themselves. Let us examine here under what conditions we can consider a solution of the eikonal equation to be a good approximation to the solution of the wave equation. In most problems, we shall be interested in solutions in terms of simple harmonic motion of a particular frequency, ω or ν, or a Fourier synthesis of this to a given initial pulse. The properties of simple harmonic motion have been discussed in Sections 1.9 and 1.10. For our present example, the amplitude may not be considered a constant, and our solution of the form of Eq. (6.108) will be in simple harmonic form

$$\psi = A(x, y, z) e^{i\omega[W(x,y,z)/c_0 - t]} \tag{6.111}$$

where, as before, the wavelength λ_0 is given by

$$\lambda_0 = \frac{2\pi}{k_0} = \frac{2\pi c_0}{\omega} = \frac{c_0}{\nu} \tag{6.112}$$

For this to be a solution of the wave equation, the conditional relations between A and W will be, substituting Eq. (6.111) into Eq. (6.105) and equating real and imaginary parts,

$$\frac{\partial^2 A}{\partial x^2} + \frac{\partial^2 A}{\partial y^2} + \frac{\partial^2 A}{\partial z^2} - A\frac{\omega^2}{c_0^2}\left[\left(\frac{\partial W}{\partial x}\right)^2 + \left(\frac{\partial W}{\partial y}\right)^2 + \left(\frac{\partial W}{\partial z}\right)^2\right] = -\frac{\omega^2}{c^2}A$$

and

$$2\left(\frac{\partial W}{\partial x}\frac{\partial A}{\partial x} + \frac{\partial W}{\partial y}\frac{\partial A}{\partial y} + \frac{\partial W}{\partial z}\frac{\partial A}{\partial z}\right) + A\left(\frac{\partial^2 W}{\partial x^2} + \frac{\partial^2 W}{\partial y^2} + \frac{\partial^2 W}{\partial z^2}\right) = 0$$

or

$$\left(\frac{\partial W}{\partial x}\right)^2 + \left(\frac{\partial W}{\partial y}\right)^2 + \left(\frac{\partial W}{\partial z}\right)^2 - n^2 - \frac{\lambda_0^2}{4\pi^2}\left[\frac{1}{A}\left(\frac{\partial^2 A}{\partial x^2} + \frac{\partial^2 A}{\partial y^2} + \frac{\partial^2 A}{\partial z^2}\right)\right] = 0 \quad (6.113)$$

and

$$\frac{\partial^2 W}{\partial x^2} + \frac{\partial^2 W}{\partial y^2} + \frac{\partial^2 W}{\partial z^2} + \frac{2}{A}\left(\frac{\partial W}{\partial x}\frac{\partial A}{\partial x} + \frac{\partial W}{\partial y}\frac{\partial A}{\partial y} + \frac{\partial W}{\partial z}\frac{\partial A}{\partial z}\right) = 0 \quad (6.114)$$

For W also to be a solution of the eikonal equation (6.109), the last expression in Eq. (6.113) must be zero. In general, the expression in the parentheses is not zero so that this condition will be met only if λ_0 equals zero, that is, in the limit of high frequencies. If W is a solution of the eikonal equation, it will be a good approximation to the wave equation if the last expression of Eq. (6.113) is small compared with the sum of the first three terms. However, it is not sufficient to say that this will be true for small λ_0; we should like to know how small λ_0 must be with regard to the physical conditions of a particular problem. Such a relation can be obtained from order-of-magnitude considerations. Let W be an exact solution of the eikonal equation. Then designating by the prime symbol a general space derivative, ψ in the form of Eq. (6.111) will be a good approximation to the solution of the wave equation if the condition from Eq. (6.113) that

$$\lambda_0^2 \frac{A''}{A} \ll (W')^2 \quad (6.115)$$

is met. From the eikonal equation (6.109), we have that

$$(W')^2 \sim n^2$$

or

$$W' \sim n \sim 1 \quad (6.116)$$

since we can choose c_0 approximately equal to c so that $n = c_0/c \sim 1$. Taking a second space derivative of Eq. (6.109), we obtain

$$W'W'' \sim nn'$$

or

$$W'' \sim n' \quad (6.117)$$

Further, from Eq. (6.114) we may write that

$$W'' \sim W' \frac{A'}{A}$$

of from conditions (6.116) and (6.117) that

$$W'' \sim \frac{A'}{A} \sim n' \qquad (6.118)$$

From condition (6.116) we may rewrite the condition (6.115) as

$$\lambda_0^2 \frac{A''}{A} \ll 1 \qquad (6.119)$$

From condition (6.115) or (6.119) we see that the unit of measurement of interest to us is the wavelength λ_0 and that the inequality relates to changes in A' with respect to A over the distance of a wavelength. For convenience, we shall rewrite condition (6.119) in the form

$$\lambda_0 \frac{\delta A'}{A} \ll 1 \qquad (6.120)$$

where $\delta A'$ is taken to mean the change in A' over a distance λ_0. From condition (6.118) this is equivalent to

$$\lambda_0 \delta W'' \ll 1 \qquad (6.121)$$

or

$$\lambda_0 \delta n' \ll 1 \qquad (6.122)$$

From Eq. (6.110) the condition (6.122) may be written as

$$\lambda_0 \frac{\delta c'}{c} \ll 1 \qquad (6.123)$$

without regard to sign. The inequality (6.123) is the desired result. It states that a solution to the eikonal equation will be a good approximation to the wave equation if the fractional change in the velocity gradient $\delta c'$ over a wavelength is small compared with the gradient c/λ_0. Thus, if in a particular problem, there is a boundary between two media of different velocities, the condition (6.123) will not be met; or if there is a region in which there is a rapid change in velocity over the dimensions of a wavelength, condition (6.123) will not be met. Such problems must be treated separately. The inequalities (6.121) and (6.120) are equivalent to condition (6.123) for the measurable quantities of the wave front W and the amplitude A. Condition (6.121) states that the change in the curvature of the wave front must be small, and condition (6.120) states that the fractional change in space rate of change of amplitude must be small over a wavelength

The eikonal equation is a first-order partial differential equation. Its solutions

$$W(x, y, z) = \text{constant} \tag{6.124}$$

represent surfaces in three-dimensional space. For a given value of W and a given instant of time t_0, the phase $\theta = W(x, y, z) - c_0 t$ of Eq. (6.108) or (6.111) will be a constant. All points along the surface will be in phase, although now in the general case not necssarily of the same amplitude. There will be a one-to-one correspondence of the motion along this surface called a *wave surface* or *wave front*. At a later time t, the motion along the surface will be in a different phase of its motion. A picture of the motion throughout space at a given time t_0 can be found by taking successive increments of the constant of Eq. (6.124). The normals to this surface define the direction of propagation, that is, the direction in which the motion of a particle on one wave surface is passed on to the next. They are called the *rays* or *ray paths*. We shall see later, as might be expected, that the rays define the direction of energy propagation so that the energy contained within a narrow bundle of rays is confined to that bundle throughout all space. The ray solution, then, is a complete solution to any particular propagation problem within the validity of the approximation of the eikonal equation to the wave equation. The variations in intensity are given by the diminutions or expansions of the ray bundles, and the time of arrival by the integral $(ds)/c$ along the ray path, where s is the coordinate distance measured along a ray.

The equation for the normals to the wave surface is

$$\frac{dx}{\frac{\partial W}{\partial x}} = \frac{dy}{\frac{\partial W}{\partial y}} = \frac{dz}{\frac{\partial W}{\partial z}} \tag{6.125}$$

where $\partial W/\partial x$, $\partial W/\partial y$, $\partial W/\partial z$ are the direction numbers of the normals. The direction cosines dx/ds, dy/ds, dz/ds are proportional to the direction numbers and satisfy the subsidiary relation

$$\left(\frac{dx}{ds}\right)^2 + \left(\frac{dy}{ds}\right)^2 + \left(\frac{dz}{ds}\right)^2 = 1 \tag{6.126}$$

This proportionality may be expressed by the three equations

$$\begin{aligned}\frac{dx}{ds} &= a\frac{\partial W}{\partial x} \\ \frac{dy}{ds} &= a\frac{\partial W}{\partial y} \\ \frac{dz}{ds} &= a\frac{\partial W}{\partial z}\end{aligned} \tag{6.127}$$

where a is a constant. Squaring and adding and using Eqs. (6.109) and (6.126), $a = 1/n$ so that Eqs. (6.127) become

$$n\frac{dx}{ds} = \frac{\partial W}{\partial x}$$

$$n\frac{dy}{ds} = \frac{\partial W}{\partial y} \qquad (6.128)$$

$$n\frac{dz}{ds} = \frac{\partial W}{\partial z}$$

Let us take a derivative d/ds along the ray path of the first of these three equations; we obtain

$$\frac{d}{ds}\left(n\frac{dx}{ds}\right) = \frac{d}{ds}\left(\frac{\partial W}{\partial x}\right) = \frac{\partial}{\partial x}\left(\frac{\partial W}{\partial x}\frac{dx}{ds} + \frac{\partial W}{\partial y}\frac{dy}{ds} + \frac{\partial W}{\partial z}\frac{dz}{ds}\right)$$

or substituting from Eqs. (6.128) and using Eq. (6.126),

$$\frac{d}{ds}\left(n\frac{dx}{ds}\right) = \frac{\partial}{\partial x}\left\{n\left[\left(\frac{dx}{ds}\right)^2 + \left(\frac{dy}{ds}\right)^2 + \left(\frac{dz}{ds}\right)^2\right]\right\} = \frac{\partial n}{\partial x}$$

so that our three equations (6.128) reduce to

$$\frac{d}{ds}\left(n\frac{dx}{ds}\right) = \frac{\partial n}{\partial x}$$

$$\frac{d}{ds}\left(n\frac{dy}{ds}\right) = \frac{\partial n}{\partial y} \qquad (6.129)$$

$$\frac{d}{ds}\left(n\frac{dz}{ds}\right) = \frac{\partial n}{\partial z}$$

These are the equations that determine the rays in terms of the index of refraction $n = n(x, y, z)$. These equations state that the variation along the ray path of the product of the index of refraction and a direction cosine is equal to the space rate of variation of the index of refraction with respect to the appropriate coordinate. They may be considered as a generalized form of *Snell's law*.

It is of interest to prove Fermat's principle. *Fermat's principle* states that a ray path between any two points P_0 and P_1, as shown in Fig. 6.4, is also a

Fig. 6.4

208 ¶ Seismology, Gravity, and Magnetism

path of stationary time between these points. We want to prove that the integral

$$F = c_0 \int_{P_0}^{P_1} dt = c_0 \int \frac{ds}{c} = \int n \, ds \qquad (6.130)$$

has an extremum value, maximum or minimum, along the ray path. It is convenient to express Eq. (6.130) in parametric form. The differential ds along the path is given by

$$ds = \sqrt{x'^2 + y'^2 + z'^2} \, d\sigma \qquad (6.131)$$

where the prime symbol indicates derivatives with respect to the parameter σ. Substituting into Eq. (6.130) gives

$$F = \int_{P_0}^{P_1} n(x, y, z) \sqrt{x'^2 + y'^2 + z'^2} \, d\sigma = \int G(x, y, z, x', y', z') \, d\sigma \qquad (6.132)$$

This is now in the form of a line integral with respect to σ of the three space variables and their derivatives with respect to σ. From the calculus of variations, this line integral will have an extremum value if Euler's equations

$$\frac{\partial G}{\partial x} - \frac{d}{d\sigma}\left(\frac{\partial G}{\partial x'}\right) = 0$$

$$\frac{\partial G}{\partial y} - \frac{d}{d\sigma}\left(\frac{\partial G}{\partial y'}\right) = 0$$

$$\frac{\partial G}{\partial z} - \frac{d}{d\sigma}\left(\frac{\partial G}{\partial z'}\right) = 0$$

are satisfied or, for this case,

$$\sqrt{x'^2 + y'^2 + z'^2} \, \frac{\partial n}{\partial x} - \frac{d}{d\sigma}\left(\frac{nx'}{\sqrt{x'^2 + y'^2 + z'^2}}\right) = 0$$

$$\sqrt{x'^2 + y'^2 + z'^2} \, \frac{\partial n}{\partial y} - \frac{d}{d\sigma}\left(\frac{ny'}{\sqrt{x'^2 + y'^2 + z'^2}}\right) = 0 \qquad (6.133)$$

$$\sqrt{x'^2 + y'^2 + z'^2} \, \frac{\partial n}{\partial z} - \frac{d}{d\sigma}\left(\frac{nz'}{\sqrt{x'^2 + y'^2 + z'^2}}\right) = 0$$

Substituting for $d\sigma$ in terms of ds from Eq. (6.131) gives

$$\frac{d}{ds}\left(n\frac{dx}{ds}\right) = \frac{\partial n}{\partial x}$$

$$\frac{d}{ds}\left(n\frac{dy}{ds}\right) = \frac{\partial n}{\partial y}$$

$$\frac{d}{ds}\left(n\frac{dz}{ds}\right) = \frac{\partial n}{\partial z}$$

which are simply the original ray equations. Thus, we have proved that the time along the ray path has an extremum value. For most physical problems, this is a minimum or least time.

It is instructive to give an alternative derivation of the eikonal equation from somewhat more physical arguments through the use of Huygens' principle. *Huygens' principle* states that the disturbance at some later time, $t = t_0 + dt$, can be obtained by considering the effect produced by each point on a wave surface at a given time, $t = t_0$, acting as a secondary source. Each point on the wave surface will produce a wavelet propagating outward with the velocity c at that point; the resultant wave front at some later time, $t = t_0 + dt$, will be the envelope of all the wavelets taken at that time. If c is a sufficiently slowly varying function over the original wave surface, the wave surface at $t_0 + dt$ will be defined approximately by the normals $c\,dt$ to the

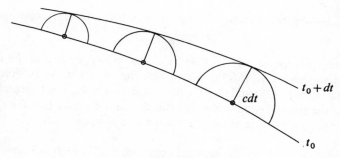

Fig. 6.5

original surface, as shown in Fig. 6.5. The wave surface, $W(x, y, z) = $ constant, is the locus of points that are in the same phase of motion at a given time t_0 so that

$$W(x, y, z) - c_0 t_0 = \text{constant}$$

where c_0 is a constant. At a later time, $t_0 + dt$, the wavelets have defined a new wave surface, which is in the same phase of motion at $t_0 + dt$ that the original wave surface was at t_0 so that

$$W(x + \alpha c\,dt,\ y + \beta c\,dt,\ z + \gamma c\,dt) - c_0(t_0 + dt) = W(x, y, z) - c_0 t_0 \quad (6.134)$$

where α, β, γ are the direction cosines of the normals to the original wave surface. Since the increments $\alpha c\,dt, \beta c\,dt, \gamma c\,dt$ are small quantities, the new wave surface can be approximated by

$$W(x + \alpha c\,dt,\ y + \beta c\,dt,\ z + \gamma c\,dt)$$
$$= W(x, y, z) + \frac{\partial W}{\partial x} \alpha c\,dt + \frac{\partial W}{\partial y} \beta c\,dt + \frac{\partial W}{\partial z} \gamma c\,dt \quad (6.135)$$

Equating Eqs. (6.134) and (6.135), we have

$$\alpha \frac{\partial W}{\partial x} + \beta \frac{\partial W}{\partial y} + \gamma \frac{\partial W}{\partial z} = \frac{c_0}{c} = n \qquad (6.136)$$

The direction numbers to a surface, $W =$ constant, are $\partial W/\partial x$, $\partial W/\partial y$, $\partial W/\partial z$. They will be proportional to the direction cosines so that we may write

$$\alpha = b \frac{\partial W}{\partial x}$$

$$\beta = b \frac{\partial W}{\partial y} \qquad (6.137)$$

$$\gamma = b \frac{\partial W}{\partial z}$$

where, as before, the direction cosines satisfy the subsidiary relation

$$\alpha^2 + \beta^2 + \gamma^2 = 1 \qquad (6.138)$$

Multiplying the first of Eqs. (6.137) by α, the second by β, and the third by γ, we obtain using Eqs. (6.136) and (6.138) that $b = 1/n$. As before, we can then obtain Eqs. (6.128) and (6.129). Squaring each of Eqs. (6.128) and adding we obtain the eikonal equation. From this derivation we see that the direction of energy propagation is along the normals to the wave surfaces, the rays.

Problem 6.4(a) Carry through the reductions to Eqs. (6.113) and (6.114).

Problem 6.4(b) Show that, for the case of a perfect liquid, the total energy per unit volume is given by $E = \rho(\dot{u}^2 + \dot{v}^2 + \dot{w}^2)/2 + p^2/2k$.

Problem 6.4(c) Derive an equation of continuity of energy flow similar to the equation of continuity of mass transport of Section 1.1 and show, using the results of Problem 6.4(b), that the net energy flow per unit area per unit time, or power, is given by $\mathbf{F} = p\dot{\mathbf{s}}$.

Problem 6.4(d) For a perfect liquid and simple harmonic motion, show that the intensity, or time average of energy flow per unit area per unit time, is transmitted along the ray paths and is given by $I = a^2/2\rho c$, where a is the pressure amplitude.

6.5 Ray Characteristics for a Flat Earth

For most physical cases of interest, the velocity is a function of only one of the space coordinates, the depth. Equations (6.129) then reduce for $n = n(z)$ to

$$n\frac{dx}{ds} = \text{constant}$$

$$n\frac{dy}{ds} = \text{constant} \tag{6.139}$$

$$\frac{d}{ds}\left(n\frac{dz}{ds}\right) = \frac{dn}{dz}$$

Combining the first and the second gives

$$\frac{\dfrac{dx}{ds}}{\dfrac{dy}{ds}} = \frac{\alpha}{\beta} = \text{constant} \tag{6.140}$$

The restriction that the ratio of the direction cosines in the x and y directions be a constant implies that the ray path is confined to a plane normal to the XY plane. This can easily be seen from Fig. 6.6, where the light lines define similar triangles in the XY plane formed by dx, dy. The trace of ds in the

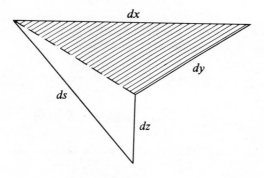

Fig. 6.6

XY plane necessarily satisfies Eq. (6.140). The trace of ds in the XY plane is confined to a line, so that ds itself is confined to a plane containing this line and normal to the XY plane. For convenience and without loss of generality, we shall choose this plane to coincide with the XZ plane. Equations (6.139) become

$$n\frac{dx}{ds} = \text{constant} \tag{6.141}$$

$$\frac{d}{ds}\left(n\frac{dz}{ds}\right) = \frac{dn}{dz} \tag{6.142}$$

212 ¶ *Seismology, Gravity, and Magnetism*

From Fig. 6.7 the direction cosines α and γ are given by

$$\alpha = \frac{dx}{ds} = \sin \theta$$

$$\gamma = \frac{dz}{ds} = \cos \theta$$

where θ is the angle the ray makes with the z direction. Equation (6.141) then becomes, using the definition of the index of refraction, Eq. (6.110),

$$\frac{\sin \theta}{c} = p \text{ (constant)} \qquad (6.143)$$

Equation (6.143) is *Snell's law* and is the basic equation of ray seismology. It states that the ratio of the sine of the angle of inclination of a ray at any given depth to the velocity at that depth is a constant along the ray path. The ray parameter p is a constant along a ray path, but varies from one ray path to the next. It is sometimes useful to state this relation in terms of the angle of inclination and velocity of the ray at its source,

$$\frac{\sin \theta_0}{c_0} = \frac{\sin \theta}{c} \qquad (6.144)$$

where c_0 is the velocity at the source and θ_0 the inclination of a particular ray at the source. Equation (6.142) reduces to

$$\frac{dn}{dz} = \frac{d}{ds}(n \cos \theta) = -n \sin \theta \frac{d\theta}{ds} + \cos \theta \frac{dn}{dz}\frac{dz}{ds}$$

$$= -n \sin \theta \frac{d\theta}{ds} + \cos^2 \theta \frac{dn}{dz}$$

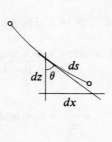

Fig. 6.7

and solving for $d\theta/ds$

$$\frac{d\theta}{ds} = -\frac{\sin\theta}{n}\frac{dn}{dz} \qquad (6.145)$$

From the definition of n, Eq. (6.110), we have upon differentiating

$$\frac{dn}{n} = -\frac{dc}{c}$$

so that Eq. (6.145) becomes

$$\frac{d\theta}{ds} = \frac{\sin\theta}{c}\frac{dc}{dz}$$

or

$$\frac{d\theta}{ds} = p\frac{dc}{dz} \qquad (6.146)$$

This is also a useful relation and states that the curvature of a ray, $d\theta/ds$, is directly proportional to the velocity gradient dc/dz. For a region in which the velocity is increasing with depth, dc/dz positive, $d\theta/ds$ will be positive, and the ray will curve upward. Similarly, for a region in which the velocity is decreasing, the ray will curve downward. The rays always curve toward a region of minimum velocity; such a region will tend to form a wave guide. This result is to be expected from a qualitative consideration of the motion of wave fronts. The portion of the wave front that is in a region of higher velocity will travel faster than that in a lower-velocity region, and the wave front will be bent toward the lower-velocity region, as shown in Fig. 6.8.

The intensity I can be given by several equivalent relations. All refer back to the fact that the energy flow is defined by the ray paths and that the energy initially confined to a narrow bundle of rays near the source will continue to be confined within that bundle throughout its propagation. One such relation for the intensity is derived here. Consider a point source at the origin, emitting P units of energy per unit time per unit solid angle of a given type, for example, dilatational or distortional. Since the rays are confined to a

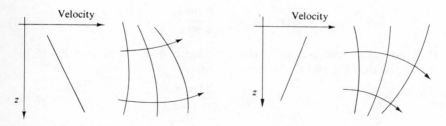

Fig. 6.8

plane with the z axis, there is cylindrical symmetry about this axis. We shall define the unit solid angle Ω with symmetry about the z axis, so that with respect to a unit sphere about the origin in Fig. 6.9,

$$d\Omega = 2\pi \sin \theta_0 \, d\theta_0$$

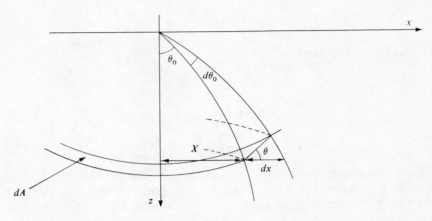

Fig. 6.9

The intensity then at any other point along the ray path will be P multiplied by the ratio of $d\Omega$ to the area dA swept out by the wave surface normal to the rays. From Fig. 6.9 this will be

$$I = P\frac{d\Omega}{dA} = P\frac{2\pi \sin \theta_0 \, d\theta_0}{2\pi x \cos \theta \, dx} = P\frac{\sin \theta_0 \, d\theta_0}{x \cos \theta \, dx}$$

The x coordinate of the ray can be expressed as a function of the initial angle θ_0 and the depth z through Eq. (6.143). At a given depth then,

$$dx = \frac{\partial x}{\partial \theta_0} d\theta_0$$

or

$$I(x, \theta_0) = \frac{P \sin \theta_0}{x \cos \theta (\partial x/\partial \theta_0)} \tag{6.147}$$

If the source and receiver are at the same depth, $c = c_0$ and from Eq. (6.144) $\theta = \theta_0$ so that Eq. (6.147) becomes

$$I(X, \theta_0) = \frac{P \tan \theta_0}{X(\partial X/\partial \theta_0)} \tag{6.148}$$

The $1/x$ term is the loss due to cylindrical spreading out from the source, and the $\sin \theta/[\cos \theta(\partial x/\partial \theta_0)]$ term expresses the decrease or increase in intensity due to spreading or contracting of the ray bundle along its path.

The travel time and horizontal range along a ray can be expressed in terms of the depth z or the angle of inclination θ by the following parametric equations. From Fig. 6.7

$$x = \int_0^z \tan\theta \, dz \qquad (6.149)$$

and

$$t = \int \frac{ds}{c} = \int_0^z \frac{dz}{c\cos\theta} \qquad (6.150)$$

where it is understood that θ is expressed as a function of z through Eq. (6.143). In terms of θ, these integrals will be, taking differentials of Eq. (6.143) and substituting into Eqs. (6.149) and (6.150),

$$x = \int_0^z \sin\theta \frac{dz}{\cos\theta} = \int_{\theta_0}^\theta \frac{\sin\theta \, d\theta}{pc'(z)} \qquad (6.151)$$

and

$$t = \int_{\theta_0}^\theta \frac{d\theta}{c'(z)\sin\theta} \qquad (6.152)$$

where it is understood that here the velocity gradient $c'(z)$ can be expressed in terms of θ through Eq. (6.143). To eliminate θ from Eqs. (6.149) and (6.150), we have from Eq. (6.143) that

$$\tan\theta = \frac{pc}{(1-p^2c^2)^{1/2}}$$

and

$$\cos\theta = (1-p^2c^2)^{1/2}$$

so that

$$x = \int_0^z \frac{pc}{(1-p^2c^2)^{1/2}} \, dz \qquad (6.153)$$

and

$$t = \int_0^z \frac{dz}{c(1-p^2c^2)^{1/2}} \qquad (6.154)$$

It is convenient to define the parameter η

$$\eta = \frac{1}{c} \qquad (6.155)$$

Substituting into Eqs. (6.153) and (6.154), we have

$$x = p \int_0^z \frac{dz}{(\eta^2 - p^2)^{1/2}} \qquad (6.156)$$

and
$$t = \int_0^z \frac{\eta^2 dz}{(\eta^2 - p^2)^{1/2}} \tag{6.157}$$

The travel time may also be expressed by
$$t = \int_0^z \left[\frac{p^2}{(\eta^2 - p^2)^{1/2}} + (\eta^2 - p^2)^{1/2} \right] dz$$

so that
$$t = px + \int_0^z (\eta^2 - p^2)^{1/2} \, dz \tag{6.158}$$

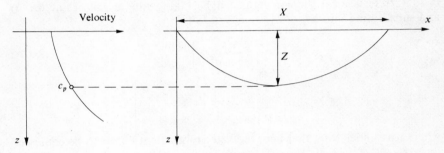

Fig. 6.10

From Fig. 6.10 the horizontal range X and travel time T for the ray to reach the depth of the source again, that is, if it does return to this depth, will be
$$X = 2p \int_0^Z \frac{dz}{(\eta^2 - p^2)^{1/2}} \tag{6.159}$$

and
$$T = 2 \int_0^Z \frac{\eta^2 \, dz}{(\eta^2 - p^2)^{1/2}} \tag{6.160}$$

where Z is the depth of the vertex, the maximum depth the ray penetrates.

We now wish to derive a useful subsidiary relation among the ray parameter p and the horizontal range x and travel time t. Consider two adjacent rays defined by the ray parameters p and $p + dp$ propagating out to a given depth z and draw the wave surfaces normal to them. They will be separated by a horizontal range interval dx and a time interval dt. From the geometry of Fig. 6.11, we have
$$\sin \theta = \frac{c \, dt}{dx}$$

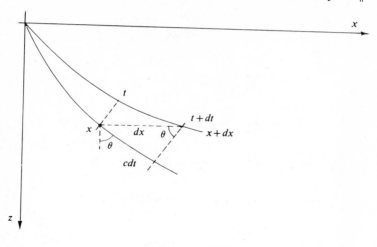

Fig. 6.11

and from Eq. (6.143) this gives

$$p = \frac{dt}{dx} \qquad (6.161)$$

Another relation for p in terms of the velocity c_p at the vertex of a ray can be given. At this point, the angle of inclination θ is $\pi/2$ so that $\sin \theta_p = 1$. Thus, from Eq. (6.143)

$$p = \frac{1}{c_p} \qquad (6.162)$$

or from Eq. (6.155)

$$p = \eta_p \qquad (6.163)$$

The velocity c_p at the vertex is the maximum velocity that the ray reaches and is attained at the greatest depth Z. If through a series of measurements from a source to a line of receivers at the surface we should obtain a plot of T versus X, the inverse slope of this graph at any given range will be, from Eqs. (6.161) and (6.162), the velocity at the vertex for the ray emergent at this range, as shown in Fig. 6.12. The validity of this relation can be seen from the following geometrical construction. A ray emergent at the surface will be inclined at the same angle it had starting out from the source. The inverse slope at this point on the T versus X plot will be simply the trace velocity $\bar{c}$, with which the wave front associated with this ray moves along the surface $z = 0$. From the geometry of Fig. 6.13, this velocity will be

$$\bar{c} = \frac{c_0}{\sin \theta_0} \qquad (6.164)$$

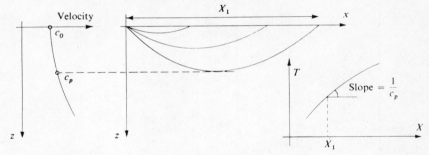

Fig. 6.12

and from Eq. (6.144), remembering that $\sin \theta_p = 1$ at the vertex,

$$\bar{c} = c_p \tag{6.165}$$

In many problems in geophysics, the measurements give a plot of T versus X, and it is desired to find from this the velocity c as a function of depth z. From the above, this reduces to the problem of finding the depth of the vertex Z from the T, X relation; for the velocity c_p at the vertex is known and is simply the inverse slope of the T, X graph. From expression (6.159) for X in terms of Z and from Eq. (6.163) we have

$$X = 2p \int_0^Z \frac{dz}{(\eta^2 - p^2)^{1/2}} = 2p \int_{\eta_0}^{\eta_p} \frac{dz/d\eta}{(\eta^2 - p^2)^{1/2}} d\eta \tag{6.166}$$

where the integration is taken along the ray path. Let us apply the operation

$$\int_{p_0}^{p_1} (p^2 - \eta_1^2)^{1/2} dp$$

to both sides of Eq. (6.166). This is an integration across the rays from the ray at zero range down to the ray whose vertex is at a depth where the velocity $c_1 = 1/\eta_1$. In this operation, the limits of integration have the relations

$$p_0 = \eta_{p0} = \eta_0 = \frac{1}{c_0}$$

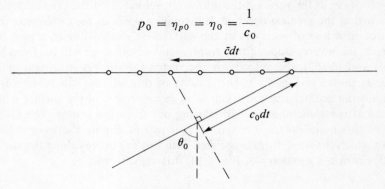

Fig. 6.13

and
$$p_\perp = \eta_{p1} = \eta_1 = \frac{1}{c_1}$$

Expression (6.166) becomes

$$\int_{\eta_0}^{\eta_1} \frac{X}{(p^2-\eta_1^2)^{1/2}} dp = \int_{\eta_0}^{\eta_1} dp \int_{\eta_0}^{\eta_p} \frac{2p(dz/d\eta)}{(p^2-\eta_1^2)^{1/2}(\eta^2-p^2)^{1/2}} d\eta \quad (6.167)$$

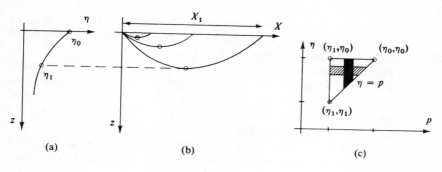

Fig. 6.14

From Fig. 6.14(c) we see that the order and limits of integration for the double integral on the right-hand side can be changed as follows:

$$\int_{\eta_0}^{\eta_1} \frac{X}{(p^2-\eta_1^2)^{1/2}} dp = \int_{\eta_0}^{\eta_1} d\eta \int_{p}^{\eta_1} \frac{2p(dz/d\eta)}{[-p^4+(\eta^2+\eta_1^2)p^2-\eta_1^2\eta^2]^{1/2}} dp$$

Integrating by parts on the left-hand side and carrying out the first integration on the right-hand side, we obtain

$$\left[X \cosh^{-1}\left(\frac{p}{\eta_1}\right)\right]_{\eta_0}^{\eta_1} - \int_{\eta_0}^{\eta_1} \frac{dX}{dp} \cosh^{-1}\left(\frac{p}{\eta_1}\right) dp$$
$$= \int_{\eta_0}^{\eta_1} \frac{dz}{d\eta} \left\{\sin^{-1}\left[\frac{2p^2-(\eta^2+\eta_1^2)}{(\eta^2-\eta_1^2)}\right]\right\}_{p=\eta}^{\eta_1} d\eta$$

The first term on the left-hand side is zero since $X = 0$ when $p = \eta_0$. We then have

$$\int_{\eta_0}^{\eta_1} \frac{dX}{dp} \cosh^{-1}\left(\frac{p}{\eta_1}\right) dp = \pi \int_{\eta_0}^{\eta_1} \frac{dz}{d\eta} d\eta$$

or
$$\int_0^X \cosh^{-1}\left(\frac{p}{\eta_1}\right) dX = \pi \int_0^{Z_1} dz = \pi Z_1 \quad (6.168)$$

and substituting for η_1 and p finally

$$Z_1 = \frac{1}{\pi} \int_0^{X_1} \cosh^{-1}\left(\frac{c_1}{c_p}\right) dX = \frac{1}{\pi} \int_0^{X_1} \cosh^{-1}\left(\frac{c_1}{\bar{c}}\right) dX \qquad (6.169)$$

This is an expression for Z_1 in terms of T and X, the quantities $1/c_1$ and $1/c_p$ being given by $(dT/dX)_1$, and (dT/dX). Generally, the integration is carried out numerically from the T versus X plot. The quantity $\bar{c}$ is the inverse slope to this graph and is determined point by point along the graph from $X = 0$ to $X = X_1$. The quantity c_1 is the inverse slope at $X = X_1$; the integration can be carried out to determine the value of Z_1 at which the velocity c_1 is reached. The velocity-depth relation $c = c(z)$ can be determined by carrying out successive integrations of Eq. (6.169) for $X = X_1, X_2, X_3,$ $\cdots$. In this derivation, it has been implicitly assumed that η was a continuously decreasing function of depth, Fig. 6.14(a). For an increasing section, $d\eta/dz$ would change signs and the integral [Eq. (6.167)] would become indeterminate. Expression (6.169) is valid only for those cases where the velocity is a continuously increasing function of depth.

It is of interest to consider some of the characteristics of various types of velocity functions. First, we need to derive an expression for dX/dp. From Eq. (6.159) this will be

$$\frac{dX}{dp} = 2\int_0^Z \frac{dz}{(\eta^2 - p^2)^{1/2}} + 2p\frac{d}{dp}\int_0^Z \frac{dz}{(\eta^2 - p^2)^{1/2}} \qquad (6.170)$$

Before carrying out the indicated differentiation on the last term, we shall integrate by parts

$$\int_0^Z \frac{dz}{(\eta^2 - p^2)^{1/2}} = \int_0^Z \frac{dz}{d\eta}\frac{d\eta/dz}{(\eta^2 - p^2)^{1/2}} dz = \int_0^Z f(\eta)\frac{d\eta/dz}{(\eta^2 - p^2)^{1/2}} dz$$

$$= \left[f(\eta)\cosh^{-1}\left(\frac{\eta}{p}\right)\right]_{\eta_0}^p - \int_0^Z f'(\eta)\cosh^{-1}\left(\frac{\eta}{p}\right)\frac{d\eta}{dz} dz$$

$$= -f(\eta_0)\cosh^{-1}\left(\frac{\eta_0}{p}\right) - \int_0^Z f'(\eta)\cosh^{-1}\left(\frac{\eta}{p}\right)\frac{d\eta}{dz} dz$$

so that Eq. (6.170) will be

$$\frac{dX}{dp} = 2\int_0^Z \frac{dz}{(\eta^2 - p^2)^{1/2}} + 2p\left[\frac{\eta_0 f(\eta_0)}{p(\eta_0^2 - p^2)^{1/2}} + \int_0^Z \frac{\eta f'(\eta)}{p(\eta^2 - p^2)^{1/2}}\frac{d\eta}{dz} dz\right]$$

$$= 2\frac{\eta_0 f(\eta_0)}{(\eta_0^2 - p^2)^{1/2}} + 2\int_0^Z [f(\eta) + \eta f'(\eta)]\frac{d\eta/dz}{(\eta^2 - p^2)^{1/2}} dz \qquad (6.171)$$

Let us now make the substitution $\zeta = \zeta(z)$, where

$$\zeta(z) = \frac{1}{c}\frac{dc}{dz} \qquad (6.172)$$

Then we shall have

$$\eta f(\eta) = \eta \frac{dz}{d\eta} = -c\frac{dz}{dc} = -\frac{1}{\zeta}$$

remembering from Eq. (6.155) that $d\eta/\eta = -dc/c$. Taking derivatives with respect to z will give

$$[f(\eta)+\eta f'(\eta)]\frac{d\eta}{dz} = \frac{1}{\zeta^2}\frac{d\zeta}{dz}$$

Substituting into Eq. (6.171) gives finally

$$\frac{dX}{dp} = -\frac{2}{\zeta_0(\eta_0^2-p^2)^{1/2}} + 2\int_0^z \frac{d\zeta/dz}{\zeta^2(\eta^2-p^2)^{1/2}} dz \qquad (6.173)$$

In many geophysical problems, the velocity increases gradually with depth so that ζ will be a moderate positive quantity and $d\zeta/dz$ small. Starting out from $z = 0$, the integrated term of Eq. (6.173) will be large compared with the integral so that dX/dp will be negative. For increasing depth, $p = \eta_p$ will decrease, and dX/dp will decrease. From Fig. 6.15(d) we can then obtain Fig. 6.15(e); since $p = dT/dX$, from Fig. 6.15(e) we can obtain the graph of T versus X. It is seen, as expected, that T increases gradually with X corresponding to penetration of the deeper, higher velocity media. Furthermore, the intensity from Eq. (6.148) is proportional to $\tan\theta_0/X(\partial X/\partial\theta_0)$. Since

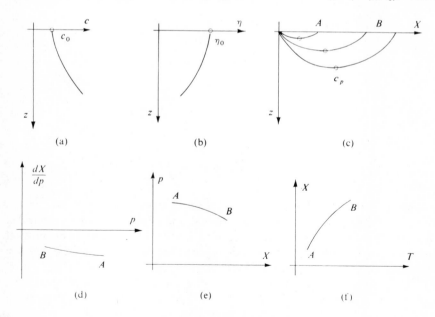

Fig. 6.15

$p = \sin \theta/c$, the ordinate of Fig. 6.15(d) or the slope of Fig. 6.15(e) can be considered to be a measure of $\partial X/\partial \theta_0$. Thus, we see that the intensity will decrease gradually with range, the rate of decrease being smaller the larger the velocity gradient.

Let us assume now that at some depth, there is a marked increase in the rate of increase in velocity, and then a return to the conditions of a gradual increase in velocity. Then ζ will become large and $d\zeta/dz$ will be at first a moderate positive number and then a moderate negative number. If this rate of increase is sufficiently large, the integral of Eq. (6.173) can become larger than the integrated term. dX/dp will become positive, and then return to its normal condition with increasing depth. As before, from the graph of dX/dp versus p, Fig. 6.16(d), the graphs of p versus X and T versus X can be

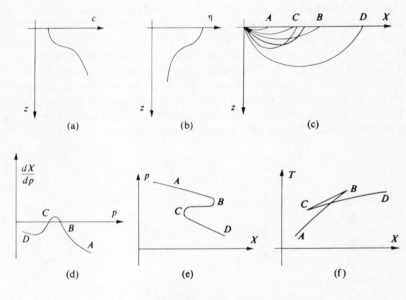

Fig. 6.16

obtained. We see here that there is a reversed segment from B to C in the travel time curve. Within this range interval, there is triplication in the travel time curve, three arrivals being received at the same range. The normal segment from A to B has a lower velocity, corresponding to the medium above the depth of the rapid increase in velocity, than the normal segment from C to D, corresponding to penetration of the higher-velocity medium below the depth of the rapid increase. At the points B and C, the expression $1/(\partial X/\partial \theta_0)$ will be zero, and these points will be bright spots of intensity. Along the reversed segment from B to C, the intensity will be low.

Seismology—Ray Theory ¶ 223

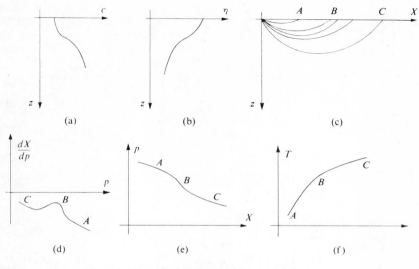

Fig. 6.17

If, on the other hand, the rate of increase is not sufficiently large for the value of dX/dp to pass through zero, the resulting ray situation will be as illustrated in Fig. 6.17. There is no triplication in the travel time curve. The travel time curve changes slope rapidly in the vicinity of B, and in this vicinity the intensity will be high.

Let us now consider the case in which at some depth there is a decrease in velocity followed at a greater depth by a return to the more normal condition of a gradual increase of velocity with depth. At the depth where the decrease in velocity begins, ζ will be zero, and the integral of Eq. (6.173) will become divergent. At depths in the regions of decreasing velocity, the value of p should increase since $p = \eta_p$. However, this is an impossibility; for a ray that is inclined at an angle θ_1, at a depth of higher velocity c_1, will from Snell's law [Eq. (6.143)] be inclined at a steeper angle θ_2 at a greater depth of lower velocity c_2. The ray cannot have a vertex until the depth is reached, at which the velocity is equal to that at the depth where it first began to decrease, as shown in Fig. 6.18(a). In this interval of forbidden depths, the quantity $(\eta^2 - p^2)^{1/2}$ in the integral of Eq. (6.173) is indeterminate.

Considered in another way we want to obtain the value of dX/dp for all rays. Starting from grazing incidence at the source, $\theta_0 = \pi/2$, and moving around to normal incidence, $\theta_0 = 0$, all rays will be covered. Since $p = \sin \theta_0/c_0$, p will be a continually decreasing quantity, so that dX/dp will be a single valued function of p. As a result, there will be a discontinuity in Z and a corresponding discontinuity in X and T.

We can get an approximation to dX/dp for those rays whose vertex is

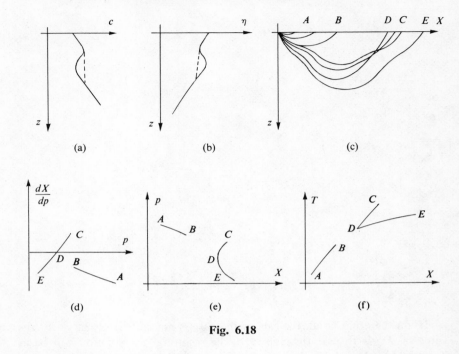

Fig. 6.18

below the negative gradient from the case of a discontinuous decrease in velocity with a normal gradual increase in velocity above and below. The integrals of Eq. (6.173) above and below the discontinuity will be small, because $d\zeta/dz$ is small, and dX/dp becomes approximately

$$\frac{dX}{dp} \approx -\frac{2}{\zeta_0(\eta_0^2-p^2)^{1/2}} + \frac{2}{\zeta_a(\eta_a^2-p^2)^{1/2}} - \frac{2}{\zeta_b(\eta_b^2-p^2)^{1/2}}$$

where η_a and η_b are the values of η just above and just below the discontinuity. For the first ray to be returned from below the discontinuity (ray C of Fig. 6.18), p will be nearly equal to η_a so that dX/dp will change from a moderate negative value at B to a large positive value at C. For increasing depth, the importance of the η_a term will diminish, and dX/dp will approach its normal value. These circumstances are illustrated in Fig. 6.18(d) from which Fig. 6.18(e) and (f) have been derived. There is a discontinuity in the travel time curve, Fig. 6.18(f), such that over the range interval from B to D, no rays are received at the surface. This is known as a *shadow zone*. Beyond the shadow zone, two rays are received out to the range of C. From Fig. 6.18(d) and (e) the intensity in this region will be higher than if there had been no negative gradient section. Physically this is understandable in that the energy that has been removed from the shadow zone must appear somewhere else. Further, in the immediate vicinity of D, $\partial X/\partial \theta_0$ is very small;

this will be a focus point of intensity. It is a characteristic of a velocity structure with a negative gradient section that there will be a shadow zone followed by a focus point and a zone of moderately high intensity.

Problem 6.5(a) Show that for the case where the velocity is a linear function of depth, all the ray paths are circles, and that for a positive gradient g, the radius of curvature of a ray is $c_0/g \sin \theta_0$ and the centers of the circular rays all lie on a line c_0/g units of distance above the source level.

Problem 6.5(b) For the case of Problem 6.5(a) determine x, t, and z along the rays.

Ans.

$$x = \frac{c_0}{g} \left(\frac{\cos \theta_0 - \cos \theta}{\sin \theta_0} \right)$$

$$t = \frac{1}{g} \log \frac{\tan \theta/2}{\tan \theta_0/2}$$

$$z = \frac{c_0}{g} \left(\frac{\sin \theta}{\sin \theta_0} - 1 \right)$$

Problem 6.5(c) Determine the ray intensity for the case of Problem 6.5(a).

Ans.

$$I = P \frac{\sin^2 \theta_0}{X}$$

6.6 Ray Characteristics for a Spherically Stratified Earth

The derivations in the above section for a flat earth apply to seismic problems over distances for which the curvature of the earth may be neglected. When we consider seismic propagation to stations distributed worldwide, which is readily obtainable with the larger earthquakes, we must include the effects of the spherical shape of the earth. We shall assume here that the earth is spherical and that the seismic velocities, either P or S, are functions of the radius vector only. Much of the derivation will follow directly from the previous section. We shall be changing from rectangular coordinates to spherical coordinates and, in particular, shall be replacing the depth coordinate z by $-r$, the radial distance from the center of the earth; $c = c(z)$ by $c = c(r)$; x by φ, the central angle measured from the source assumed to be at or near the earth's surface to any point along the ray; and X by Δ, the central angle from the source to a receiver also assumed to be at the earth's surface.

With the above in mind and with reference to Fig. 6.19, we can derive

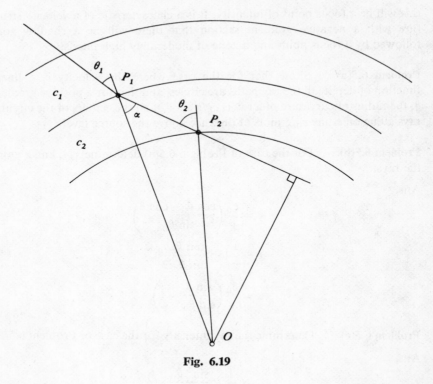

Fig. 6.19

our ray equation. Let us consider for the moment that our earth model is composed of an indefinitely large number of thin concentric homogeneous shells across which there is a discrete velocity change. Then for a ray crossing the spherical boundary at P_1, we shall have

$$\frac{\sin \theta_1}{c_1} = \frac{\sin \theta_2}{c_2}$$

or

$$\frac{OP_1 \sin \theta_1}{c_1} = \frac{OP_1 \sin \alpha}{c_2} \qquad (6.174)$$

From the geometry of Fig. 6.19, we shall also have

$$OP_1 \sin \alpha = OP_2 \sin \theta_2 \qquad (6.175)$$

Combining Eqs. (6.174) and (6.175), we obtain

$$\frac{OP_1 \sin \theta_1}{c_1} = \frac{OP_2 \sin \theta_2}{c_2}$$

or, in general, reducing our problem back to a continuously varying velocity, $c = c(r)$, we get

$$\frac{r \sin \theta}{c} = p \text{ (constant)} = \frac{r_0 \sin \theta_0}{c_0} \qquad (6.176)$$

corresponding to expressions (6.143) and (6.144).

Let us next consider two adjacent rays as shown in Fig. 6.20, the para-

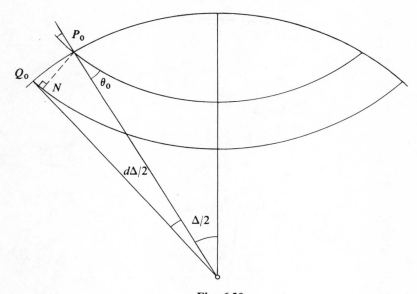

Fig. 6.20

meters of the ray starting at P_0 being p, Δ, T and those of the ray starting at Q_0 being $p+dp$, $\Delta+d\Delta$, $T+dT$. From the geometry of Fig. 6.20, we shall have

$$\sin \theta_0 = \frac{NQ_0}{P_0Q_0} = \frac{c_0 dT/2}{r_0 d\Delta/2}$$

or from Eq. (6.176)

$$p = \frac{dT}{d\Delta} \qquad (6.177)$$

corresponding to Eq. (6.161).

As before, it will be useful to introduce a ray parameter η corresponding to Eq. (6.155), but here defined as

$$\eta = \frac{r}{c} \qquad (6.178)$$

Since again at the vertex of the ray, θ will be $\pi/2$ or $\sin \theta_p = 1$, we shall have from Eqs. (6.176) and (6.178) that

$$p = \eta_p = \frac{r_p}{c_p} \qquad (6.179)$$

corresponding to Eqs. (6.162) and (6.163).

We are now in a position to derive the integral expressions for Δ and T. With reference to Fig. 6.21, let the coordinates of a point P along the ray be

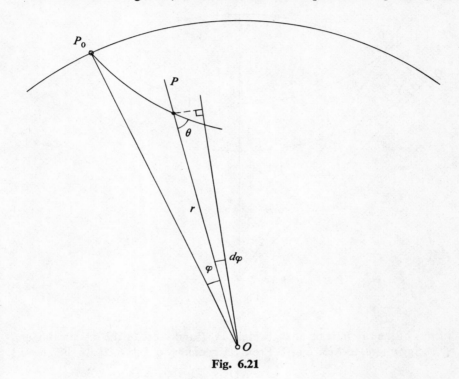

Fig. 6.21

given by r and φ. Then the arc distance along the ray will be simply

$$\overline{ds}^2 = \overline{dr}^2 + r^2 \overline{d\varphi}^2 \qquad (6.180)$$

and from the geometry of Fig. 6.21, we shall also have

$$\sin \theta = \frac{r d\varphi}{ds}$$

or from Eq. (6.176)

$$p = \frac{r^2}{c} \frac{d\varphi}{ds} \qquad (6.181)$$

Eliminating ds from Eqs. (6.180) and (6.181), we obtain

$$\frac{r^4}{c^2p^2}\overline{d\varphi}^2 = \overline{dr}^2 + r^2\overline{d\varphi}^2$$

or

$$d\varphi = \frac{dr}{r\left(\dfrac{r^2}{c^2p^2}-1\right)^{1/2}} = \frac{p\,dr}{r(\eta^2-p^2)^{1/2}} \qquad (6.182)$$

using Eq. (6.178). The total arc distance Δ will then be simply

$$\Delta = 2p\int_{r_p}^{r_0} \frac{dr}{r(\eta^2-p^2)^{1/2}} \qquad (6.183)$$

corresponding to Eq. (6.159). Eliminating $d\varphi$ from Eqs. (6.180) and (6.181), we obtain

$$\overline{ds}^2 = \overline{dr}^2 + \frac{p^2c^2}{r^2}\overline{ds}^2$$

or

$$ds = \frac{dr}{\left(1-\dfrac{p^2c^2}{r^2}\right)^{1/2}} = \frac{\eta\,dr}{(\eta^2-p^2)^{1/2}} \qquad (6.184)$$

using Eq. (6.178) again. The total travel time T will then be simply

$$T = 2\int_{r_p}^{r_0}\frac{ds}{c} = 2\int_{r_p}^{r_0}\frac{\eta^2\,dr}{r(\eta^2-p^2)^{1/2}} \qquad (6.185)$$

corresponding to Eq. (6.160). We may also express T as

$$T = 2\int_{r_p}^{r_0}\left[\frac{p^2}{r(\eta^2-p^2)^{1/2}} + \frac{(\eta^2-p^2)^{1/2}}{r}\right]dr$$

or

$$T = p\Delta + 2\int_{r_p}^{r_0}\frac{(\eta^2-p^2)^{1/2}}{r}\,dr \qquad (6.186)$$

corresponding to Eq. (6.158).

As in the previous section, our measurements will usually give us a plot of T versus Δ, from which it is desired to determine the velocity c as a function of the radial distance r. We see that our integral [Eq. (6.183)] is similar to the integral [Eq. (6.166)] with the replacement of $(dz/d\eta)$ by $(1/r)(\partial r/\partial \eta)$ and reversal of the limits of integration. The derivation can be followed through exactly the same ending with the equation

$$\int_0^{\Delta_1}\cosh^{-1}\left(\frac{p}{\eta_1}\right)d\Delta = \pi\int_{r_1}^{r_0}\frac{dr}{r} \qquad (6.187)$$

corresponding to Eq. (6.168). Our final expression will then be

$$\log\left(\frac{r_0}{r_1}\right) = \frac{1}{\pi}\int_0^{\Delta_1} \cosh^{-1}\left(\frac{p}{\eta_1}\right) d\Delta \qquad (6.188)$$

corresponding to Eq. (6.169). This is an expression for r_1 in terms of quantities measured from the T versus Δ plot. From Eq. (6.177) the quantity $\eta_1 = (dT/d\Delta)_1$ is the slope of the T versus Δ plot at Δ_1. From Eq. (6.179) the quantity $p = (dT/d\Delta)$ is the slope measured sequentially from $\Delta = 0$ to $\Delta = \Delta_1$. As before, the integration is usually carried out numerically for r_1. The value of c_1 is given simply by Eq. (6.178) from $\eta_1 = r_1/c_1$.

We have been considering so far only the first P and first S arrivals. There will be many other arrivals following both the initial P and initial S arrivals. Some of these will be surface waves, to be considered in the next chapter, and some of them will be reflections and refractions from velocity discontinuities within the earth and at the earth's outer surface. In general, we can recognize the following gross features for the internal constitution of the earth—a *crust* extending from the earth's outer surface down to a depth of 10 to 50 km, a *mantle* extending from the bottom of the crust to a depth of about 2900 km, and a *core* extending from the base of the mantle to the earth's center, 6400 km. The seismic discontinuity between the crust and the mantle is known as the *Mohorovicic discontinuity*. A boundary is also observed within the core at a depth of about 5100 km, separating it into an inner and an outer core. In the crust and mantle, both P and S waves are observed; in the core, only P waves are observed, leading to the conclusion that the core is liquid. By ignoring the thin crust, the following types of secondary waves may be observed. A P or S wave incident at the earth's outer surface will produce both a reflected P and a reflected S wave. A P or S wave incident on the core-mantle boundary will produce a reflected P, reflected S, and a refracted P wave. A P wave incident on the inner core–outer core boundary will produce a reflected P and a refracted P wave. Also, since there is a decrease in P wave velocity across the mantle-core boundary, there will be, from the discussion of Section 6.5, a shadow zone followed by a focusing region.

From the experimentally determined T versus Δ plots, Eq. (6.188) can be used sequentially to obtain plots of the P and S velocities versus depth. It is of particular interest to us here to see what we can determine concerning the physical properties of the earth from such a plot. It should be obvious at the outset that since we have only two known quantities α and β defined by three unknown quantities λ, μ, and ρ, we cannot obtain a unique solution. We must impose some additional conditions in order to be able to obtain some estimate of λ, μ, and ρ.

We shall assume a spherical earth with spherical symmetry and designate by r the radial distance out from the center of the earth. Then the density ρ, the incompressibility k, the total hydrostatic pressure p, and the gravitational

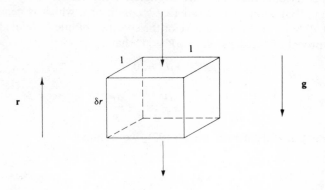

Fig. 6.22

attraction g will be functions of r only. We may assume that the stress in the interior of the earth is essentially hydrostatic. The increase in pressure from one depth to the next will then be simply related to the increase in mass, from Fig. 6.22, approximately by

$$\frac{dp}{dr}\delta r = -g\rho\,\delta r$$

or

$$\frac{dp}{dr} = -g\rho \tag{6.189}$$

The scalar value of g at any distance r from the center of a spherical earth is given simply as

$$g = \gamma\frac{m}{r^2} \tag{6.190}$$

where m is the total mass within the sphere of radius r. Combining Eqs. (6.189) and (6.190), we obtain

$$\frac{dp}{dr} = -\frac{\gamma m \rho}{r^2} \tag{6.191}$$

From Eqs. (6.29) and (6.25) we have for the incompressibility k—remembering that our pressure p used here is the total hydrostatic pressure so that the differential pressure quantity related to wave propagation in an elastic medium in Eq. (6.29) will be dp—

$$k = -\frac{dp}{\frac{dV}{V}} \tag{6.192}$$

232 ¶ Seismology, Gravity, and Magnetism

If we further assume that there are negligible temperature effects associated with wave propagation, that is, adiabatic conditions, which is generally the case, we shall have for the derivative of the specific volume V defined as usual as the reciprocal of the density, or $\rho V = 1$, that

$$\rho dV - V d\rho = 0$$

or

$$\frac{dV}{V} = -\frac{d\rho}{\rho} \qquad (6.193)$$

Substituting Eq. (6.193) into Eq. (6.192), we obtain

$$k = \frac{dp}{\dfrac{d\rho}{\rho}}$$

or

$$\frac{dp}{d\rho} = \frac{k}{\rho} = \frac{3\lambda + 2\mu}{3\rho} = \alpha^2 - \frac{4}{3}\beta^2 \qquad (6.194)$$

substituting also from Eqs. (6.29), (6.37), and (6.39). This expression will be a valid relation for $dp/d\rho$ in terms of α and β for the elastic medium at any particular depth r. We shall now make one final assumption, which is the most critical, that the material is of uniform composition as a function of depth. We are assuming in effect that $dp/d\rho$ or k/ρ is a function of depth only and not of changes in composition as well. We know that this assumption cannot be valid for the entire earth and, in particular, in the most gross aspects across the core boundaries and the Mohorovicic discontinuity and in the transition regions of rapid change in seismic properties. Nevertheless, with this assumption, we may write from Eqs. (6.191) and (6.194) that

$$\frac{d\rho}{dr} = \frac{d\rho}{dp}\frac{dp}{dr} = -\frac{\gamma m \rho}{r^2(\alpha^2 - \frac{4}{3}\beta^2)} \qquad (6.195)$$

Equation (6.195) is an expression for the density variation within any region of the earth for which the constitution is essentially uniform.

Equation (6.195) gives us an expression for the density variation in the earth in terms of known measured quantities. We see that conceivably we can apply Eq. (6.195) sequentially from the earth's outer surface with appropriate adjustment across regions where there are known gross changes in composition to obtain an approximate determination of ρ as a function of r. From Eqs. (6.191) and (6.190) we can then determine immediately p and g as a function of r, and from Eqs. (6.39), (6.37), (6.29), and (6.27), the elastic parameters μ, λ, k, and E as a function of r.

Problem 6.6(a) Derive the expression for $d\Delta/dp$ corresponding to Eq. (6.173), using the quantities $f(\eta) = (1/r)(dr/d\eta)$ and $\zeta(r) = (r/c)(dc/dr)$.

Problem 6.6(b) Determine the radius of curvature for the rays and find the velocity function that will produce circular rays.

Ans.

$$\rho = [-r/p][1/(dc/dr)], \quad c = a - br^2$$

Problem 6.6(c) Examine and discuss the validity and limitations of some reasonably recent seismic determinations of the variation of the earth's physical properties as a function of depth.

References

Båth, M. 1968. *Mathematical Aspects of Seismology*. New York: Elsevier.
Bullen, K. E. 1947. *An Introduction to the Theory of Seismology*. Cambridge: Cambridge University Press.
Byerly, P. 1942. *Seismology*. Englewood Cliffs, N.J.: Prentice-Hall.
Dobrin, M. B. 1952. *Introduction to Geophysical Prospecting*. New York: McGraw-Hill.
Ewing, W. M., W. S. Jardetsky, and F. Press. 1957. *Elastic Waves in Layered Media*. New York: McGraw-Hill.
Garland, G. D. 1971. *Introduction to Geophysics*. Philadelphia: W. B. Saunders.
Grant, F. S., and G. F. West. 1965. *Interpretation Theory in Applied Geophysics*. New York: McGraw-Hill.
Gutenberg, B., and C. F. Richter. 1954. *Seismicity of the Earth and Associated Phenomena*. Princeton, N.J.: Princeton University Press.
Jeffreys, H. 1952. *The Earth*. Cambridge: Cambridge University Press.
Lamb, H. 1932. *Hydrodynamics*. New York: Dover.
Love, A. E. H. 1927. *A Treatise on the Mathematical Theory of Elasticity*. New York: Dover.
Nettleton, L. L. 1940. *Geophysical Prospecting for Oil*. New York: McGraw-Hill.
Officer, C. B. 1958. *Introduction to the Theory of Sound Transmission*. New York: McGraw-Hill.
Page, L. 1935. *Introduction to Theoretical Physics*. New York: Van Nostrand.
Rayleigh, L. 1894. *The Theory of Sound*. New York: Dover.
Stacey, F. D. 1969. *Physics of the Earth*. New York: Wiley.
Takeuchi, H. 1966. *Theory of the Earth's Interior*. Waltham: Blaisdell.

CHAPTER 7

SEISMOLOGY—WAVE THEORY

7.1 Normal Modes

Many of the theoretical problems encountered in seismology do not permit a simple and easily understood type of solution directly in terms of rays as given in the previous chapter. This is because certain constraints, or boundary conditions, are imposed on the propagation. This, then, involves interference effects from multiple reflections and refractions and conversion effects from P to SV, or vice versa, which at a distance from the source determine the characteristics of the propagation. We arrive at *normal mode* effects, that is, preferred modes of vibration for the system and at *wave guide* effects, that is, propagation constrained for certain frequencies and incident angles to a given layer by the boundary conditions imposed by the upper and lower surfaces of the layer.

Let us consider the simple harmonic motion solution of the one-dimensional wave equation in the form of Eqs. (1.183) and (1.184) or (1.185). From the familiar trigonometric identities, we may write

$$\cos(kx \pm \omega t) = \cos(kx)\cos(\omega t) \mp \sin(kx)\sin(\omega t) \tag{7.1}$$

or

$$\cos(kx)\cos(\omega t) = \tfrac{1}{2}[\cos(kx-\omega t)+\cos(kx+\omega t)]$$
$$\sin(kx)\sin(\omega t) = \tfrac{1}{2}[\cos(kx-\omega t)-\cos(kx+\omega t)] \tag{7.2}$$

and similarly for $\sin(kx-\omega t)$ in terms of $\sin(kx)\cos(\omega t)$ and $\cos(kx)\sin(\omega t)$. The expressions (7.2) are of interest. They state that two simple harmonic waves of equal amplitude and the same frequency combine to produce a stationary vibration. This would occur for the combined effect of an incident wave and its reflection from a boundary for which the amplitudes of the incident and reflected waves are the same. Solutions of the form of the left-hand side of Eqs. (7.2) are known as *standing waves*. Points along the X coordinate for which the motion is zero for all times are referred to as *nodes*; for a standing wave of the form $\cos(kx)\cos(\omega t)$, this corresponds simply to

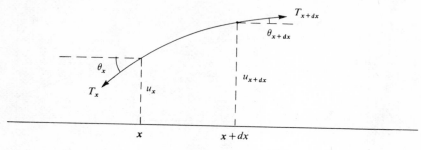

Fig. 7.1

the zeros of cos (kx). We may then write, as a form of solution of the one-dimensional wave equation,

$$\psi = {\sin \atop \cos}(kx) {\sin \atop \cos}(\omega t) \qquad (7.3)$$

where, by the notation, we mean any of the four combinations of sines and cosines.

Let us extend our consideration of the one-dimensional wave equation to two boundaries. It will be convenient to illustrate this through the use of a physical example to which these conditions apply. We shall take the example of a vibrating string fixed at both ends. We shall let ρ be the mass per unit length of the string and T be its tension. We shall neglect gravity, that is, the weight of the string, and shall assume that the displacements u are sufficiently small that we may approximate $\sin \theta$ by $\tan \theta$, which is simply $\partial u/\partial x$. Then from Fig. 7.1 we shall have for Newton's second law

$$T_{x+dx} \sin \theta_{x+dx} - T_x \sin \theta_x = \rho dx \frac{\partial^2 u}{\partial t^2}$$

$$\left(T\frac{\partial u}{\partial x}\right)_{x+dx} - \left(T\frac{\partial u}{\partial x}\right)_x = \rho dx \frac{\partial^2 u}{\partial t^2}$$

$$\frac{\partial}{\partial x}\left(T\frac{\partial u}{\partial x}\right) dx = \rho dx \frac{\partial^2 u}{\partial t^2}$$

$$T\frac{\partial^2 u}{\partial x^2} = \rho \frac{\partial^2 u}{\partial t^2}$$

$$c^2 \frac{\partial^2 u}{\partial x^2} = \frac{\partial^2 u}{\partial t^2} \qquad (7.4)$$

where

$$c^2 = \frac{T}{\rho} \qquad (7.5)$$

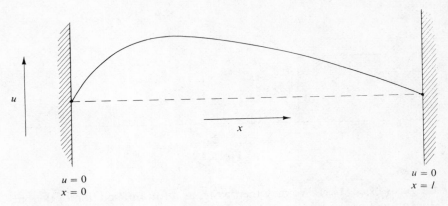

Fig. 7.2

and where we have also assumed that T and ρ are constant. Equation (7.4) is, of course, simply the one-dimensional wave equation. With reference to Fig. 7.2, we can now see the usefulness of a solution in the form of Eq. (7.3). The boundary condition that $u = 0$ at $x = 0$ is met if only terms in $\sin(kx)$ are in the solution. The boundary condition that $u = 0$ at $x = l$ is met if

$$\sin(kl) = 0$$

or

$$k = \frac{n\pi}{l} \quad (n = 1, 2, 3, \ldots) \tag{7.6}$$

From Eqs. (1.182) and (1.188) we have the corresponding expressions

$$\omega_n = kc = \frac{n\pi c}{l} \tag{7.7}$$

and

$$\nu_n = \frac{\omega_n}{2\pi} = \frac{nc}{2l} \tag{7.8}$$

for the frequency. We see that only discrete values of the wave number k and correspondingly of the frequency ω or ν are permitted. Our final solution, satisfying the boundary conditions, will then be

$$u(x,t) = \sum_{n=1}^{\infty} a_n \sin\left(\frac{n\pi x}{l}\right) \cos(\omega_n t) + b_n \sin\left(\frac{n\pi x}{l}\right) \sin(\omega_n t) \tag{7.9}$$

This is a normal mode solution. It satisfies both boundary conditions and states that the motion is composed of a set of discrete frequencies ω_1, ω_2, $\omega_3, \ldots$, whose amplitude components vary in a sinusoidal manner.

From Eq. (1.186) we have alternatively for the wavelength λ, in terms of the wave number k,

$$\lambda = \frac{2\pi}{k} = \frac{2l}{n} \tag{7.10}$$

We see that each mode starting with the first corresponds to the length of the string being successively, $\frac{1}{2}$, 1, $1\frac{1}{2}$, ... times the wavelength, as illustrated in Fig. 7.3.

The arbitrary constants of Eq. (7.9) are determined by the initial conditions, that is, by the initial displacement $u(x,0)$ and the initial velocity $(\partial u/\partial t)(x,0)$ of the string. This may easily be done through the Fourier series analysis of Section 1.8. From the given initial conditions and Eq. (7.9), we have

$$u(x,0) = \sum_{n=1}^{\infty} a_n \sin\left(\frac{n\pi x}{l}\right)$$

and

$$\frac{\partial u}{\partial t}(x,0) = \sum_{n=1}^{\infty} b_n \omega_n \sin\left(\frac{n\pi x}{l}\right)$$

Then performing the integration,

$$\int_0^l u(x,0) \sin\left(\frac{m\pi x}{l}\right) dx = \sum_{n=1}^{\infty} a_n \int_0^l \sin\left(\frac{n\pi x}{l}\right) \sin\left(\frac{m\pi x}{l}\right) dx$$

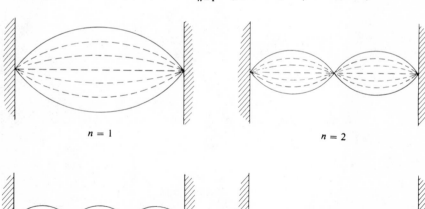

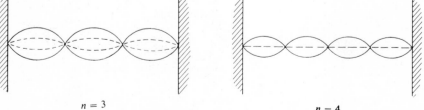

Fig. 7.3

we obtain

$$a_m = \frac{2}{l}\int_0^l u(x,0) \sin\left(\frac{m\pi x}{l}\right) dx \qquad (7.11)$$

Similarly, we obtain

$$b_m = \frac{2}{\omega_m l}\int_0^l \frac{\partial \dot{u}}{\partial t}(x,0) \sin\left(\frac{m\pi x}{l}\right) dx \qquad (7.12)$$

The solution of the more complicated normal mode problems that one encounters in nature is quite similar to this simple case, with the exception that the waves progress outward in a second space coordinate. The normal mode patterns exist in the depth coordinate z because of the plane parallel boundaries perpendicular to z. The propagation is outward in a radial direction.

Let us consider one of the simplest examples of such three-dimensional propagation in the vicinity of a boundary. We shall take the case of a point source in a liquid bounded by a free surface, for example, the ocean. We shall also take the velocity of propagation c to be a constant, so that our solution for waves diverging from a point source will be of the form of Eq. (1.177) for the displacement potential, or

$$\varphi_1 = \frac{1}{r} f\left(t - \frac{r}{c}\right) \qquad (7.13)$$

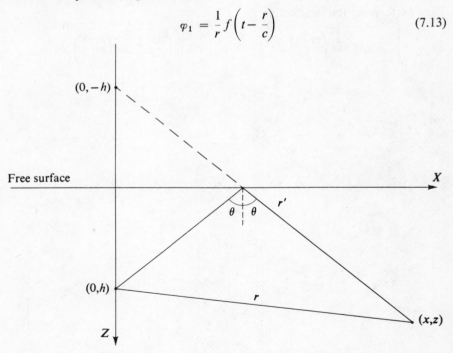

Fig. 7.4

where we have, for convenience, changed the argument by a constant factor. For the reflection from a free surface, we have from Eq. (6.88) and Section 1.10 that the reflected wave is the negative of the incident wave, or

$$\varphi_2 = -\frac{1}{r'} f\left(t - \frac{r'}{c}\right) \tag{7.14}$$

where r' is the distance between the source and receiver via the reflection path. The total received signal will then be simply

$$\varphi = \varphi_1 + \varphi_2 = \frac{1}{r} f\left(t - \frac{r}{c}\right) - \frac{1}{r'} f\left(t - \frac{r'}{c}\right) \tag{7.15}$$

where from the geometry of Fig. 7.4, r and r' are given by

$$r = \sqrt{x^2 + (z-h)^2}$$

and $\tag{7.16}$

$$r' = \sqrt{x^2 + (z+h)^2}$$

From Eq. (7.15) we see that we might consider our problem as the interference effect between two sources, one our original source located a distance h below the reflecting surface and the other the negative of the original source located a distance $-h$ above the surface. The effect of the free surface can be replaced by an *image source* located an equal distance above the surface. The geometry of Fig. 7.5 attests to this reasoning.

For a simple harmonic source, $f = a \cos(\omega t)$, the steady-state solution will be

$$\varphi = a \left\{ \frac{1}{r} \cos\left[\omega\left(t - \frac{r}{c}\right)\right] - \frac{1}{r'} \cos\left[\omega\left(t - \frac{r'}{c}\right)\right] \right\} \tag{7.17}$$

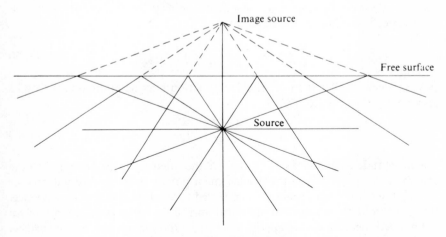

Fig. 7.5

We are interested in the case where z and h are both small with respect to x. Then r and r' may be approximated by

$$r = x\left[1 + \frac{(z-h)^2}{x^2}\right]^{1/2} = x\left[1 + \tfrac{1}{2}\frac{(z-h)^2}{x^2} + \cdots\right]$$

and similarly (7.18)

$$r' = x\left[1 + \tfrac{1}{2}\frac{(z+h)^2}{x^2} + \cdots\right]$$

We may then write for φ in Eq. (7.17)

$$\begin{aligned}\varphi &= \frac{a}{x}\left\{\cos\left[\omega\left(t - \frac{r}{c}\right)\right] - \cos\left[\omega\left(t - \frac{r'}{c}\right)\right]\right\} \\ &= \frac{a}{x}\left\{-2\sin\left[\omega\left(t - \frac{r+r'}{2c}\right)\right]\sin\left[\omega\left(\frac{r'-r}{2c}\right)\right]\right\} \\ &= -\frac{2a}{x}\sin\left(\frac{\omega h z}{cx}\right)\sin\left[\omega\left(t - \frac{x}{c}\right)\right]\end{aligned}$$ (7.19)

using the approximations from Eqs. (7.18) that

$$\frac{r+r'}{2} = x$$

and

$$\frac{r'-r}{2} = \frac{hz}{x}$$

Under these conditions, the combined effect of the direct and surface reflected waves is a wave propagating in the x direction near the surface whose amplitude varies according to the function

$$\frac{2a}{x}\sin\left(\frac{\omega h z}{cx}\right) = \frac{2a}{x}\sin\left(2\pi\frac{hz}{\lambda x}\right)$$ (7.20)

If hz is much smaller than λx, the sine function may be approximated by its argument giving for Eq. (7.20)

$$\frac{4\pi a\, hz}{\lambda x^2}$$ (7.21)

The amplitude decreases as x^{-2}; or since the intensity is proportional to the square of the amplitude of the displacement potential, the intensity decreases as x^{-4}. Such a range dependence is referred to as a *Lloyd mirror* dependence. It is greater than the normal spherical spreading for a point source in an unbounded medium and is applicable only to the region near a free surface. We also note from expression (7.21), as we might expect, that the amplitude approaches zero as either h or z approaches zero. This is sometimes referred to as a *pressure release* effect and is dependent on the wavelength. For long wavelengths (low frequencies), the pressure release effect will become apparent at greater distances away from the boundary than for short wavelengths (high frequencies). We also note, referring to Eq. (7.20) for farther distances h and z away from the boundary and for shorter horizontal ranges x, that φ will vanish for increments of

$$2\frac{hz}{\lambda x} = 0, 1, 2, 3 \ldots$$

Problem 7.1(a) Derive the expressions for the preferred wavelengths and frequencies and discuss the normal mode effects for vertical propagation in a liquid bounded by (1) free surface at the top and bottom, (2) free surface at the top and rigid (no displacement) surface at the bottom, and (3) free surface at the top and a second fluid medium at the bottom.

Problem 7.1(b) Solve the problem for the vibration of a uniform string of length l, which has been pulled aside a distance d at $x = l/2$ and then released from rest. Assume that the initial shape is made up of two straight lines. Discuss.

7.2 Dispersion

As we have seen in the previous chapter, the wave equation for many problems will be more complicated than the simple form discussed in Section 1.10 either through the variation of the elastic parameters, density, or velocity with one or more space coordinates, usually the depth, or through the introduction of additional terms to the equations of motion for gravity effects, absorption, or the like, or through a combination of both. This condition, of course, also applies in principle to problems in which there are layers of constant elastic parameters and density separated by boundaries across which there is a discontinuity in these quantities. In such cases, it cannot be assumed that the solutions developed in Section 1.10 will be solutions to these more complex equations.

Let us consider the one-dimensional wave equation in such a more complex form. One of the more useful methods of solution is to try a simple harmonic solution of the form

$$\psi = Ae^{i[k(Vt-x)]} = Ae^{i(\omega t - kx)} \tag{7.22}$$

where

$$\omega = Vk \tag{7.23}$$

We cannot assume that the velocity of wave propagation V is equal to c as was true in the simple case of constant coefficients. In general, expression (7.22) will be a solution to the more complex wave equations, but upon substitution of Eq. (7.22) into the wave equation, a conditional equation between k and V results, which also must be satisfied. This conditional equation is referred to as the *characteristic equation*, and we may write it in the form

$$G(k,V) = 0 \tag{7.24}$$

Equation (7.24) states that the velocity of wave propagation V, also referred to as the *phase velocity*, is a function of the wave number, or through Eq. (7.23), of the frequency. This phenomenon is known as *dispersion*. Each frequency will travel with a different velocity. An initial disturbance will gradually become a long train of waves, whose frequency will vary in a regular manner, those frequencies associated with the higher velocities being near the head of the train and those associated with the lower velocities being near the rear.

Let us now generalize this solution to arbitrary initial conditions. One of the easier methods is to follow an analysis similar to that for the Fourier series generalization of the vibrating string problem. However, in this case, at present, there are no boundary conditions so that an integration over all wave numbers, or frequencies, rather than a set of discrete wave numbers is necessary. This is achieved through our Fourier integral transforms of Section 1.9. From Eq. (1.172) we have

$$f(x) = \int_{-\infty}^{\infty} g(k)e^{ikx}\, dx$$
$$g(k) = \frac{1}{2\pi}\int_{-\infty}^{\infty} f(\chi)e^{-ik\chi}\, d\chi \tag{7.25}$$

for the initial conditions, $f(\chi)$, given as a particular distribution of stress and/or displacement at time $t = 0$, or

$$f(t) = \int_{-\infty}^{\infty} g(\omega)e^{i\omega t}\, d\omega$$
$$g(\omega) = \frac{1}{2\pi}\int_{-\infty}^{\infty} f(\tau)e^{-i\omega\tau}\, d\tau \tag{7.26}$$

for the initial conditions, $f(\tau)$, given as a particular stress and/or displacement time function at $x = 0$. In Eqs. (7.25) $g(k)$ is, in general, a complex number and represents the amplitude and phase of the *spectrum* of wave numbers which, when added together, give the desired initial displacement $f(x)$. Similarly, $g(\omega)$ represents the amplitude and phase of the spectrum of frequencies, which are required to produce the initial disturbance $f(t)$.

From this formulation, we may write the complete solution for the given initial conditions. Using the initial conditions of Eqs. (7.26) and the simple harmonic solution [Eq. (7.22)], where $g(\omega)$ is now used instead of A, we obtain

$$\Psi(x,t) = \int_{-\infty}^{\infty} g(\omega)e^{i[\omega t - k(\omega)x]} \, d\omega \tag{7.27}$$

Similarly, for the initial conditions of Eqs. (7.25), we obtain

$$\Psi(x,t) = \int_{-\infty}^{\infty} g(k)e^{i[kx - \omega(k)t]} \, dk \tag{7.28}$$

We can be sure that either Eq. (7.27) or (7.28) is the correct solution, for in general an integration of a simple harmonic solution of the wave equation with respect to a parameter not in the original wave equation will also be a solution, and at $x = 0$, Eq. (7.27) reduces to

$$\Psi(0,t) = \int_{-\infty}^{\infty} g(\omega)e^{i\omega t} \, d\omega = f(t)$$

and similarly for Eq. (7.28). The problem reduces to the determination of the Fourier transform $g(\omega)$ of the initial conditions $f(t)$ and the evaluation of the integral [Eq. (7.27)]. In most problems involving normal modes, certain boundary conditions also have to be met, which often lead to an additional summation or integration over the wave number k, such an integration usually being taken prior to the integration [Eq. (7.27)].

Evaluation of an integral such as Eq. (7.27) is generally facilitated through the approximation of the method of *stationary phase*. If we have an integral of the form

$$u = \int_a^b \varphi(z)e^{if(z)} \, dz \tag{7.29}$$

it represents integration of a rapidly oscillating function $e^{if(z)}$ of amplitude $\varphi(z)$. If $\varphi(z)$ is a slowly changing function with respect to an oscillation of $e^{if(z)}$, the various contributions to the integral will cancel out by annulling plus and minus values except in the neighborhood of those values of z for which $f(z)$ is stationary. Here, and only here, will a nonoscillating contribution exist. Let us designate by α the value of z at such a stationary position of $f(z)$. Then we may write

$$z = \alpha + \xi$$

and

$$f'(\alpha) = 0 \tag{7.30}$$

so that in the neighborhood of α, by a series expansion,

$$f(z) = f(\alpha) + \tfrac{1}{2}\xi^2 f''(\alpha) + \cdots$$

Since the integral has an appreciable value in this neighborhood only, the limits of integration of ξ may be extended to $-\infty$ and ∞, so that

$$u = \varphi(\alpha)e^{if(\alpha)} \int_{-\infty}^{\infty} e^{i(\xi^2/2)f''(\alpha)} \, d\xi \tag{7.31}$$

This integral may be evaluated giving

$$u = \frac{\sqrt{2\pi}\,\varphi(\alpha)}{|f''(\alpha)|^{1/2}}\, e^{i[f(\alpha)\pm\pi/4]} \qquad (7.32)$$

where | | indicates the absolute value, and the plus or minus sign in the exponential is taken as $f''(\alpha)$ is positive or negative. Only the second term of the series expansion for $f(\alpha)$ has been used. The evaluation [Eq. (7.32)], then, is valid only under the condition that the third and succeeding terms in the series expansion are much smaller than the second. There are some cases for which this condition does not hold, in particular at such values of α for which $f''(\alpha)$ is also zero, for which a further evaluation is necessarily required.

The solution for Eq. (7.27), and similarly for Eq. (7.28), when the above conditions are met is then

$$\Psi(x,t) = \sqrt{\frac{2\pi}{x}}\,\frac{\varphi(\omega)}{|d^2k/d\omega^2|^{1/2}}\, e^{i[\omega t - kx \mp \pi/4]} \qquad (7.33)$$

where

$$f(\omega) = \omega t - kx$$

$$f'(\omega) = t - \frac{dk}{d\omega}x = 0 \qquad (7.34)$$

$$f''(\omega) = -\frac{d^2k}{d\omega^2}x$$

and where the value of ω in Eq. (7.33) is determined by Eqs. (7.34). The particular value of ω and also of k from Eq. (7.23) that is used in Eq. (7.33) is determined as a function of x and t by Eqs. (7.34). Thus, at a particular value of x and t, the dominant energy contribution will have a particular frequency and wave number. Rewriting Eqs. (7.34) in the following form and defining a new velocity U by

$$U = \frac{x}{t} = \frac{d\omega}{dk} \qquad (7.35)$$

we see that a particular value of frequency and wave number found at a space-time location (x_1, t_1) will be found again at (x_2, t_2) such that

$$\frac{x_1}{t_1} = \frac{x_2}{t_2}$$

It will have traveled with a velocity U given by Eq. (7.35). Thus, the energy associated with a particular frequency group will travel with a velocity U, known as the *group velocity*, which is different from the velocity of wave propagation or phase velocity V.

The existence and significance of these two velocities can be demonstrated physically through a consideration of the mechanisms that occur in such propagation. First, however, it is desirable to write down some alternative forms of Eq. (7.35). From Eq. (7.23) we have

$$U = \frac{d\omega}{dk} = \frac{d(Vk)}{dk} = V + k\frac{dV}{dk} \qquad (7.36)$$

From Eq. (1.186) we have

$$k = \frac{2\pi}{\lambda} \qquad dk = -\frac{2\pi}{\lambda^2}d\lambda$$

so that

$$U = V - \lambda\frac{dV}{d\lambda} \qquad (7.37)$$

Again from Eq. (7.23) we have

$$\frac{1}{U} = \frac{d\left(\frac{\omega}{V}\right)}{d\omega} = \frac{1}{V} - \frac{\omega}{V^2}\frac{dV}{d\omega} \qquad (7.38)$$

Consider a dispersive wave train in which the phase velocity V increases with frequency. Then at a given distance, the wave train may resemble Fig. 7.6 as a function of time. At some slightly greater distance, A will arrive at a later time A', having traveled with a velocity V_A. B travels more slowly than A so that the time between B and B' will be greater than that between A and A', and the relative separation $A-B$ will increase. The same is true in a corresponding manner for every other point on the wave train. The wave train is lengthened as it progresses out, and the crest between AB, which was associated with a frequency $1/[2(B-A)]$ at the near distance, is associated with a lower frequency $1/[2(B'-A')]$ at a greater distance. The frequency associated with a particular phase of the motion, such as $A-A'$, is variable with distance. In the example of Fig. 7.6, the increase in distance was chosen such that the frequency of the crest BC at the nearer distance is found at the

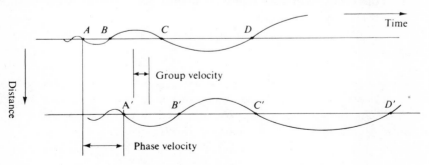

Fig. 7.6

trough $A'B'$ at the greater distance. The group velocity being that of a particular frequency group is the velocity from BC to $A'B'$, whereas the phase velocity is that from AB to $A'B'$. In this example, the group velocity is greater than the phase velocity.

The expression for the group velocity in terms of the phase velocity can be derived from a simple diagram such as that shown above. For simplicity in arithmetic, the coordinates of Fig. 7.6 will be reversed, and we shall consider

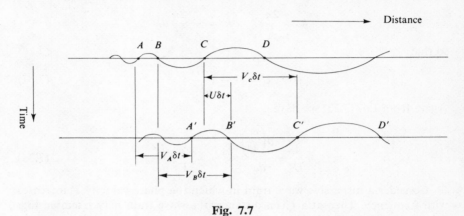

Fig. 7.7

the X coordinate motion at a specific instant of time and then at some slightly later time. Also, in this case, we shall assume that the phase velocity V decreases with frequency. In this case (Fig. 7.7), it is assumed that the time interval δt between the two examinations is such that the wavelength originally associated with C is found at B' at the later time. Then

$$U = \frac{B'-C}{\delta t} = \frac{C'-C}{\delta t} - \frac{C'-B'}{\delta t} = V_C - \tfrac{1}{2}\frac{\lambda_{B'C'}}{\delta t} \qquad (7.39)$$

Now δt has been chosen such that $C'-B' = D-C$. $C-B$ becomes $C'-B'$ in this time interval, so that

$$\delta t = \frac{\tfrac{1}{2}(\lambda_{CD} - \lambda_{BC})}{V_C - V_B} \qquad (7.40)$$

where $\tfrac{1}{2}(\lambda_{CD} - \lambda_{BC})$ is the difference in distance between CD and BC expressed in wavelengths. It is the relative distance that BC must increase in passing to $B'C'$, and $V_C - V_B$ is the relative velocity at which C is spreading away from B. Substituting Eq. (7.40) into Eq. (7.39) gives

$$U = V_C - \lambda_{CD}\frac{V_C - V_B}{\lambda_{CD} - \lambda_{BC}}$$

or in the limit

$$U = V - \lambda \frac{dV}{d\lambda}$$

which is simply Eq. (7.37).

The phase velocity is determined from the physical characteristics of the medium. The group velocity is determined from the net effect of adding up all the little elements of different frequency in the wave train traveling with different velocities. In the case where the dispersion is brought about through the variation of velocity with one of the space coordinates, usually depth, either in a continuous manner or discontinuously across boundaries, it is referred to as *geometrical dispersion*. In the case where the dispersion is brought about by a dependence of frequency on the physical parameters determining the wave velocity, it is referred to as *material dispersion*.

Another useful concept in visualizing the mechanism of dispersive wave propagation is to consider a group of records that might be obtained at a number of receiving stations away from the source. In Fig. 7.8 the records are aligned at time $t = 0$. Each frequency of the wave train will propagate through the train with its appropriate group velocity, and this group velocity will be constant for a given frequency and different for different frequencies. The location, then, of a given frequency in the wave trains from successive receiving stations will plot on a straight line passing through the origin. The slope of this line will be the group velocity. Such lines have been drawn for three different frequencies ω_1, ω_2, and ω_3 in Fig. 7.8. If we had chosen to follow a particular phase of the wave train from record to record, such as the crest A, it would plot on a curve passing through the origin. The phase A is associated with a different frequency from one record to the next. In Fig. 7.8 the frequency decreases with distance for a particular phase of the wave train.

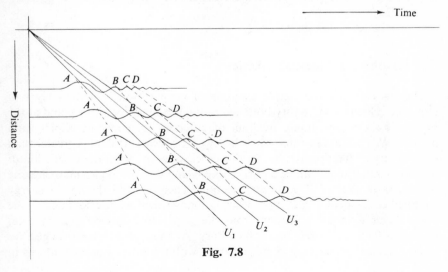

Fig. 7.8

The slope of this curve at any point will be the phase velocity for the particular frequency associated with that point. Such curves have been drawn for the crests A, B, C, and D in Fig. 7.8.

As an example of dispersion, we may take the previous example of the vibrating string, which we shall now subject to an additional restoring force proportional to the displacement. The equation of motion becomes

$$\frac{\partial^2 \psi}{\partial t^2} = c^2 \frac{\partial^2 \psi}{\partial x^2} - a^2 \psi \tag{7.41}$$

Substituting Eq. (7.22) into Eq. (7.41) gives

$$k^2 V^2 = k^2 c^2 + a^2 \tag{7.42}$$

corresponding to Eq. (7.24). For the phase velocity, we then have

$$V = \left(c^2 + \frac{a^2}{k^2}\right)^{1/2} \tag{7.43}$$

or from Eq. (7.23) in terms of ω

$$V = c\left(\frac{\omega^2}{\omega^2 - a^2}\right)^{1/2} \tag{7.44}$$

For the group velocity, we have from Eqs. (7.36) and (7.42)

$$U = \frac{d(Vk)}{dk} = (k^2 c^2 + a^2)^{-1/2} c^2 k = \frac{c^2}{V} \tag{7.45}$$

For small wave numbers, low frequencies greater than $\omega = a$, the phase velocity approaches ∞, and the group velocity approaches zero. For large wave numbers, high frequencies, both the phase and the group velocities approach the value c, the solution for the simple wave equation.

Problem 7.2(a)   Evaluate Eq. (7.31).

7.3 Dispersive Normal Modes

Let us consider the simple example of propagation in a fluid layer of constant velocity and density bounded at the top by a free surface and at the bottom by a second liquid medium of different but constant velocity and density. We shall assume that the velocity of the lower medium is greater than the velocity of the upper layer. In this case, we know from the discussion of Section 6.3 that there will be a critical angle, and that for angles more grazing than critical, there will be total reflection. Therefore, all the sound energy confined to incident angles more grazing than critical will be confined to the upper layer by total reflection from the upper free surface and by total reflection from the lower liquid boundary. At moderate to long ranges, we have a normal mode type of propagation with the normal mode patterns in

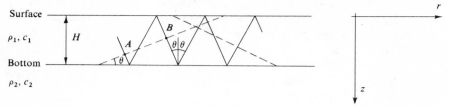

Fig. 7.9

the vertical and the propagation outward in a radial direction. The solution is not that of each ray path considered as a separate entity, but of the combined interference effects from all ray paths. This problem and its solution are applicable to shallow water sound propagation in the ocean, such as on the continental shelf.

At long ranges from a source, we may consider the incident wave fronts to be plane, and we may consider the propagation to consist of a large number of multiply reflected plane waves covering all angles of incidence. Let us consider one set of multiply reflected waves reflected from the bottom and the surface at an angle θ, as shown in Fig. 7.9. The density and velocity of the ocean and subbottom material are designated by the constant values ρ_1, c_1 and ρ_2, c_2, respectively, and the depth of water by H. There will be a set of downgoing and upgoing multiply reflected waves, whose direction is defined by the angle θ. The description of the interference effects for the set of multiply reflected rays will depend on the frequency. At those frequencies for which the phase difference between the downgoing wave associated with the point A and the downgoing wave that has undergone one additional bottom and surface reflection associated with the point B is an integral number of 2π, there will be constructive interference. Further, if this condition is met, there will be constructive interference for all the downgoing and upgoing waves associated with this set of rays for all time and for all r and z throughout the layer. For all other frequencies, the waves will interfere out of phase, and their net effect will be small compared with those frequencies for which there is always constructive interference.

These frequencies may be determined simply as follows. We desire the total phase change for the ray path from A to B minus the phase change for reflection from the free surface π, minus the phase change for reflection from the bottom ϵ of Eq. (6.101) to be equal to $2(n-1)\pi$, where n is an integer and can take successively the values $n = 1, 2, 3, \ldots$. From Fig. 7.9 this will be

$$\frac{2\pi}{\lambda_n}\left[\frac{H}{\cos\theta} + \frac{H}{\cos\theta}\cos(2\theta)\right] - \epsilon - \pi = 2(n-1)\pi$$

$$\frac{2\pi\nu_n H}{c_1}\cos\theta - \frac{\epsilon}{2} = (2n-1)\frac{\pi}{2} \qquad (7.46)$$

where λ_n and ν_n are the wavelength and frequency for the solutions to Eq. (7.46) for the respective value of n. The phase change upon reflection from the bottom for angles more grazing than the critical angle is given from Eq. (6.102) as

$$\tan\frac{\epsilon}{2} = \frac{\rho_1\sqrt{\sin^2\theta - (c_1^2/c_2^2)}}{\rho_2\sqrt{1-\sin^2\theta}} \tag{7.47}$$

The spatial interference between the downgoing and upgoing waves, both at an angle θ with the horizontal, is such as to produce a standing-wave pattern in the z direction and a net propagation in the r direction only. The velocity with which the wave progresses in the r direction is dependent on the angle θ and will always be greater than the velocity c_1 of each upgoing or downgoing ray as measured along the ray direction. From Fig. 7.9 it is given simply as

$$V = \frac{c_1}{\sin\theta} \tag{7.48}$$

The velocity V corresponds to the phase velocity of the previous section. It should be noted that the wave number k_n, associated with the wavelength λ_n of Eq. (7.46) measured along the ray direction, is different from the wave number κ_n, corresponding to Eq. (7.23) measured in the radial direction of propagation of the interference effect. From Eqs. (7.48) and (7.23) the relation is simply

$$\kappa_n = \frac{\omega_n}{V} = \frac{\omega_n \sin\theta}{c_1} = k_n \sin\theta \tag{7.49}$$

Substituting Eqs. (7.47) and (7.48) into Eq. (7.46), we obtain

$$\frac{2\pi\nu_n H}{c_1}\sqrt{1-\frac{c_1^2}{V^2}} = (2n-1)\frac{\pi}{2} + \frac{\epsilon}{2}$$

$$\tan\left(\frac{2\pi\nu_n H}{c_1}\sqrt{1-\frac{c_1^2}{V^2}}\right) = \tan\left[(2n-1)\frac{\pi}{2} + \frac{\epsilon}{2}\right] = -\cot\left(\frac{\epsilon}{2}\right)$$

$$\tan\left(\frac{2\pi\nu_n H}{c_1}\sqrt{1-\frac{c_1^2}{V^2}}\right) = -\frac{\rho_2\sqrt{1-(c_1^2/V^2)}}{\rho_1\sqrt{(c_1^2/V^2)-(c_1^2/c_2^2)}}$$

$$\tan\left(\frac{2\pi\nu_n H}{Vc_1}\sqrt{V^2-c_1^2}\right) = -\frac{\rho_2 c_2\sqrt{V^2-c_1^2}}{\rho_1 c_1\sqrt{c_2^2-V^2}}$$

$$\tan\left(\frac{2\pi\nu_n H}{V}\sqrt{\frac{V^2}{c_1^2}-1}\right) = -\frac{\rho_2\sqrt{V^2/c_1^2-1}}{\rho_1\sqrt{1-(V^2/c_2^2)}} \tag{7.50}$$

Equation (7.46) is a relation between the angle of incidence of a particular group of multiply reflected rays and the set of discrete frequencies that will be prominent. For each angle of incidence, there is a particular set of frequencies. Equation (7.50) is a relation between the velocity of propagation along the layer, or channel, and the frequency. It states that the phase velocity is a function of the frequency. The interference effects from the multiple reflections have led to a dispersive wave propagation.

The formal solution to this problem, which will not be repeated here, following Eqs. (7.22) to (7.24) gives the same phase velocity relation as derived above. The derivation given here shows somewhat more directly the physical principles that are involved.

Returning to Eq. (7.50), we see that it is multiple valued in v_n. For each value of V, as for each value of θ, there will be an infinite set of values of v_n. The dispersion curve, phase velocity versus frequency, for $n = 1$ is referred to as the first mode; the dispersion curve for $n = 2$ is referred to as the second mode, and so on. Equations (7.46) and (7.50) are valid only for angles of incidence more grazing than critical. For angles of incidence steeper than critical, there will be no phase change upon reflection from the bottom. Moreover, there will be energy refracted into the bottom upon each bottom reflection, so that at long ranges, the net contributions from incidence steeper than critical will be small compared with incidence more grazing than critical and may be neglected. For unattenuated propagation, then, there will be a low-frequency cutoff for each mode corresponding to critical angle reflection. From Eqs. (6.99) and (7.48) we shall then have for V_c

$$V_c = \frac{c_1}{\sin \theta_c} = c_2 \tag{7.51}$$

and for the corresponding cutoff frequency v_{cn} from Eq. (7.50),

$$\frac{2\pi v_{cn} H}{c_1} \sqrt{1 - \frac{c_1^2}{c_2^2}} = (2n-1)\frac{\pi}{2}$$

or

$$v_{cn} = \frac{(2n-1)c_1}{4H\sqrt{1-(c_1^2/c_2^2)}} \tag{7.52}$$

As has been discussed in the previous section, a complete description of dispersive wave propagation from an arbitrary initial disturbance requires a second evaluation to describe the change in shape of the dispersive wave train as it progresses outward in space and time. This evaluation led to the concept of group velocity U, which is the velocity of propagation for a particular frequency in the train as contrasted with the phase velocity V of a particular point or phase of the train. From Eq. (7.38) we have for U

$$\frac{1}{U} = \frac{1}{V} - \frac{v_n}{V^2}\frac{dV}{dv_n} \tag{7.53}$$

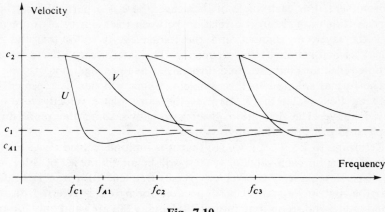

Fig. 7.10

For this problem, the evaluation for U is most easily done numerically from the V-ν_n plot. A sketch of the form of the phase and group velocities for the first three modes is given in Fig. 7.10.

From Fig. 7.10 it is possible to comprehend qualitatively the form of the dispersive wave train. Considering the first mode only—the train will start with the frequency f_{c1} having traveled with the velocity c_2. With increasing time, the wave train will increase in frequency down to the part that has traveled with the velocity c_1. From there on, there will be two superimposed wave trains to the termination traveling with the velocity c_{A1}. The first wave train is a continuation of the original with increasing frequency down to f_{A1}; the second starts at very high frequencies with the velocity c_1 and decreases in frequency to f_{A1}. The resultant wave train would resemble that shown in Fig. 7.11. It should be noted that at f_{A1}, $dU/d\omega = 0$, and our

Fig. 7.11

condition for the evaluation [Eq. (7.32)] is not met. The evaluation of Eq. (7.29) under this condition leads to a constant frequency train of decreasing amplitude, referred to as the *Airy phase*.

Problem 7.3(a) Derive the amplitude dependence as a function of depth for the normal modes and discuss the dependence.

Problem 7.3(b) Derive the expressions for V and U for the attenuated modes and discuss the propagation.

Problem 7.3(c)   Calculate V/c_1 and U/c_1 for the unattenuated first mode for $c_2 = 1.5c_1$ and graph as a function of the dimensionless quantities V_1/c_1 and U/c_1 versus $\nu H/c_1$.

Problem 7.3(d)   Show that for $dU/d\omega = 0$ that $d^2k/d\omega^2 = 0$ and the condition for the evaluation [Eq. (7.33)] is not met.

7.4 Rayleigh Waves

In the previous sections, we have developed the concepts of normal modes, dispersion, and wave guide propagation and have discussed, in a physical sense, how such wave guide propagation occurs. We shall proceed to examine in this and the following section the possibility of and conditions for such waves being propagated along the boundary surface of an elastic medium or in the vicinity of a series of boundary surfaces between elastic media. We shall assume a wave propagating only in a direction parallel to the boundary and shall see if such a wave can satisfy the boundary conditions. Such waves are referred to as *surface waves* in seismology.

With reference to Fig. 6.2 of Section 6.3, consider the possibility of a wave propagating in the X direction in such a manner that (1) the disturbance is largely confined to the neighborhood of the boundary and (2) at any instant the particle motion in any line parallel to the Y axis is the same. On account of (1), the wave is a surface wave, and on account of (2), it is analogous to the plane waves of Section 6.3.

We may then use the displacement potentials defined by Eqs. (6.48) and their corresponding wave equations (6.49), (6.50), and (6.51). We shall then assume simple harmonic solutions of the form

$$\varphi = f(z)e^{i\kappa(x-ct)} \qquad (7.54)$$

$$\psi = g(z)e^{i\kappa(x-ct)} \qquad (7.55)$$

$$v = h(z)e^{i\kappa(x-ct)} \qquad (7.56)$$

where c is the velocity of propagation of the surface wave and κ its associated wave number, defined by the usual relation

$$c = \frac{\omega}{\kappa} \qquad (7.57)$$

Similar forms of solution are assumed for the upper medium with the functions f, g, h replaced by f_1, g_1, h_1.

Substituting any of the above forms of solution into their corresponding wave equation will lead to the same type of solution. Let us consider the substitution of Eq. (7.56) into its corresponding wave equation (6.51), obtaining

$$-\kappa^2 c^2 h = \beta^2 \frac{d^2 h}{dz^2} - \kappa^2 \beta^2 h$$

$$\frac{d^2 h}{dz^2} + \kappa^2 s^2 h = 0 \tag{7.58}$$

where, for convenience, we have introduced the quantities r, s, r_1, s_1, defined by

$$\begin{aligned} r &= \left(\frac{c^2}{\alpha^2} - 1\right)^{1/2} \\ s &= \left(\frac{c^2}{\beta^2} - 1\right)^{1/2} \\ r_1 &= \left(\frac{c^2}{\alpha_1^2} - 1\right)^{1/2} \\ s_1 &= \left(\frac{c^2}{\beta_1^2} - 1\right)^{1/2} \end{aligned} \tag{7.59}$$

and wnere we shall take these quantities as defining only the positive square root. The solution of the ordinary differential equation (7.58) for h is simply

$$h(z) = Ce^{i\kappa s z} + Fe^{-i\kappa s z} \tag{7.60}$$

For the effect to be essentially confined to the surface and for the propagation to be in the X direction only, the exponential of Eq. (7.60) must be real and negative. Hence, the quantities r, s, r_1, s_1 must be imaginary; the constant F must vanish for the solution $h(z)$, and the constant C_1 for the solution $h_1(z)$. Our final forms of solution are then in medium M

$$\varphi = A e^{i\kappa(rz + x - ct)} \tag{7.61}$$

$$\psi = B e^{i\kappa(sz + x - ct)} \tag{7.62}$$

$$v = C e^{i\kappa(sz + x - ct)} \tag{7.63}$$

and in medium M_1

$$\varphi_1 = D_1 e^{i\kappa(-r_1 z + x - ct)} \tag{7.64}$$

$$\psi_1 = E_1 e^{i\kappa(-s_1 z + x - ct)} \tag{7.65}$$

$$v_1 = F_1 e^{i\kappa(-s_1 z + x - ct)} \tag{7.66}$$

where A, B, C, D_1, E_1, F_1 are constants, and r, s, r_1, s_1 are all positive imaginaries.

We shall now proceed to examine if such a surface wave can exist in the vicinity of a single, plane boundary between two elastic media. We must apply our usual boundary conditions of continuity of displacement and continuity of stress across the boundary surface. From Eqs. (6.48) the first set of boundary conditions for u, w, and v are

$$A + sB = D_1 - s_1 E_1 \tag{7.67}$$

$$rA - B = -r_1 D_1 - E_1 \tag{7.68}$$

$$C = F_1 \tag{7.69}$$

From Eqs. (6.53), (6.54), and (6.55), the second set of boundary conditions are

$$\lambda(1+r^2)A + 2\mu(r^2 A - sB) = \lambda_1(1+r_1^2) + 2\mu_1(r_1^2 D_1 + s_1 E_1)$$

$$\mu[2rA - (1-s^2)B] = \mu_1[-2r_1 D_1 - (1-s_1^2)E_1]$$

$$\mu sC = -\mu_1 s_1 F_1$$

or from Eqs. (6.37) and (6.39) in terms of the seismic velocities, α and β,

$$\{[\alpha^2(1+r^2) - 2\beta^2]A - 2\beta^2 sB\}\rho = \{[\alpha_1^2(1+r_1^2) - 2\beta_1^2]D_1 + 2\beta_1^2 s_1 E_1\}\rho_1 \tag{7.70}$$

$$[2rA - (1-s^2)B]\beta^2 \rho = [-2r_1 D_1 - (1-s_1^2)E_1]\beta_1^2 \rho_1 \tag{7.71}$$

$$s\beta^2 \rho C = -s_1 \beta_1^2 \rho_1 F_1 \tag{7.72}$$

Since both s and s_1 are positive imaginaries, we see immediately that C and F_1 are zero. There can be no SH type surface wave for a single, plane boundary.

The solutions of these boundary conditions for the P and SV type motions are laborious. We shall examine here only the relatively simpler case of a free surface. Following the discussion of Section 6.3, the coefficients D_1 and E_1 vanish, and the boundary conditions [Eqs. (6.53) and (6.54)] reduce to zero stress on the free surface. From Eqs. (7.70) and (7.71) we shall then have

$$[\alpha^2(1+r^2) - 2\beta^2]A - 2\beta^2 sB = 0$$

$$2rA - (1-s^2)B = 0 \tag{7.73}$$

For Eqs. (7.73) to have other than vanishing solutions for A and B, the determinant of the coefficients must be zero, or

$$[\alpha^2(1+r^2) - 2\beta^2](1-s^2) - 4rs\beta^2 = 0 \tag{7.74}$$

Substituting for r and s from Eqs. (7.59), we obtain

$$(c^2 - 2\beta^2)\left(2 - \frac{c^2}{\beta^2}\right) = -4\beta^2 \left(1 - \frac{c^2}{\alpha^2}\right)^{1/2}\left(1 - \frac{c^2}{\beta^2}\right)^{1/2}$$

$$\left(2 - \frac{c^2}{\beta^2}\right)^2 = 4\left(1 - \frac{c^2}{\alpha^2}\right)^{1/2}\left(1 - \frac{c^2}{\beta^2}\right)^{1/2} \tag{7.75}$$

Rationalizing Eq. (7.75), we finally obtain

$$\frac{c^6}{\beta^6} - 8\frac{c^4}{\beta^4} + c^2\left(\frac{24}{\beta^2} - \frac{16}{\alpha^2}\right) - 16\left(1 - \frac{\beta^2}{\alpha^2}\right) = 0 \qquad (7.76)$$

from which c can be determined.

It is in order to examine whether Eq. (7.76) has a solution that will satisfy our restrictions that both r and s are imaginary. If we substitute into the left-hand side of Eq. (7.76) the values $c = \beta$ and $c = 0$, we obtain unity and $-16(1 - \beta^2/\alpha^2)$. This latter value is always negative since β is less than α. Hence Eq. (7.76) will always have a solution c less than β, which will meet the conditions that r and s are imaginary. A surface wave of this type is referred to as a *Rayleigh wave*. The term Rayleigh wave is also used in the more general sense for P and SV type surface waves involving a sequence of boundary surfaces. It may seem unusual from the discussion of the previous section that any surface wave can exist at all for a single, plane boundary since no multiple reflections are involved to maintain the horizontal propagation. A part of the explanation is that for the Rayleigh wave, there is conversion of P and SV type motions associated with the interface. There is no such conversion for SH type motion, and no surface wave exists for SH motion along a single, plane boundary. It should also be noted for this limiting case of a single boundary that the Rayleigh wave propagation is nondispersive.

Problem 7.4(a) Carry through the rationalization of Eqs. (7.75) to (7.76).

Problem 7.4(b) For the case where Poisson's relation is valid, solve Eq. (7.76) for c in terms of β.

Ans. $c = 0.92\beta$

Problem 7.4(c) For the conditions of Problem 7.4(b), solve for the expressions for the particle displacements u and w and show that the motion is retrograde elliptical.

7.5 Love Waves

When a layer rather than a single boundary is considered, surface waves of the SH type can exist. Such surface waves are referred to as *Love waves*. Let us consider the simple case of a layer M bounded above by a free surface and bounded below by a second medium M_1, as shown in Fig. 7.12. For the form of solution for the displacement v in the layer M, we can have both of the expressions on the right-hand side of Eq. (7.60); for v_1 in M_1 we shall have only the form of Eq. (7.63), giving

$$\begin{aligned} v &= Ce^{i\kappa(sz+x-ct)} + Fe^{i\kappa(-sz+x-ct)} \\ v_1 &= C_1 e^{i\kappa(s_1 z+x-ct)} \end{aligned} \qquad (7.77)$$

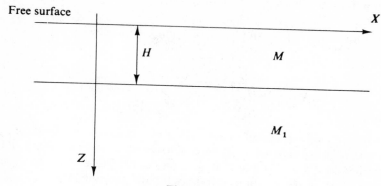

Fig. 7.12

Our boundary conditions are that the stress vanish at the free surface and that the displacement and stress be continuous across the boundary between M and M_1, or

$$\frac{\partial v}{\partial z} = 0 \qquad (z = 0)$$

$$v = v_1 \qquad (z = H) \qquad (7.78)$$

$$\mu \frac{\partial v}{\partial z} = \mu_1 \frac{\partial v_1}{\partial z} \qquad (z = H)$$

Substituting Eqs. (7.77) into Eqs. (7.78), we obtain

$$C - F = 0$$
$$Ce^{i\kappa sH} + Fe^{-i\kappa sH} = C_1 e^{i\kappa s_1 H}$$
$$\mu s C e^{i\kappa sH} - \mu s F e^{-i\kappa sH} = \mu_1 s_1 C_1 e^{i\kappa s_1 H}$$

or

$$2\cos(\kappa sH)C - e^{i\kappa s_1 H} C_1 = 0$$
$$2is\mu \sin(\kappa sH)C - \mu_1 s_1 e^{i\kappa s_1 H} C_1 = 0 \qquad (7.79)$$

As before, for Eqs. (7.79) to have a solution for C and C_1, the determinant of the coefficients must vanish, giving

$$-\mu_1 s_1 + i\mu s \tan(\kappa sH) = 0 \qquad (7.80)$$

We have, as a requirement still, that the displacement diminish away from the boundary into the lower medium, or that s_1 be imaginary. Under this restriction, we can have a solution of Eq. (7.80) if s is real. This condition will be met only if the layer velocity β is less than the velocity β_1 in the underlying medium. Substituting for s and s_1 from Eqs. (7.59), Eq. (7.80) becomes

$$\mu_1 \left(1 - \frac{c^2}{\beta_1^2}\right)^{1/2} - \mu \left(\frac{c^2}{\beta^2} - 1\right)^{1/2} \tan\left[\kappa H \left(\frac{c^2}{\beta^2} - 1\right)^{1/2}\right] = 0$$

or

$$\tan\left(\frac{\omega H}{c}\sqrt{\frac{c^2}{\beta^2} - 1}\right) = \frac{\mu_1 \sqrt{1 - (c^2/\beta_1^2)}}{\mu \sqrt{(c^2/\beta^2) - 1}} \qquad (7.81)$$

The similarity of expression in Eq. (7.81) with Eq. (7.50) should be noted. We see that our surface wave is now dispersive with its phase velocity c being determined by Eq. (7.81) and that we have, as before, a sequence of modes corresponding to the multiple solutions for the tangent in Eq. (7.81). The group velocity is given, also as before, by Eq. (7.36).

If we were to examine the problem of Rayleigh wave propagation for the example of Fig. 7.12, we would obtain a conditional equation similar to Eq. (7.80), defining a dispersive wave train. If we were to continue further and examine the problems of either Rayleigh or Love wave propagation for a sequence of layers, a conditional equation would result defining a more complex dispersive wave propagation.

As one might expect, surface waves do occur and are recorded on seismograms from distant earthquakes. The Rayleigh and Love waves that are recorded generally have propagated around the earth in the outer portions, crustal layers and upper mantle. From a comparison of the recorded dispersive properties of these surface waves with the group velocity calculations from Eqs. (7.36) and (7.81), or other more complex models, useful properties of the crust and mantle can be determined. As one might also expect, these surface waves are more substantially developed from shallow focus earthquakes occurring within the layers rather than from deep focus earthquakes. In general, then, a seismogram from a distant earthquake will consist of the discrete arrivals defined by the rays followed because of the longer paths and lower velocities by the Rayleigh and Love surface wave trains.

Problem 7.5(a) Using the methods of Section 7.3, derive the dispersion equation (7.81).

References

Båth, M. 1968. *Mathematical Aspects of Seismology*. New York: Elsevier.
Bullen, K. E. 1947. *An Introduction to the Theory of Seismology*. Cambridge: Cambridge University Press.
Ewing, W. M., W. S. Jardetsky, and F. Press. 1957. *Elastic Waves in Layered Media*. New York: McGraw-Hill.
Garland, G. D. 1971. *Introduction to Geophysics*. Philadelphia: W. B. Saunders.
Jeffreys, H. 1952. *The Earth*. Cambridge: Cambridge University Press.

Lamb, H. 1932. *Hydrodynamics*. New York: Dover.
Morse, P. M. 1948. *Vibration and Sound*. New York: McGraw-Hill.
Officer, C. B. 1958. *Introduction to the Theory of Sound Transmission*. New York: McGraw-Hill.
Pekeris, C. L. 1948. "Theory of propagation of explosive sound in shallow water," *Geol. Soc. Am.*, Memoir 27, 1948.
Stacey, F. D. 1969. *Physics of the Earth*. New York: Wiley.
Takeuchi, H. 1966. *Theory of the Earth's Interior*. Waltham: Blaisdell.
Wangsness, R. K. 1963. *Introduction to Theoretical Physics*. New York: Wiley.

Chapter 8

GRAVITY

8.1 Fundamental Relations

With reference to Section 6.1, we have for a conservative field of force

$$-dV = F_x dx + F_y dy + F_z dz$$

from which it follows that

$$F_x = -\frac{\partial V}{\partial x} \quad F_y = -\frac{\partial V}{\partial y} \quad F_z = -\frac{\partial V}{\partial z}$$

or, in vector notation,

$$\mathbf{F} = -\nabla V \tag{8.1}$$

This is a fundamental relation in the consideration of problems in gravitational theory. It states that we may determine the vector quantity $\mathbf{F}$, the gravitational attraction, by first solving for the scalar potential V and then taking the derivative. For most complex gravitational problems, this is often the simplest method of solution.

It was stated in Section 1.1 that a vector function of position in space can be represented by the gradient of a scalar function of the coordinates when, and only when, the curl of the vector function vanishes. Therefore, the necessary and sufficient condition for the existence of a potential in a field of force is that the curl of the force be zero. This is the case for gravitational attraction and, in general, whenever the force between particles is an attraction or repulsion, depending only on the distance between them. Assuming the force to be of the form $f(r)\mathbf{r}$, we have

$$\begin{aligned}
\nabla \times [f(r)\mathbf{r}] &= \nabla \times [f(r)(\mathbf{i}x + \mathbf{j}y + \mathbf{k}z)] \\
&= \mathbf{i}\left\{\frac{\partial}{\partial y}[zf(r)] - \frac{\partial}{\partial z}[yf(r)]\right\} + \cdots \\
&= \mathbf{i}\frac{df}{dr}\left[\frac{zy}{r} - \frac{yz}{r}\right] + \cdots = 0 \tag{8.2}
\end{aligned}$$

since $\partial r/\partial x = x/r$, and so on.

If m_1 is the mass of a small body and m_2 that of a second small body, Newton's law of gravitation states that the force of attraction exerted by m_2 on m_1, or vice versa, is

$$F = \gamma \frac{m_1 m_2}{r^2} \tag{8.3}$$

where γ is the universal constant of gravitation. Since we shall usually be interested in the *gravity* or *gravitational attraction* at a point, which is defined as the force per unit mass observed at a measuring point due to an attracting body's gravitational field, Eq. (8.3) becomes

$$g = \gamma \frac{m}{r^2} \tag{8.4}$$

where g is gravity. In vector notation Eq. (8.4) is simply

$$\mathbf{g} = \gamma \frac{m}{r^2} \mathbf{r}_1 \tag{8.5}$$

where $\mathbf{r}_1$ is the unit vector measured in a direction from the observing point toward the attracting mass. If we designate by Φ the potential due to gravity, the relation between it and the gravitational attraction is simply, from Eq. (8.1),

$$\mathbf{g} = -\nabla \Phi \tag{8.6}$$

From Newton's second law of motion, which states that the force on a body is equal to its mass multiplied by its acceleration, we see that the gravitational attraction is the same as the acceleration that a particle would experience at the observing point in the particular gravitational field. In CGS units, g can be given in either dyn/g or in its equivalent cm/sec^2; this unit is called a *gal*, named for Galileo. Because of the numerical size of this unit, most earth gravity measurements and calculations are quoted in *milligals* (mgal).

Problem 8.1(a) Find the gravitational potential at a distance r from (1) a mass m, (2) a distributed mass of density ρ with a volume τ, and (3) a surficial mass of density σ on a surface s.

Ans. (1) $\gamma \dfrac{m}{r}$, (2) $\gamma \displaystyle\int_\tau \dfrac{\rho \, d\tau}{r}$, (3) $\gamma \displaystyle\int_s \dfrac{\sigma \, ds}{r}$

Problem 8.1(b) Find the gravitational potential at a point P (1) outside and (2) inside a spherical shell of uniform density and radius a.

Ans. (1) $\gamma \dfrac{M}{r}$, (2) $\gamma \dfrac{M}{a}$

Problem 8.1(c) Show that the gravitational potential outside a sphere in which the density is a function of the distance from the center only is the same as if the entire mass of the sphere were concentrated at its center. Prove that two such spheres attract each other as if their masses were concentrated at their centers.

8.2 Gauss' Law and Green's Equivalent Layer

To proceed to the derivation of Gauss' law, we first define the *gravitational flux dN*, through an element of surface *d*s, as the product of the component of the gravitational attraction normal to the surface by the area of the surface, that is,

$$dN = \mathbf{g} \cdot d\mathbf{s} \tag{8.7}$$

Consider the flux through the closed surface S (Fig. 8.1) due to a small mass

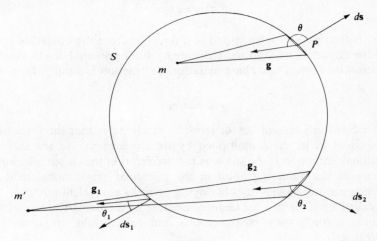

Fig. 8.1

m inside. The flux through the element of surface ds at P is

$$dN = g\,ds\cos\theta = \gamma\frac{m\,ds\cos\theta}{r^2}$$

But the solid angle subtended at m by ds is

$$d\Omega = -\frac{ds\cos\theta}{r^2}$$

Therefore,

$$dN = -\gamma m\,d\Omega$$

Summing up, the entire flux through the closed surface is

$$N = -4\pi\gamma m \tag{8.8}$$

On the other hand, the total flux through the surface due to a small mass m' outside the surface is zero; for if we divide the surface into pairs of elements subtending the same solid angle at m', such as $d\mathbf{s}_1$ and $d\mathbf{s}_2$, the flux through $d\mathbf{s}_2$ is annulled by the equal and opposite flux through $d\mathbf{s}_1$.

Therefore, the total gravitational flux through any closed surface is equal to $-4\pi\gamma$ times the mass enclosed by the surface. This is *Gauss' law*. We may rewrite this for an extended mass of density ρ within the surface S

$$N = -4\pi\gamma \int_\tau \rho \, d\tau$$

By definition, the flux N through the closed surface is the surface integral

$$N = \int_s \mathbf{g} \cdot d\mathbf{s}$$

Equating these two expressions, we obtain

$$\int_s \mathbf{g} \cdot d\mathbf{s} = -4\pi\gamma \int_\tau \rho \, d\tau \tag{8.9}$$

Applying Gauss' theorem to the left-hand member, we obtain

$$\int_\tau \nabla \cdot \mathbf{g} \, d\tau = -4\pi\gamma \int_\tau \rho \, d\tau$$

As this is true for every volume τ, however small, the integrands must be equal. Therefore, the differential form of Gauss' law corresponding to the integral form of Eq. (8.9) is

$$\nabla \cdot \mathbf{g} = -4\pi\gamma\rho \tag{8.10}$$

Furthermore, we have from Section 8.1 and Eq. (8.2) that $\mathbf{g}$ is expressible as the gradient of a scalar potential or, in other words, that the curl of $\mathbf{g}$ vanishes

$$\nabla \times \mathbf{g} = 0 \tag{8.11}$$

By integrating Eq. (8.11) over an arbitrary surface and converting the surface integral into a line integral around the periphery l of the surface by means of Stokes' theorem, we obtain the integral form of Eq. (8.11)

$$\oint \mathbf{g} \cdot d\mathbf{l} = 0 \tag{8.12}$$

the path of integration being *any* closed curve.

The differential equations (8.10) and (8.11), or the equivalent integral forms of Eqs. (8.9) and (8.12), completely describe a gravitational field. Equation (8.10) tells us that the gravitational attraction is a solenoidal vector at all points, where no mass is present. In regions containing mass, the gravitational attraction has a divergence equal to $-4\pi\gamma$ times the density. Furthermore, Eq. (8.11) shows that the gravitational attraction is also an irrotational vector everywhere, whether mass is present or not.

Combining Eqs. (8.6) and (8.10), we obtain *Poisson's equation*

$$\nabla \cdot \nabla \Phi = 4\pi\gamma\rho \tag{8.13}$$

which must be satisfied everywhere. In regions where no mass is present, this reduces to *Laplace's equation*

$$\nabla \cdot \nabla \Phi = 0 \tag{8.14}$$

Gauss' law has provided us with two extremely useful equations, Eqs. (8.13) and (8.14), for the gravitational potential. It also provides us with a powerful method in itself for solving certain simple gravitational problems, particularly those in which symmetry allows us to construct a geometrically simple surface normal to **g**, at all points of which the gravitational attraction has the same magnitude. Suppose, for instance, that we wish to find the gravitational attraction at a point P (Fig. 8.2), distant r from the center O of a spherical mass distribution in which the density is a function of the radius vector only. Construct a spherical surface of radius r about O as center. It is clear from symmetry that **g** is everywhere normal to this surface and has the same magnitude at all points on it. Therefore, the flux through it is $-4\pi r^2 g$, and according to Gauss' law, Eq. (8.8),

$$4\pi r^2 g = 4\pi\gamma M$$

where M is the total mass inside the sphere. Hence, the gravitational attraction is

$$g = \gamma \frac{M}{r^2} \tag{8.15}$$

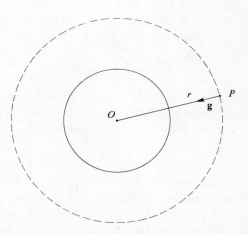

Fig. 8.2

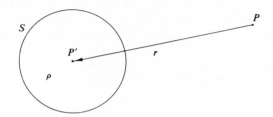

Fig. 8.3

the same as if the mass M were concentrated at point O. If the sphere of radius r lies inside the region of the mass, the quantity M in Eq. (8.15) represents only the portion of the mass inside the sphere. The portion of the mass outside the sphere gives rise to no field at P.

We may make an interesting and useful extension of Gauss' law by the use of Green's theorem. In Fig. 8.3 let S be an *equipotential surface* enclosing all of an extended mass of density ρ. Let P' be a point inside S, and P a point outside, and let r be the distance between P and P'.

Green's theorem is an integral relation between any two scalar functions. We shall let the first function be Φ, the potential at P' due to all the mass enclosed within S. We shall let the second function be $1/r$. From the results of Problem 8.1(a), the function $1/r$ represents the potential at P' due to a mass of $1/\gamma$ at P. From Eq. (1.40) we have

$$\int_\tau \left[\Phi \nabla \cdot \nabla \left(\frac{1}{r} \right) - \frac{1}{r} \nabla \cdot \nabla \Phi \right] d\tau = \int_s \left[\Phi \nabla \left(\frac{1}{r} \right) - \frac{1}{r} \nabla \Phi \right] \cdot d\mathbf{s} \quad (8.16)$$

For the first term on the left-hand side of Eq. (8.16), we have

$$\int_\tau \Phi \nabla \cdot \nabla \left(\frac{1}{r} \right) = 0$$

since $1/r$ is the potential due to a point mass at P and we have from Laplace's equation (8.14) that $\nabla \cdot \nabla (1/r)$ is uniquely zero except in the vicinity of P. For the second term on the left-hand side, we have

$$\int_\tau \frac{1}{r} \nabla \cdot \nabla \Phi \, d\tau = 4\pi \gamma \int_\tau \frac{\rho \, d\tau}{r}$$

since Φ is the potential due to the extended mass at a point P' within the mass and we have from Poisson's equation (8.13) that $\nabla \cdot \nabla \Phi$ is equal to $4\pi\gamma\rho$. For the first term on the right-hand side, we have

$$\int_s \Phi \nabla \left(\frac{1}{r} \right) \cdot d\mathbf{s} = \Phi_s \int_s \frac{\partial}{\partial n} \left(\frac{1}{r} \right) ds = 0$$

since by definition Φ_s is a constant on the equipotential surface and since the reduced integral is the flux across S due to a point mass outside S, which from Gauss' law is uniquely zero. For the second term on the right-hand side, we have

$$\int_s \frac{1}{r} \nabla\Phi \cdot d\mathbf{s} = -\int_s \frac{1}{r}\frac{\partial\Phi}{\partial n} ds$$

Combining, we obtain for Eq. (8.16)

$$\gamma \int_\tau \frac{\rho\, d\tau}{r} = -\frac{1}{4\pi}\int_s \frac{\dfrac{\partial\Phi}{\partial n} ds}{r}$$

From the results of Problem 8.1(a), the left-hand side of this equation is simply an expression for the potential Φ at P due to the extended mass within S. Also from the results of Problem 8.1(a), the right-hand side is an expression for the potential at P due to a surface distribution of mass on S of density $(-1/4\pi\gamma)(\partial\Phi/\partial n)$. Since S is an equipotential surface, gravity is normal to S and from Eq. (8.6) given by $-\partial\Phi/\partial n$ at S. Let us designate by g_0 the value of the gravitational attraction at S due to the extended mass within S. We then have, finally,

$$\Phi(P) = \frac{1}{4\pi}\int_s \frac{g_0}{r} ds \tag{8.17}$$

Such a surface layer is known as a *Green's equivalent layer*. Our proof shows that if the surface is an equipotential surface, the contained mass may be replaced by an equivalent surface layer of density $g_0/4\pi\gamma$ without changing the potential or attraction at any external point.

The fiction of a surface layer, particularly in the form of a Green's equivalent layer, is a useful concept to employ in the solution of certain types of potential theory problems. Let us examine now the boundary conditions that must be met for Φ and its derivative $\partial\Phi/\partial n$, normal to the surface, as the boundary is crossed. Consider for the moment the surface layer to have a small thickness t (Fig. 8.4). From the definition of the potential, we know that it is equal to the gravitational attraction force times a distance. For the slab of Fig. 8.4, the gravitational attraction will remain finite as $t \to 0$. The in-

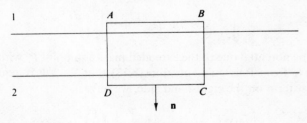

Fig. 8.4

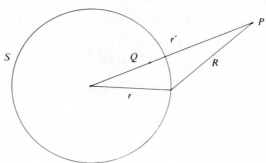

Fig. 8.5

cremental work done, then, in passing through the slab approaches zero as $t \to 0$, or our potential is continuous in passing the boundary

$$\Phi_1 = \Phi_2 \tag{8.18}$$

This is our first boundary condition. For our second boundary condition, let us consider the integral form of Gauss' law, Eq. (8.9), over a box of unit area $ABCD$ in Fig. 8.4. The two integrals over the sides BC and AD will approach zero as $t \to 0$, and we shall have

$$\left(\frac{\partial \Phi}{\partial n}\right)_2 - \left(\frac{\partial \Phi}{\partial n}\right)_1 = 4\pi\gamma\sigma \tag{8.19}$$

where the outward drawn normal **n** for side 1 is the opposite of that for side 2 and where we have substituted for the mass density ρ the surface density σ given by $\sigma = \rho t$.

Let us now consider such an equivalent layer over a spherical surface, which we shall assume to be an equipotential surface. We wish to find the potential Φ at a point outside and inside the spherical shell. From Fig. 8.5 we have for the potential at a point P due to the shell

$$\Phi = \gamma \int \frac{g_0}{4\pi\gamma R} \, ds$$

From our discussion of Legendre polynomials in Section 1.6, we have for R in terms of r and r' at a point outside S, where $r < r'$,

$$\frac{1}{R} = \frac{1}{r'} \sum_{n=0}^{\infty} \left(\frac{r}{r'}\right)^n P_n(\cos\theta)$$

and for a point inside S, where $r' < r$,

$$\frac{1}{R} = \frac{1}{r} \sum_{n=0}^{\infty} \left(\frac{r'}{r}\right)^n P_n(\cos\theta)$$

Substituting into the equation for the potential, we obtain

$$\Phi(P) = \frac{1}{4\pi r'} \sum \left(\frac{r}{r'}\right)^n \int P_n g_0 \, ds$$

$$= \frac{r}{4\pi} \sum \left(\frac{r}{r'}\right)^{n+1} \int P_n g_0 \, d\omega \quad (8.20)$$

and

$$\Phi(Q) = \frac{1}{4\pi r} \sum \left(\frac{r'}{r}\right)^n \int P_n g_0 \, ds$$

$$= \frac{r}{4\pi} \sum \left(\frac{r'}{r}\right)^n \int P_n g_0 \, d\omega \quad (8.21)$$

where r and r' are constant over the surface integration and where we have substituted for $ds = r^2 d\omega$, the incremental solid angle. Taking the partial derivatives of Eqs. (8.20) and (8.21) with respect to r' to find the radial component of the gravitational attraction, we obtain

$$-\frac{\partial \Phi(P)}{\partial r'} = \frac{1}{4\pi} \sum (n+1) \left(\frac{r}{r'}\right)^{n+2} \int P_n g_0 \, d\omega \quad (8.22)$$

and

$$-\frac{\partial \Phi(Q)}{\partial r'} = \frac{-1}{4\pi} \sum n \left(\frac{r'}{r}\right)^{n-1} \int P_n g_0 \, d\omega \quad (8.23)$$

As we approach the surface S of the mass distribution from a point on either side of S, we see that there is, as expected, a discontinuity in the radial component of the gravitational attraction. We may determine the value of this discontinuity from Eq. (8.19),

$$\left(\frac{\partial \Phi(Q)}{\partial r'}\right)_{r'=a} - \left(\frac{\partial \Phi(P)}{\partial r'}\right)_{r'=a} = 4\pi\gamma\sigma = g_0 \quad (8.24)$$

where we have substituted for σ from our initial assumed surface density. Subtracting Eq. (8.22) from Eq. (8.23) and equating to Eq. (8.24), we have finally

$$g_0 = \sum_{n=0}^{\infty} \frac{2n+1}{4\pi} \int P_n g_0 \, d\omega \quad (8.25)$$

It is to be noticed that this expression, as expected, is the same as the combination of Eqs. (1.109) and (1.111).

Problem 8.2(a) By applying Gauss' law, find the gravitational attraction (1) at a distance $r > a$ from the axis of an infinite cylinder of radius a with uniform density λ per unit length and (2) in the neighborhood of an infinite plane with a uniform density σ per unit area.

Ans.   (1) $\gamma \dfrac{2\lambda}{r}$,   (2) $2\pi\gamma\sigma$

Problem 8.2(b)   Find the gravitational attraction at a distance $r < a$ (1) from the center of a uniform density sphere of radius a with total mass M and (2) from the axis of a uniform density infinite cylinder of radius a with total density λ per unit length.

Ans.   (1) $\gamma \dfrac{M}{a^3} r$,   (2) $\gamma \dfrac{2\lambda}{a^2} r$

Problem 8.2(c)   Show from Gauss' law that the gravitational field vanishes inside a uniform density spherical shell.

8.3 Gravitational Potential and Attraction for an Ellipsoidal Earth

If the earth were a nonrotating spherical mass in which the density were a function of the radius only, the gravitational attraction at its outer surface would be a constant, given by Eq. (8.15), and its potential would be given by the results of Problem 8.1(a).

To proceed to the next approximation and the one that is used for most theoretical gravity determinations for the earth, we shall assume that we have a rotating earth. We shall make no assumption as to the internal density distribution within the earth, but shall assume that the equipotential surface for the outer boundary of the earth may be approximated by an *ellipsoid of revolution*, the axis of revolution being the earth's polar axis.

At the outer boundary of the earth, the actual equipotential surface over the oceanic areas is simply *mean sea level*. Over the continental areas, it is a referenced sea level surface. This actual equipotential surface is known as the *geoid*. It is important to repeat here that we are approximating this surface by an ellipsoid of revolution. Subsequent measurements of gravity and the reductions of such measurements will determine how well our theoretical approximation coincides with the geoid.

The potential on this outer equipotential surface is referred to as the *geopotential* Ψ and is made up of two parts. One is the gravitational potential Φ due to the attracting masses of the earth. The other is the kinetic potential due to the rotation of the earth. This latter term is simply the kinetic energy per unit mass for a point at the earth's outer surface as given by Eq. (6.7), or

$$\tfrac{1}{2}\omega^2 r^2 \cos^2 \varphi' \qquad (8.26)$$

where r is the radius measured from the center of the earth and φ' the geocentric latitude. We may then write for Ψ

$$\Psi = \Phi + \tfrac{1}{2}\omega^2 r^2 \cos^2 \varphi' \tag{8.27}$$

where Φ must satisfy Laplace's equation (8.14).

The equation for an elliptical section of the ellipsoid of revolution, measured in a plane containing the polar axis, is

$$r^2 \left[\frac{\cos^2 \varphi'}{a^2} + \frac{\sin^2 \varphi'}{a^2(1-e)^2} \right] = 1 \tag{8.28}$$

where a is the semimajor axis, b the semiminor axis, and e the ellipticity defined by

$$e = \frac{a-b}{a} \tag{8.29}$$

For our calculations of Ψ, we are interested in obtaining an expansion of $1/r$ in the dimensionless form a/r and of r to terms of the order e^2. Our final expressions for the geopotential Ψ and the gravity g on the assumed equipotential surface will then be valid to second-order terms in the ellipticity. From Eq. (8.28) we can then obtain

$$\frac{a}{r} = \left[\cos^2 \varphi' + \frac{\sin^2 \varphi'}{(1-e)^2} \right]^{1/2}$$

$$= 1 + e \sin^2 \varphi' + \frac{3}{2} e^2 \sin^2 \varphi' - \frac{1}{2} e^2 \sin^4 \varphi' + \cdots \tag{8.30}$$

and

$$r = \left[\frac{a^2(1-e)^2}{(1-e)^2 \cos^2 \varphi' + \sin^2 \varphi'} \right]^{1/2}$$

$$= \frac{a(1-e)}{(1-e \cos^2 \varphi')} [1 - \tfrac{1}{2} e^2 \cos^2 \varphi' \sin^2 \varphi' + \cdots] \tag{8.31}$$

using in each case the binomial expansion twice.

Our gravitational potential Φ must satisfy Laplace's equation. Since we are dealing with a problem in spherical coordinates and one with symmetry about the polar axis, or no dependence on azimuth, our solution will be of the form of Eq. (1.128). Further, we are dealing with a problem that has symmetry across the equatorial plane, so that from the discussion of Section 1.6, it is clear that the odd-order Legendre polynomials cannot enter our solution, and by choice we shall consider an expansion in terms of $1/r$, so that only the expressions represented by the second term in the brackets of Eq. (1.128) will enter our solution. Our solution for Φ, then, satisfying Laplace's equation can be written as

$$\Phi = \frac{\gamma M}{a}\left[\frac{a}{r} - \frac{2}{3}J_2\left(\frac{a}{r}\right)^3 P_2 + \frac{8}{35}J_4\left(\frac{a}{r}\right)^5 P_4 + \cdots\right] \quad (8.32)$$

We are at some liberty in choosing the exact form of our undetermined coefficients J_2 and J_4. First, we know that the zero-order solution for Φ, that is, for a sphere, is given by the result of Problem 8.1(a) as $\gamma M/r$. Second, we have expressed our expansion in a/r rather than $1/r$ since the powers of a/r are nearly unity. Third, we have included the arithmetic coefficients to cancel out the arithmetic coefficients of P_2 and P_4. We hope, then, that we have chosen a solution that will give a uniformly convergent series and that our undetermined coefficient J_2 will be of the order of e, and our undetermined coefficient J_4 of the order of e^2. We shall see from the subsequent discussions that this is indeed the case.

For the geopotential, we have first for the kinetic potential term of Eq. (8.26), using Eq. (8.31),

$$\frac{1}{2}\omega^2 r^2 \cos^2 \varphi' = \frac{1}{2}\frac{\gamma Mm}{a}\frac{(1-e)}{(1-e\cos^2\varphi')^2}\left(1 - \frac{1}{2}e^2 \cos^2\varphi' \sin^2\varphi'\right)^2 \cos^2\varphi'$$

$$= \frac{1}{2}\frac{\gamma Mm}{a}[1 + e - (1+3e)\sin^2\varphi' + 2e\sin^4\varphi' + \cdots] \quad (8.33)$$

where we have used the binomial expansion again to terms of the order e^2 or em and have made the substitution for m defined by

$$m = \frac{\omega^2 a^3(1-e)}{\gamma M} \quad (8.34)$$

To zero-order terms, the quantity m is simply the ratio of the centrifugal acceleration at the equator to the zero-order gravity term; for the earth constants ω, a, and M, the quantity m is small and of the order e. We, then, have for the geopotential, substituting for a/r from Eq. (8.30) into Eq. (8.32), substituting for the Legendre polynomials P_2 and P_4 from the results of Problem 1.6(a), remembering that the angle φ' is the complement of θ used in Section 1.6, and substituting for the kinetic potential from Eq. (8.33), to terms of the order e^2, em, $J_2 e$, or J_4,

$$\Psi = \frac{\gamma M}{a}[1 + (e + \tfrac{3}{2}e^2)\sin^2\varphi' - \tfrac{1}{2}e^2\sin^4\varphi'$$

$$-J_2(1+3e\sin^2\varphi')(\sin^2\varphi' - \tfrac{1}{3}) + J_4(\sin^4\varphi' - \tfrac{6}{7}\sin^2\varphi' + \tfrac{3}{35})$$

$$+ \tfrac{1}{2}m(1+e-(1+3e)\sin^2\varphi' + 2e\sin^4\varphi')] \quad (8.35)$$

Now, since by definition the geopotential Ψ is a constant on the assumed equipotential surface of an ellipsoid of revolution, the coefficients of the $\sin^2 \varphi'$ and $\sin^4 \varphi'$ terms in Eq. (8.35) must be uniquely zero, or

$$e + \tfrac{3}{2}e^2 - J_2(1-e) - \tfrac{6}{7}J_4 - \tfrac{1}{2}m(1+3e) = 0 \tag{8.36}$$

and

$$-\tfrac{1}{2}e^2 - 3J_2 e + J_4 + me = 0 \tag{8.37}$$

We can then solve these two equations for the appropriate values of J_2 and J_4, which we do by successive approximations obtaining first a first-order approximation for J_2, using that to obtain a value for J_4 and, in turn, using the determined value of J_4 to obtain a second-order approximation for J_2, giving

$$J_2 = e - \tfrac{1}{2}m + e(-\tfrac{1}{2}e + \tfrac{1}{7}m) \tag{8.38}$$

and

$$J_4 = \tfrac{7}{2}e^2 - \tfrac{5}{2}me \tag{8.39}$$

We see from our final result that, as expected, J_2 is of the order e and J_4 of the order e^2.

We are now in a position to determine the value of gravity g on our equipotential surface. From Eq. (8.6) we have that gravity is the negative of the gradient of the geopotential, which in our coordinate system will give, for the scalar magnitude of g,

$$\begin{aligned} g &= -\left[\left(\frac{\partial \Psi}{\partial r}\right)^2 + \left(\frac{\partial \Psi}{r \partial \varphi'}\right)^2\right]^{1/2} \\ &= -\frac{\partial \Psi}{\partial r}\left[1 + \tfrac{1}{2}\frac{\left(\frac{\partial \Psi}{r \partial \varphi'}\right)^2}{\left(\frac{\partial \Psi}{\partial r}\right)^2} + \cdots\right] \\ &= -\frac{\partial \Psi}{\partial r}\left[1 + \tfrac{1}{2}\left(\frac{a^2}{\gamma M}\frac{\partial \Psi}{r \partial \varphi'}\right)^2\right] \end{aligned} \tag{8.40}$$

remembering that the tangential component of g is small compared with the radial component and where the zero-order expression of $\partial \Psi / \partial r$ has been substituted as a multiplying factor for $\partial \Psi / r \partial \varphi'$ in the final expression. From Eqs. (8.32) and (8.26) for $\partial \Psi / \partial r$ to second-order terms, we obtain

$$\begin{aligned} \frac{\partial \Psi}{\partial r} &= -\frac{\gamma M}{a^2}\left[\left(\frac{a}{r}\right)^2 - 2J_2\left(\frac{a}{r}\right)^4 P_2 + \tfrac{8}{7}J_4\left(\frac{a}{r}\right)^6 P_4\right] + \omega^2 r \cos^2 \varphi' \\ &= -\frac{\gamma M}{a^2}\{1 + e - \tfrac{3}{2}m + e(e - \tfrac{27}{14}m) + [-e + \tfrac{5}{2}m - e(e - \tfrac{39}{14}m)]\sin^2 \varphi' \\ &\qquad\qquad - \tfrac{1}{8}e(11e - 15m)\sin^2 2\varphi'\} \end{aligned} \tag{8.41}$$

after a certain amount of algebra and using the values of J_2 and J_4 from Eqs. (8.38) and (8.39). Now since the tangential component of gravity is small compared with the radial component, we may compute $\partial \Psi / \partial \varphi'$ to first-order terms as compared with the calculation of $\partial \Psi / \partial r$ to second-order terms, obtaining again from Eqs. (8.32) and (8.26)

$$\frac{\partial \Psi}{\partial \varphi'} = \frac{\gamma M}{a} [-2J_2 \sin \varphi' \cos \varphi'] - \omega^2 r^2 \cos \varphi' \sin \varphi'$$

$$= -\frac{\gamma M}{a} (e - \tfrac{1}{2}m) \sin 2\varphi' - \tfrac{1}{2} \frac{\gamma M m}{a} \sin 2\varphi'$$

$$= -\frac{\gamma M}{a} e \sin 2\varphi' \qquad (8.42)$$

Substituting Eqs. (8.41) and (8.42) into Eq. (8.40), we then obtain finally

$$g = \frac{\gamma M}{a^2} \{1 + e - \tfrac{3}{2}m + e(e - \tfrac{27}{14}m) + [-e + \tfrac{5}{2}m - e(e - \tfrac{39}{14}m)] \sin^2 \varphi' - \tfrac{1}{8}e(7e - 15m) \sin^2 2\varphi'\} \qquad (8.43)$$

If we designate by g_0 the value of gravity on the equator, we shall have from Eq. (8.43) simply

$$g_0 = \frac{\gamma M}{a^2} [1 + e - \tfrac{3}{2}m + e(e - \tfrac{27}{14}m)] \qquad (8.44)$$

and substituting back into Eq. (8.43) for g

$$g = g_0 [1 + (-e + \tfrac{5}{2}m + \tfrac{15}{4}m^2 - \tfrac{17}{14}em) \sin^2 \varphi' - \tfrac{1}{8}e(7e - 15m) \sin^2 2\varphi'] \qquad (8.45)$$

This is the usual form in which gravity is expressed for the assumed ellipsoidal equipotential surface. It is often referred to as the theoretical gravity formula for the earth. It states simply that the gravity may be expressed by the sum of a constant term g_0 plus a first-order correction term times the sine squared of the geocentric latitude minus a second-order correction term times the sine squared of twice the geocentric latitude.

Problem 8.3(a) Carry out the indicated reductions to obtain Eqs. (8.30) and (8.31).

Problem 8.3(b) Carry out the indicated reduction to obtain Eq. (8.33).

Problem 8.3(c) Solve Eqs. (8.36) and (8.37) for the indicated values of J_2 and J_4 in Eqs. (8.38) and (8.39).

Problem 8.3(d) Carry through the reduction to obtain Eq. (8.41).

Problem 8.3(e) Carry through the reduction for Eq. (8.45).

8.4 Geocentric and Geographic Latitude

In the last section, we used the *geocentric latitude*, defined as the angle between the radius vector from the center of the earth to a point P on the earth's surface and the equatorial plane as measured in a plane normal to the equatorial plane. For most earth locations, we shall be dealing with *geographic latitude*, defined as the angle between a vertical at the earth's surface and the equatorial plane as measured in a plane normal to the equatorial plane. By definition, this vertical will be the normal to the reference ellipsoid.

We wish to find the relation between geocentric latitude φ' and geographic latitude φ. From Fig. 8.6 the angles φ and φ' are as shown. The angle $\varphi - \varphi'$ is the angle between the tangent to the reference ellipse and a circular arc at P. It is the acute angle in the triangle shown from which we have

$$\tan(\varphi - \varphi') = -\frac{dr}{r d\varphi'} \tag{8.46}$$

From Eq. (8.28) we obtain for $dr/d\varphi'$

$$2r\frac{dr}{d\varphi'}\left(\frac{\cos^2 \varphi'}{a^2} + \frac{\sin^2 \varphi'}{a^2(1-e)^2}\right) + r^2\left(\frac{-2\cos\varphi'\sin\varphi'}{a^2} + \frac{2\sin\varphi'\cos\varphi'}{a^2(1-e)^2}\right) = 0$$

or

$$\frac{dr}{d\varphi'} = -\frac{r^3}{2}\frac{2e\sin 2\varphi'}{a^2(1-e)^2} = -re\sin 2\varphi'$$

to first-order terms. Since $\varphi - \varphi'$ is a small angle, we can write valid to first-order terms from Eq. (8.46)

$$\varphi - \varphi' = e \sin 2\varphi' \tag{8.47}$$

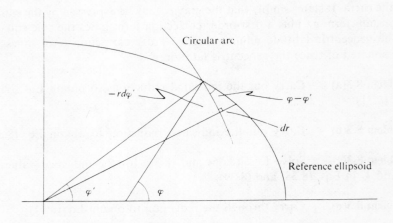

Fig. 8.6

Since φ and φ' are nearly equal, this difference may be equally well expressed, to first-order terms, as $e \sin 2\varphi$.

For our gravity formula (8.45), we wish to express $\sin^2 \varphi'$ in terms of $\sin^2 \varphi$. We can obtain from the trigonometric relation for the sum of two angles, remembering again that $\varphi-\varphi'$ is small

$$\sin \varphi = \sin[\varphi'+(\varphi-\varphi')] = \sin \varphi' \cos(\varphi-\varphi') + \cos \varphi' \sin(\varphi-\varphi')$$

$$= \sin \varphi' + (\varphi-\varphi') \cos \varphi'$$

or

$$\sin^2 \varphi = \sin^2 \varphi' + 2(\varphi-\varphi') \cos \varphi' \sin \varphi' + \cdots$$

$$= \sin^2 \varphi' + e \sin^2 2\varphi'$$

to first-order terms. Since the coefficient of $\sin^2 2\varphi'$ in Eq. (8.45) is already to the second order, we need make no correction there for changing from geocentric to geographic latitudes. We then obtain for Eq. (8.45)

$$g = g_0[1 + (-e + \tfrac{5}{2}m + \tfrac{15}{4}m^2 - \tfrac{17}{14}em) \sin^2 \varphi$$

$$+ (\tfrac{1}{8}e^2 - \tfrac{5}{8}em) \sin^2 2\varphi] \qquad (8.48)$$

Since the quantity m depends on M, the mass of the earth, which is not determinable to sufficient accuracy by other measurements for our purposes, one method that has been used to evaluate the coefficients of $\sin^2 \varphi$ and $\sin^2 2\varphi$ in Eq. (8.48) is to determine experimentally from measured values of g at various latitudes the coefficient of $\sin^2 \varphi$ and, in turn, m to a first approximation. Then this value of m is used to determine the coefficient of $\sin^2 2\varphi$ and, in turn, to recorrect the coefficient of the $\sin^2 \varphi$ term and the value of m. Several determinations of these coefficients have been made as more gravity measurements and more accurate gravity measurements are obtained. Typical of the values so obtained are g_0 equal to 978.049 in gals, the coefficient of the $\sin^2 \varphi$ term equal to 5.2884×10^{-3}, and the coefficient of the $\sin^2 2\varphi$ term equal to -5.9×10^{-6}.

Problem 8.4(a) Derive the expressions for Ψ and g in Sections 8.3 and 8.4 to first-order terms and show that the result for g is

$$g = g_0[1 + (\tfrac{5}{2}m - e) \sin^2 \varphi]$$

where m is now simply the ratio of centrifugal acceleration at the equator to the zero-order gravity term. The coefficient of the $\sin^2 \varphi$ term in g is known as *Clairaut's formula*.

8.5 Deviations of the Geoid from a Reference Ellipsoid

In general, the earth geoid will not conform with the reference ellipsoid of revolution, which we have developed in Section 8.3, or with any spheroidal surface of reference that we may choose to develop. We have taken into account in our previous derivation, by the inclusion of ellipticity and the rotation of the earth, the two major factors affecting gravity from an ideal, nonrotating, spherical earth. The other factors, such as mass differences between continents and oceans, mass attractions of mountains, deep sea trenches, and the buried mass differences which may be associated with each of these, are not of a sufficiently symmetrical nature over the earth that they can be treated in any simple theoretical manner.

We shall choose, instead, to develop a formula that will allow us at each point on the earth's surface to determine from gravity measurements over the earth the distance between the geoid and the reference ellipsoid. This formula will include the difference in gravity, or *gravity anomaly*, between the measured gravity reduced to the reference ellipsoid and the theoretical gravity on this surface from Eq. (8.48), or similar derived equation. As might be expected, the formula is principally sensitive to gravity anomalies in the vicinity of the point at which the determination is to be made. Since our theoretical gravity derivation has been carried out to second-order terms and since we may assume that the determined coefficients for this equation are also valid only to second-order terms, we may carry out this derivation to first-order terms. Stated another way, the observed gravity anomalies are small compared to the ellipticity and rotational terms.

On the geoid itself, the geopotential Ψ is a constant. Let us designate by Ψ_1 the potential on our reference ellipsoid, as determined from Eq. (8.35). In actuality, of course, the potential on the reference ellipsoid is not a constant, and the ellipsoid itself is not an equipotential surface. For our calculations, nevertheless, we shall assume Ψ_1 to be constant and determine a potential difference Ψ_2 from the gravity anomalies, thus making Ψ variable. This would give the same results for Ψ_2 if we had been able to follow the correct procedure making Ψ constant and calculating a variable Ψ_1. Then

$$\Psi = \Psi_1 + \Psi_2 \tag{8.49}$$

Correspondingly, g is the measured gravity *as reduced to the reference ellipsoid*; g_1, theoretical gravity on the reference ellipsoid at the same latitude;

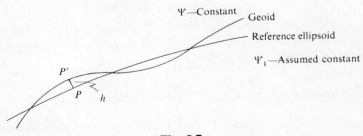

Fig. 8.7

and g_2, the gravity anomaly, or difference between the two. Then

$$g = g_1 + g_2 \tag{8.50}$$

From Fig. 8.7 we may develop an expression for Ψ_2 in terms of g_2. From Eq. (8.6) we have directly that g_2 on the geoid is $-\partial \Psi_2 / \partial r$. This is not, however, the value of the gravity anomaly that is observed on the reference ellipsoid in Eq. (8.50). We have an additional effect due to the change in total gravity from the geoid to the reference ellipsoid. Since the distance h between the two surfaces is small and of the same order as similar terms developed from the ellipticity, we have from Eq. (8.15) to zero order that

$$\frac{\partial g}{\partial r} = -2\gamma \frac{M}{r^3} = -2\frac{g}{a} \tag{8.51}$$

or

$$\Delta g = -\frac{2gh}{a} \tag{8.52}$$

Since Ψ_2 is the difference in potential over the small distance h from the geoid to the ellipsoid, we have again from Eq. (8.6) and to the first order

$$g = -\frac{\Delta \Psi}{\Delta r} = -\frac{\Psi_2}{(-h)} = \frac{\Psi_2}{h} \tag{8.53}$$

remembering from Fig. 8.7 that h is the elevation of the geoid above the reference ellipsoid. Substituting Eq. (8.53) into Eq. (8.52) and combining with $-\partial \Psi_2 / \partial r$, we have

$$g_2 = -\frac{\partial \Psi_2}{\partial r} - \frac{2\Psi_2}{a} \tag{8.54}$$

Our problem now reduces to finding a solution for Ψ_2, or h, in terms of g_2.

If we had carried out the solution of Laplace's equation in spherical coordinates in Section 1.7 to include terms in longitude as well as latitude, we would have obtained a solution similar to Eq. (1.128) with P_n being replaced by similar polynomials S_{ns} in both latitude and longitude. If Ψ_2 were sufficiently well known, it would be possible to express it in terms of such a solution

$$\Psi_2 = \sum A_{ns} \frac{S_{ns}}{r^{n+1}}$$

We could also express g_2 by an expansion in the same polynomials

$$g_2 = \sum g_{ns} S_{ns}$$

Substituting these two expressions into Eq. (8.54) and setting $r = a$ in the

278 ¶ *Seismology, Gravity, and Magnetism*

vicinity of our reference surface, we have, to the first order,

$$\sum (n-1) \frac{A_{ns} S_{ns}}{a^{n+2}} = \sum g_{ns} S_{ns} \tag{8.55}$$

or equating coefficients for each polynomial for the equality to hold in general

$$A_{ns} = \frac{a^{n+2}}{n-1} g_{ns} \tag{8.56}$$

We notice from Eq. (8.55) that there are no terms for g_{ns} for $n = 1$ and that for $n = 0$ the $g_{ns} S_{ns}$ is a constant, which contribution can be considered to have been included in the previous solution, so that our series begins at $n = 2$.

We shall consider, to the first order, our gravity anomalies g_2 spread as a Green's equivalent layer on a sphere of radius $r = a$ with a surface density $g_2/4\pi\gamma$. Then each of the terms on the right-hand side of Eq. (8.25) will correspond with a $g_{ns} S_{ns}$ term summed up over longitude. We have, then, for Ψ_2 at a distance r

$$\Psi_2(P') = \sum A_{ns} \frac{S_{ns}}{r^{n+1}} = \sum_{n=2}^{\infty} \frac{a^{n+2}}{n-1} g_{ns} \frac{S_{ns}}{r^{n+1}}$$

$$= \frac{1}{4\pi} \sum_{n=2}^{\infty} \int \frac{a^{n+2}}{r^{n+1}} \frac{(2n+1)}{(n-1)} P_n (\cos \theta) g_2(Q) \, d\omega \tag{8.57}$$

Looking at the series expansion in Eq. (8.57), we may write

$$\sum_{n=2}^{\infty} \frac{a^{n+2}}{r^{n+1}} \frac{(2n+1)}{(n-1)} P_n (\cos \theta)$$

$$= a^2 \left[2 \sum \frac{a^n}{r^{n+1}} P_n + 3 \sum \frac{a^n}{(n-1)r^{n+1}} P_n \right] \tag{8.58}$$

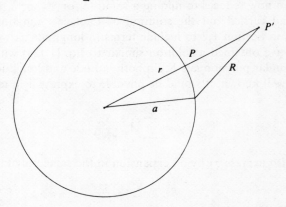

Fig. 8.8

With reference to Fig. 8.8, we have

$$R^2 = r^2 - 2ar \cos \theta + a^2$$

and from Section 1.6 we obtain

$$\frac{1}{R} = \frac{1}{r} \sum_{n=0}^{\infty} \left(\frac{a}{r}\right)^n P_n = \frac{1}{r}\left[1 + \frac{a}{r}\cos\theta + \sum_{n=2}^{\infty}\left(\frac{a}{r}\right)^n P_n\right]$$

or

$$\sum_{n=2}^{\infty} \frac{a^n}{r^{n+1}} P_n = \frac{1}{R} - \frac{1}{r} - \frac{a\cos\theta}{r^2} \qquad (8.59)$$

We may write the second term in Eq. (8.58) in integral form, substitute from Eq. (8.59), and integrate

$$\sum_{n=2}^{\infty} \frac{a^n}{(n-1)r^{n+1}} P_n = \frac{1}{r^2}\int_r^\infty \sum \frac{a^n}{r^n} P_n\, dr$$

$$= \frac{1}{r^2}\int_r^\infty r \sum \frac{a^n}{r^{n+1}} P_n\, dr$$

$$= \frac{1}{r^2}\left[-R - a\cos\theta \log\frac{R+r-a\cos\theta}{2r} + r - a\cos\theta\right] \qquad (8.60)$$

Substituting Eqs. (8.60) and (8.59) into Eqs. (8.58) and (8.57), we obtain

$$\Psi_2(P') = \frac{a^2}{4\pi}\int\left[2\left(\frac{1}{R} - \frac{1}{r} - \frac{a\cos\theta}{r^2}\right)\right.$$
$$\left. + \frac{3}{r^2}\left(-R - a\cos\theta \log\frac{R+r-a\cos\theta}{2r} + r - a\cos\theta\right)\right]g_2\, d\omega$$

$$= \frac{a^2}{4\pi r^2}\int\left[\frac{2r^2}{R} + r - 3R - 5a\cos\theta - 3a\cos\theta \log\frac{r-a\cos\theta+R}{2r}\right]g_2\, d\omega \qquad (8.61)$$

For our final result, we want $\Psi_2(P)$. Hence, we put $r = a$ and from Fig. 8.9, we have

$$R = 2a\sin\frac{\theta}{2}$$

and

$$\frac{r - a\cos\theta + R}{2r} = \frac{1 - \cos\theta + 2\sin\frac{\theta}{2}}{2} = \sin\frac{\theta}{2} + \sin^2\frac{\theta}{2}$$

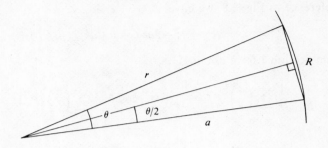

Fig. 8.9

so that Eq. (8.61) becomes

$$\Psi_2(P) = \frac{a}{4\pi}\int F(\theta)g_2(Q)\,d\omega \tag{8.62}$$

where

$$F(\theta) = \csc\frac{\theta}{2} + 1 - 6\sin\frac{\theta}{2} - 5\cos\theta - 3\cos\theta\log\left(\sin\frac{\theta}{2} + \sin^2\frac{\theta}{2}\right) \tag{8.63}$$

From Eq. (8.53) the height of the geoid above the ellipsoid is then

$$h = \frac{a}{4\pi g}\int F(\theta)g_2(Q)\,d\omega \tag{8.64}$$

Equation (8.64) is known as *Stokes' integral* and is the formula that permits us to calculate the deviation of the geoid from our assumed reference ellipsoid through the use of the gravity anomalies g_2. The value of $F(\theta)$ decreases rapidly with θ as expected. For actual computations, it is usually convenient and sufficiently accurate to make an approximate determination of the integral [Eq. (8.64)] by a series of zonal components, taking average values of g_2 over each zone and using tabulated values of the integral of $F(\theta)$ over each zone.

Problem 8.5(a)   Carry through the reduction for Eq. (8.60).

8.6 Deflection of the Vertical

As we have already discussed, the vertical at any station on the earth's surface is the normal to the geoid. From the equation for our reference ellipsoid, we may calculate the vertical to this surface at any known point on the earth's surface. For precise astronomical and surveying measurements, it is necessary to know the deviation of the vertical from the ellipsoid to the geoid. For this purpose, we shall want to calculate the angle between the

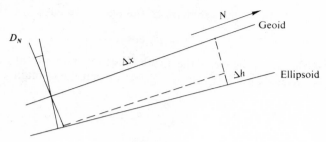

Fig. 8.10

normal to the geoid and the normal to the ellipsoid; this angle is known as the *deflection of the vertical*.

From Fig. 8.10 we see that, as drawn, the vertical to the ellipsoid is a greater angle as measured with the equatorial plane than the vertical to geoid or that the north component of the deflection of the vertical, D_N, is a negative quantity. From the geometry of Fig. 8.10, we have

$$D_N = -\frac{\partial h}{\partial x} \qquad (8.65)$$

Similarly, we would find for east component of the deflection of the vertical, D_E,

$$D_E = -\frac{\partial h}{\partial y} \qquad (8.66)$$

where x is a coordinate distance measured on the surface of the ellipsoid in a north direction and y in an east direction. From the previous section we have an expression for h, the distance between the geoid and ellipsoid, in terms of an integral of the polar angle θ measured from the observing point P to the point of the gravity anomaly Q. If we can find a relation among x and y and θ, we can take the derivative of Eq. (8.64) with respect to x and y to find D_N and D_E, respectively. In Fig. 8.11 we have P and a nearby point P' in a north

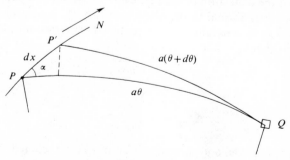

Fig. 8.11

direction from P. From the geometry of the figure, we have

$$a d\theta = dx \cos \alpha \tag{8.67}$$

where α is the angle between the north direction and the line along the ellipsoid between P and Q. From Eq. (8.64) we can then obtain

$$D_N = -\frac{\partial h}{\partial x} = -\frac{a}{4\pi g} \int g_2(Q) \frac{\partial F}{\partial \theta} \frac{\partial \theta}{\partial x} d\omega$$

$$= -\frac{K}{4\pi g} \int g_2(Q) \frac{\partial F}{\partial \theta} \cos \alpha \, d\omega \tag{8.68}$$

and similarly for the east component of the deflection of the vertical

$$D_E = -\frac{\partial h}{\partial y} = -\frac{K}{4\pi g} \int g_2(Q) \frac{\partial F}{\partial \theta} \sin \alpha \, d\omega \tag{8.69}$$

where we have added the constant $K = 2\pi/360 \times 60 \times 60$ to convert the deflection in radians to seconds of degrees. We can easily determine $\partial F/\partial \theta$ from Eq. (8.63) and can treat these two integrals in much the same manner as we did Eq. (8.64) for calculation by a series of zonal components with the added complication of dividing the zonal components into compartments for the calculation of $g_2(\theta) \cos \alpha$ or $g_2(\theta) \sin \alpha$.

Conversely, we may wish for precision location determinations to correct the vertical to the geoid to that of the reference ellipsoid to which latitude and longitude are referred. We see immediately that the north component of the deflection of the vertical is measured in the same plane as the latitude, that is, a plane normal to the equator and containing the polar axis and, further, that it is the same as the difference between the latitude angles determined for the geoid and the ellipsoid at the particular station. Thus,

$$\Delta_{\text{lat}} = D_N \tag{8.70}$$

This is not the case for the longitude as we see from Fig. 8.12. The east component of the deflection of the vertical is measured in a plane through the center of the earth normal to the plane in which D_N is measured. Longitude is measured in a plane parallel to the equatorial plane and normal to the polar axis. We have then

$$a D_E = -a \cos \varphi \, \Delta_{\text{long}}$$

or

$$\Delta_{\text{long}} = -\frac{D_E}{\cos \varphi} \tag{8.71}$$

where φ is the latitude, and the minus sign has been included since longitude is measured to the west.

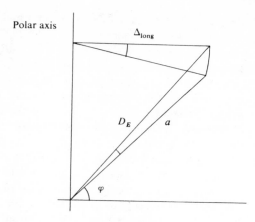

Fig. 8.12

8.7 Gravity Corrections and Gravity Anomalies

In the previous sections, we were concerned with the gravitational field of the earth as a whole. Here we shall be interested in the effects of the inhomogeneities in density and irregularities in elevation of the outer portions of the earth. We shall find that there are certain corrections related to these differences, which it is convenient and useful to make to gravity observations before using these measurements for the determination of a theoretical gravity formula or for other geodetic or geologic purposes.

There are three such corrections which are usually made to gravity observations—elevation, topographic, and isostatic corrections. The *elevation* or *free air correction* simply takes account of the difference in gravity between the elevation of the station and the reference surface. For a station above sea level, this correction will be positive, and for a station below sea level, negative. The expression free air correction derives from the fact that this correction takes account of elevation only and not that of any nearby masses below or adjacent to the station.

In the derivation of our reference ellipsoid, we have taken account of the major portions of the earth, which are below our reference surface. To compare our gravity observations with this theoretical derivation, we should make a correction for the perturbing effects of the topographic masses above our reference surface. In other words, we wish to eliminate or subtract the gravity effects of the additional masses above our reference surface. This correction is known as the *topographic* or *Bouguer correction*. For a station below sea level, such as for gravity measurements in a submarine, we do not want to eliminate the gravity effects of masses between the station and sea level, but to account for the difference in attraction due to this mass for a station below and at sea level.

In a broad sense, we may consider the earth to be in hydrostatic equilibrium. We know, however, from experience that the outer regions of the earth do have some strength. We may then expect that any topographic mass is *compensated* over some region at depth by an equal, balancing mass. If we take, for example, the rather simple analogy of a floating iceberg, we know that the mass above sea level is compensated by an equal displaced mass, defined by the mass of the iceberg minus the mass of the displaced water, below sea level. The correction for this compensating mass is known as the *isostatic correction*. We do not know a priori how this compensation is affected at depth. If we should assume that the earth's outer portions had negligible strength and that therefore compensation for each vertical column of topographic mass was compensated in an extension of that column at depth, the correction would be referred to as a *local isostatic correction*. If, on the other hand, we should assume some form of strength in the earth's outer portions, compensation for any vertical column of topographic mass will be compensated over an extended region at depth, and the correction would be referred to as a *regional isostatic correction*.

In general discussions of the earth's outer regions, there are two sets of descriptive terms, one from gravity and one from seismology, which are used; it is important to keep in mind the distinctions between these sets of terms and not use them indiscriminately or interchangeably. From gravity we have the term *lithosphere*, or more properly lithospheric shell, which is defined as the shell or region that can sustain long-term stress differences. Underlying the lithosphere is the *asthenosphere*, or asthenospheric shell, which exhibits no strength. The depth of compensation, then, for a local or regional isostatic correction would be the boundary between the lithosphere and the asthenosphere. From seismological observations, we observe major seismic discontinuities at depth, that is, abrupt changes in seismic velocity. The region above the first major discontinuity is called the *crust*, and the region below, the *mantle*.

When we make our topographic and isostatic corrections, we necessarily change the shape of the geoid itself. Stated another way, the geoid is no longer an equipotential surface after we have eliminated the gravity effects of the topographic and compensating masses. We shall define the term *fictitious geoid* as the equipotential surface after the topographic and isostatic corrections have been made and reserve the term *real geoid* for the equipotential surface of the actual earth. We must then make a final correction for the difference in elevations of the real geoid as compared with the fictitious geoid. This correction is referred to as the *indirect effect*. Following this correction, we may then make the correction of Section 8.5 from the fictitious geoid to the reference ellipsoid.

In the above paragraphs, we have discussed the application of various corrections to gravity measurements for geodetic purposes. Conversely, we may also use these same gravity measurements, or more particularly the gravity anomalies resulting from such measurements, for the determination of geologic structure, or more specifically for the determination of geologic mass distributions at depth. If we should make only an elevation correction to the observations, the resulting anomaly is referred to as a *free air anomaly*. If we should make both an elevation and a topographic correction, the resulting anomaly is referred to as a *Bouguer anomaly*. If we should make an elevation, topographic and isostatic correction, the resulting anomaly is referred to as an *isostatic anomaly*.

When we wish to determine underlying mass distributions, usually the Bouguer anomaly is used, since for such an anomaly, no assumption has been made as to the mass distributions at depth. When we do use isostatic anomalies, the problem is often one of determining how nearly a region may approach local isostatic equilibrium or how nearly a region may approach the particular strength characteristics for a given form of regional isostatic equilibrium. It should also be apparent to the reader that when use is made of a free air anomaly, we may consider it to represent an extreme form of local isostatic equilibrium in which the topographic and compensating masses are superimposed.

To continue, with the calculations for these corrections, we have that since the elevation correction g_e is usually a small quantity, we may to the first order use zero-order terms for the multiplying factors in its derivation. Using the zero-order gravity term of the earth, we have from Eq. (8.15) that

$$\frac{dg}{dr} = -\gamma \frac{2M}{r^3} = -\frac{2g}{r} \tag{8.72}$$

or

$$g_e = \frac{2g}{r} h \tag{8.73}$$

where h is the elevation of the station above sea level and is a quantity small compared with r. The numerical value of the coefficient in Eq. (8.73) is 0.307 mgal/m.

For the topographic correction, let us begin by considering first the effect of an infinite slab of material between our station elevation and sea level. This can often be taken as a reasonable approximation for the topographic effect for stations in mountain regions, where the elevation of the mountain range is usually small compared with its lateral extent. From the results of Problem 8.2(a) and remembering that the topographic correction g_t expresses the subtraction of the gravity effect of this slab, we have

$$g_t = -2\pi\gamma\sigma = -2\pi\gamma\rho h \tag{8.74}$$

where we have substituted for σ its equivalent in terms of ρ and thickness of the slab h. From Eq. (8.15) we have, for the zero-order gravity term of the earth in terms of the mean density of the earth $\bar{\rho}$,

$$g = \gamma \frac{M}{r^2} = \frac{4}{3} \pi \gamma \bar{\rho} r \tag{8.75}$$

Substituting Eq. (8.75) into Eq. (8.74) to eliminate γ, we obtain

$$g_t = -\frac{3g}{2r} \frac{\rho}{\bar{\rho}} h \tag{8.76}$$

Combining the elevation correction of Eq. (8.73) with the topographic correction [Eq. (8.76)], we obtain for the sum of these two corrections

$$g_e + g_t = \frac{2g}{r} \left(1 - \frac{3}{4} \frac{\rho}{\bar{\rho}}\right) h \tag{8.77}$$

We see that the effect of the topographic correction is to reduce the effect of the elevation correction. Using typical values of 5.52 for $\bar{\rho}$ and 2.80 for ρ, the coefficient of the topographic correction of Eq. (8.76) will be 0.117 mgal/m. In general, of course, we cannot simplify the topographic correction to that of an infinite slab, and we must calculate in some numerical form an integral of the form

$$g_t = -\gamma \int \frac{dm}{d^2} \cos \alpha \tag{8.78}$$

where d is the distance from the station to the disturbing mass, and α is the angle between the line defined by d and the vertical. The $\cos \alpha$ term must be included, since the topographic correction g_t will always be a small quantity as compared with normal gravity, and we shall, therefore, want the component of the gravity effect of the disturbing topographic mass in the direction of $\mathbf{g}$.

For the isostatic correction, we shall consider here only two overly simplified types of local isostatic compensation. One, known as the *Pratt* hypothesis of local compensation, assumes that the depth of compensation is

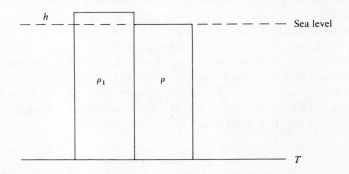

Fig. 8.13

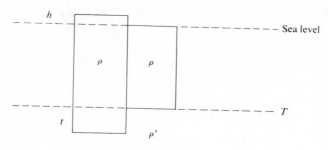

Fig. 8.14

constant everywhere and that the compensation is affected by a variation in density from one local column to the next. The other, known as the *Airy* hypothesis of local compensation, assumes that the lithospheric density remains constant and that the compensation is affected by extensions of the lithospheric shell into the asthenosphere. For the Pratt hypothesis, we then assume that the compensation is affected above a fixed depth T by local variations in the lithospheric density so that from Fig. 8.13 we shall have

$$(T+h)\rho_1 = T\rho$$

$$\rho_1 = \frac{T}{T+h}\rho \qquad (8.79)$$

or for the density difference $\Delta\rho$ of an elevated column over a sea level column

$$\Delta\rho = \rho_1 - \rho = -\frac{h}{T+h}\rho \qquad (8.80)$$

We shall then have, corresponding to Eq. (8.78), evaluations of the form

$$g_i = \gamma\rho \int \frac{h}{T+h} \frac{dv}{d^2} \cos\alpha \qquad (8.81)$$

to make where the integral is taken over the whole column. For the Airy hypothesis, we assume that the lithospheric density remains constant and that the compensation is affected locally by extensions of the lithosphere into the asthenosphere so that from Fig. 8.14 we shall have

$$t(\rho' - \rho) = h\rho$$

$$t = \frac{\rho}{\rho' - \rho} h \qquad (8.82)$$

We shall then have, again corresponding to Eq. (8.78), evaluations of the form

$$g_i = \gamma(\rho' - \rho) \int \frac{dv}{d^2} \cos\alpha \qquad (8.83)$$

where the integral is now taken only over the portion of the column t extending below the depth T.

It is important to make one further general observation on isostatic corrections. We know a priori that for every topographic mass, there will be an equal and opposite compensating mass. Therefore, as we go further away from our observing station, the distances and angles of the topographic and compensating masses will approach each other, and the combined effect of the topographic and isostatic corrections, on any assumption of isostatic compensation, will approach zero. This is of tremendous importance in geodetic problems. It reduces the effects of distant topography and permits one to carry out the combined topographic and isostatic corrections to a limited distance away from the station.

In the process of making the topographic and isostatic corrections, masses have been subtracted from or added to the earth. This will necessarily change the shape of the geoid surface. The old surface, or real geoid, is the equipotential surface for the actual earth. The new surface, or fictitious geoid, is the equipotential surface for our reduced earth after the perturbing topographic and compensating masses have been removed.

Our previous measurements of elevation and gravity have been necessarily referred to sea level of the real geoid. We wish them now to be referred to the new, fictitious geoid. Our problem, then, is to determine the difference in elevation between the two geoids and the effect this change has on gravity.

In Fig. 8.15, G_r is the trace of the real geoid and G_f that of the fictitious geoid. By definition, the geopotential Ψ is constant on G_r, and the same value of the geopotential is constant on the reduced surface G_f. We shall let d be the vertical distance between G_r and G_f with P and Q the corresponding points on G_r and G_f. Designating by V' potentials referred to the reduced earth and by V potentials referred to the original earth, we may write

$$V'_P = V'_Q + \frac{\partial V'}{\partial n} d$$

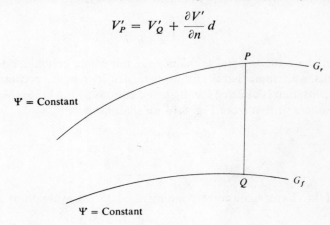

Fig. 8.15

to the first order. Since $V_P = V'_Q$ and by definition $g' = \partial V'/\partial n$, we may write

$$d = \frac{(V' - V)_P}{g'}$$

Further, since the indirect effect correction is a small quantity, the substitution of g, the value of gravity on the real geoid, for g', the value of gravity on the fictitious geoid, will only produce a second-order effect. We then have

$$d = \frac{\delta V}{g} \tag{8.84}$$

The distance between the fictitious and the real geoid is equal to the difference in potential caused by eliminating the topographic and compensating masses as measured at a point on the real geoid divided by gravity at that point.

The values of the potential for the topographic and compensating masses could be obtained in the same manner, incrementally, as the values of the gravitational attraction in Eqs. (8.78) and (8.81) or (8.83). Companion columns for the indirect effect are usually included in topographic-isostatic correction tables.

To obtain the gravity effect due to this change in distance d, one should go through the whole sequence of elevation, topographic, and isostatic corrections. However, since the gravity effect of the indirect effect is always a small quantity and since the mass deviations between the two geoid surfaces will be small and of broad extent, one usually assumes isostatic compensation with zero depth of compensation, that is, include an elevation correction only. From Eqs. (8.73) and (8.84), we then have

$$g_{ie} = \frac{2\delta V}{r} \tag{8.85}$$

As has been implied previously, the interpretation of gravity anomalies to the determination of subsurface geologic mass distributions is a somewhat subjective practice. There is no unique mass distribution to satisfy a given gravity anomaly distribution. The interpreter needs, in addition to a knowledge of gravity and potential theory, a fundamental knowledge of geology. The calculation of the gravity anomalies for a given subsurface mass distribution is usually a simple procedure; the geologic synthesis leading to the assumed mass distribution is the difficult part of the interpretation. Sometimes first or second derivative maps of the gravity anomaly field are prepared in order to assist in recognizing the type of anomalies present. We shall examine here only the form of some simple types of gravity anomalies.

Let us examine first a horizontal bed of infinite extent. We have from Eq. (8.74) that its gravity effect will be $2\pi\gamma\rho h$, where ρ and h are the density and thickness of the bed. Using for ρ a typical value of 2.80, we find that it requires a thickness h of 9 m to cause a gravity anomaly of 1 mgal. Stated another way, if we had an extended horizontal bed intruded into an area and if it had a substantial density contrast of say, 0.30, with the surrounding rocks, it would require a thickness of 80 m to cause a local gravity anomaly of 1 mgal. Since the accuracy of most gravity measuring instruments is to the order of a few tenths of milligals, this gives us an index of the order of magnitude of geologic structures that we can determine with gravity methods.

Let us examine next the effects that may be caused by bodies of a two-dimensional aspect, that is, by bodies one of whose horizontal dimensions is large compared with the other horizontal dimension and its vertical dimen-

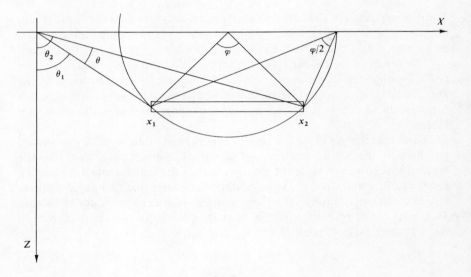

Fig. 8.16

sion. Since again we want only the vertical component of the gravitational attraction from Eq. (8.78), where z is the vertical coordinate, we have

$$g = \gamma\rho \int z \, dx \, dz \int_{-\infty}^{\infty} \frac{dy}{(x^2+y^2+z^2)^{3/2}} = 2\gamma\rho \int \frac{z \, dx \, dz}{x^2+z^2} \qquad (8.86)$$

We see, then, that for a two-dimensional mass whose dimensions Δx and Δz are small, the gravity anomaly Δg will have its maximum directly over the mass and Δg will be reduced to half its maximum value at a horizontal distance $x = z$, that is, at a horizontal distance equal to the depth of the buried mass.

Let us next consider a mass of infinite horizontal extent in the Y direction, limited extent in the X direction, and of a small thickness Δz. The resultant gravity anomaly will then be from Eq. (8.86) and Fig. 8.16.

$$\Delta g = 2\gamma\rho z\Delta z \int_{x_1}^{x_2} \frac{dx}{x^2+z^2} = 2\gamma\rho z\Delta z \left[\frac{1}{z}\tan^{-1}\frac{x}{z}\right]_{x_1}^{x_2}$$

$$= 2\gamma\rho\Delta z(\theta_2 - \theta_1) = 2\gamma\rho\Delta z\theta \qquad (8.87)$$

The horizontal extent of the buried mass is a chord of a circle, whose radius is the horizontal distance from the point at which the anomaly is a maximum to a point at which it is half the maximum value. If we should let the distance x_2 in Fig. 8.16 be extended to infinity, we would have the gravity expression for a fault, or

$$\Delta g = 2\gamma\rho\Delta z\left(\frac{\pi}{2} - \theta_1\right) \qquad (8.88)$$

We see, then, that the gravity anomaly for a fault approaches its maximum value as x increases in a positive direction, has its half value directly over the fault, and approaches zero as x increases in a negative direction.

Problem 8.7(a) Find $\Delta\rho$ for the Pratt hypothesis and t for the Airy hypothesis for an oceanic column of water depth d and density ρ_w.

Ans. $\quad \Delta\rho = \dfrac{d}{T-d}(\rho - \rho_w)$

$\quad t = \dfrac{\rho - \rho_w}{\rho' - \rho}d$

Problem 8.7(b) Determine the gravity anomaly of a buried sphere of radius a.

Ans. $\quad \frac{4}{3}\pi\gamma\rho a^3(z/r^3)$

Problem 8.7(c) Determine the gravity anomaly of a buried horizontal cylinder of radius a.

Ans. $\quad 2\pi\gamma\rho a^2(z/r^2)$

Problem 8.7(d) Calculate the first and second derivatives of the gravity anomaly for Problems 8.7(b) and 8.7(c). Graph the gravity anomaly and its first and second derivatives for both cases and discuss their similarities and differences.

Problem 8.7(e) Calculate the first and second derivatives of the gravity anomaly for a fault. Graph the gravity anomaly and its first and second derivative and compare with the results of Problem 8.7(d).

8.8 Internal Mass Distribution of the Earth

It is of interest at this point to see what, if anything, we can determine about the internal mass distribution of the earth. If we take an ellipsoid of revolution of uniform density, it should be apparent to the reader that the outer surface of this ellipsoid, unlike the outer surface of a sphere of uniform density, is not an equipotential surface. For the earth, we can make this an equipotential surface by the inclusion of the centrifugal force of rotation in our kinetic potential term. In other words, we shall assume that the flattening of the earth for a rotating body in equilibrium is due to the centrifugal force of rotation. Further, we shall assume that the earth acts as a rotating liquid, that is, the stresses are hydrostatic. This is a reasonable assumption since except for the relatively thin outer crust of the earth, the earth does act as a liquid to long-term stresses. This, then, implies that the geoid and all internal equipotential surfaces are also surfaces of constant density as well as pressure.

We shall first want to derive an expression for the external and the internal gravitational potentials for a nearly spherical body which is a figure of revolution and whose surface is given by

$$r = a\left(1 + \sum_{n=1}^{\infty} c_n P_n\right) \tag{8.89}$$

For a sphere of radius a, we have from the results of Problem 8.1(b) for the potential outside a sphere U_0 and the potential inside a sphere U_i

$$U_0 = \gamma \int_0^a \frac{4\pi\rho x^2}{r} dx = \frac{4}{3}\pi\gamma\rho a^3 \frac{1}{r} \tag{8.90}$$

and

$$U_i = \gamma \int_0^r \frac{4\pi\rho x^2}{r} dx + \gamma \int_r^a \frac{4\pi\rho x^2}{x} dx$$

$$= 4\pi\gamma\rho \left[\frac{r^2}{3} + \frac{a^2 - r^2}{2}\right]$$

$$= \frac{4}{3}\pi\gamma\rho a^3 \frac{3a^2 - r^2}{2a^3} \tag{8.91}$$

where ρ is the density. From Eq. (1.128) we can, then, write for the form of the solution for the external and internal potentials of the nearly spherical body

$$\Phi_0 = \frac{4}{3}\pi\gamma\rho a^3 \frac{1}{r} + \sum_{n=1}^{\infty} A_n \left(\frac{a}{r}\right)^{n+1} P_n \tag{8.92}$$

and

$$\Phi_i = \frac{4}{3}\pi\gamma\rho a^3 \frac{3a^2 - r^2}{2a^3} + \sum_{n=1}^{\infty} B_n \left(\frac{r}{a}\right)^n P_n \tag{8.93}$$

Now, to first-order terms, we may consider the excess mass of the nearly spherical body over that of a sphere of radius a to be spread on the outer surface of the sphere as a layer with surface density

$$\sigma = (r-a)\rho = a\rho \sum_{n=1}^{\infty} c_n P_n \tag{8.94}$$

For the surface layer, we have from Eq. (8.18) that the potential is continuous across the boundary. Referring to the second terms of Eqs. (8.92) and (8.93) and remembering the orthogonality relations of Legendre polynomials, we have that

$$A_n = B_n \tag{8.95}$$

From the second boundary condition [Eq. (8.19)], we have for the first derivatives from the second terms of Eqs. (8.92) and (8.93)

$$\sum_{n=1}^{\infty} nA_n \frac{r^{n-1}}{a^n} P_n + \sum_{n=1}^{\infty} (n+1)A_n \frac{a^{n+1}}{r^{n+2}} P_n = 4\pi\gamma\sigma$$

$$\sum_{n=1}^{\infty} n \frac{A_n}{a}\left(\frac{r}{a}\right)^{n-1} P_n + \sum_{n=1}^{\infty} (n+1)\frac{A_n}{a}\left(\frac{a}{r}\right)^{n+2} P_n = 4\pi\gamma a\rho \sum_{n=1}^{\infty} c_n P_n$$

or, using the orthogonality relations of the Legendre polynomials,

$$A_n = \frac{4\pi\gamma\rho a^2 c_n}{2n+1} \tag{8.96}$$

Substituting back for Φ_0 and Φ_i, we obtain

$$\Phi_0 = \frac{4}{3}\pi\gamma\rho a^3 \frac{1}{r} + 4\pi\gamma\rho a^2 \sum_{n=1}^{\infty} \frac{1}{2n+1}\left(\frac{a}{r}\right)^{n+1} c_n P_n$$

$$= \frac{4}{3}\pi\gamma\rho a^3 \left[\frac{1}{r} + \sum_{n=1}^{\infty} \frac{3}{2n+1} \frac{a^n c_n}{r^{n+1}} P_n\right] \tag{8.97}$$

and

$$\Phi_i = \frac{4}{3}\pi\gamma\rho a^3 \frac{3a^2-r^2}{2a^3} + 4\pi\gamma\rho a^2 \sum_{n=1}^{\infty} \frac{1}{2n+1}\left(\frac{r}{a}\right)^n c_n P_n$$

$$= \frac{4}{3}\pi\gamma\rho a^3 \left[\frac{3a^2-r^2}{2a^3} + \sum_{n=1}^{\infty} \frac{3}{2n+1} \frac{r^n c_n}{a^{n+1}} P_n\right] \tag{8.98}$$

For our rotating, hydrostatic earth, we have a density ρ varying in some manner to be determined from the center of the earth to its outer surface. We shall also have that for any surface

$$r = x(1 + \sum_{n=1}^{\infty} c_n P_n) \tag{8.99}$$

internal to the earth that ρ is a constant over this surface. We may think of the potential as being built up from shells whose surfaces are given by Eq. (8.99). A shell, which is at a distance given by $x = y$ and a thickness dy, will make a contribution to the potential equal to that of the difference of that due to two homogeneous bodies of density $\rho = \rho(y)$ and surfaces y and $y+dy$. Thus, we shall have for Φ_0 from Eq. (8.97)

$$\Phi_0 = \frac{4}{3}\pi\gamma \int_0^a \rho \frac{\partial}{\partial y}\left(\frac{y^3}{r} + \sum_{n=1}^{\infty} \frac{3}{2n+1} \frac{y^{n+3}c_n}{r^{n+1}} P_n\right) dy \qquad (8.100)$$

At internal points, the potential must be separated into two parts. From 0 to x, we shall have a contribution of the form of Eq. (8.97), and from x to a, a contribution of the form of Eq. (8.98). Thus,

$$\Phi_i = \frac{4}{3}\pi\gamma \int_0^x \rho \frac{\partial}{\partial y}\left(\frac{y^3}{r} + \sum_{n=1}^{\infty} \frac{3}{2n+1} \frac{y^{n+3}c_n}{r^{n+1}} P_n\right) dy$$

$$+ \frac{4}{3}\pi\gamma \int_x^a \rho \frac{\partial}{\partial y}\left(\frac{3}{2}y^2 + \sum_{n=1}^{\infty} \frac{3}{2n+1} \frac{r^n c_n}{y^{n-2}} P_n\right) dy \qquad (8.101)$$

where the second part of the first term in Eq. (8.98) has been eliminated since r is not a function of y.

To obtain the geopotential Ψ, we must add the kinetic potential term [Eq. (8.26)]

$$\tfrac{1}{2}\omega^2 r^2 \cos^2 \varphi = \tfrac{1}{3}\omega^2 r^2 + \tfrac{1}{2}\omega^2 r^2(\tfrac{1}{3}-\sin^2 \varphi)$$

$$= \tfrac{1}{3}\omega^2 r^2 - \tfrac{1}{3}\omega^2 r^2 P_2 \qquad (8.102)$$

We see that the kinetic potential contains only the Legendre polynomial P_2. As the kinetic potential represents the balancing force to the gravitational potential and because of the orthogonality relations of the Legendre polynomials, we may conclude that at least to the first-order hydrostatic theory, the surfaces of constant density, and consequently the equipotential surfaces, contain no harmonics other than P_2.

We may now go back to Eq. (8.89) and write

$$r = a(1+c_2 P_2) = a[1+\tfrac{3}{2}(\sin^2 \varphi - \tfrac{1}{3})c_2]$$

In terms of the ellipticity e, we have from Eq. (8.29)

$$e = \frac{r(0)-r\left(\frac{\pi}{2}\right)}{r(0)} = \frac{-\tfrac{3}{2}c_2}{1-\tfrac{1}{2}c_2} = -\tfrac{3}{2}c_2$$

Gravity ¶ 295

to the first order. In terms of e, Eqs. (8.99), (8.100), and (8.101) become

$$r = x(1 - \tfrac{2}{3} e P_2) \tag{8.103}$$

$$\Phi_0 = \frac{4}{3} \pi \gamma \int_0^a \rho \frac{\partial}{\partial y} \left(\frac{y^3}{r} - \frac{2}{5} \frac{y^5 e}{r^3} P_2 \right) dy \tag{8.104}$$

and

$$\Phi_i = \frac{4}{3} \pi \gamma \int_0^x \rho \frac{\partial}{\partial y} \left(\frac{y^3}{r} - \frac{2}{5} \frac{y^5 e}{r^3} P_2 \right) dy + \frac{4}{3} \pi \gamma \int_x^a \rho \frac{\partial}{\partial y} \left(\frac{3}{2} y^2 - \frac{2}{5} r^2 e P_2 \right) dy \tag{8.105}$$

Now we may express the total mass M of the earth simply by adding up all the shells, which is, to the first order,

$$M = 4\pi \int_0^a \rho y^2 \, dy = \frac{4}{3} \pi a^3 \bar{\rho}$$

or

$$\bar{\rho} = \frac{3}{a^3} \int_0^a \rho y^2 \, dy \tag{8.106}$$

where $\bar{\rho}$ is the average density of the earth. We may also express similarly the average density ρ_0 out to a surface at a distance r as

$$\rho_0 = \frac{3}{x^3} \int_0^x \rho y^2 \, dy \tag{8.107}$$

The condition that we must now meet for our hydrostatic earth is that these surfaces of constant density are also equipotential surface, that is, that Ψ is a function of x only. The geopotential Ψ is the sum of Eq. (8.105) and (8.102). In the first term of the first integral of Eq. (8.105), we may replace $1/r$ by

$$\frac{1}{r} = \frac{1}{x}(1 + \tfrac{2}{3} e(x) P_2)$$

to the first order. The second term of the first integral is already a small quantity, as are the second term of the second integral and the kinetic potential term, so that we may replace r by x in these terms, from which we obtain

$$\Psi = \frac{4}{3} \pi \gamma \left[\frac{1}{x} \left(1 + \frac{2}{3} e P_2 \right) \int_0^x 3\rho y^2 \, dy - \frac{2 P_2}{5 x^3} \int_0^x \rho \frac{d}{dy} (y^5 e) \, dy \right.$$
$$\left. + \int_x^a 3 \rho y \, dy - \frac{2}{5} x^2 P_2 \int_x^a \rho \frac{d}{dy} (e) \, dy \right.$$
$$\left. + \frac{1}{3} \omega^2 x^2 - \frac{1}{3} \omega^2 x^2 P_2 \right.$$

296 ¶ Seismology, Gravity and Magnetism

The first term of the first integral, the third integral, and the first term of the kinetic potential are already functions of x only. To meet our condition, we must then have that the coefficients of the P_2 terms vanish, which gives

$$\frac{4}{3}\pi\gamma\left[\frac{2e}{3x}\int_0^x 3\rho y^2\,dy - \frac{2}{5x^3}\int_0^x \rho\frac{d}{dy}(y^5 e)\,dy\right.$$
$$\left. - \frac{2x^2}{5}\int_x^a \rho\frac{d}{dy}(e)\,dy\right] - \frac{1}{3}\omega^2 x^2 = 0$$

or

$$\frac{e}{x}\int_0^x \rho y^2\,dy - \frac{1}{5}\left[\frac{1}{x^3}\int_0^x \rho\frac{d}{dy}(y^5 e)\,dy\right.$$
$$\left. + x^2\int_x^a \rho\frac{d}{dy}(e)\,dy\right] = \frac{\omega^2 x^2}{8\pi\gamma} \qquad (8.108)$$

Multiplying Eq. (8.108) by x^3 and differentiating with respect to x, we obtain, remembering that the derivative of an integral of a function $f(y)$ with respect to one of its limits of integration x is simply $f(x)$,

$$\left(x^2\frac{de}{dx} + 2xe\right)\int_0^x \rho y^2\,dy + (ex^2)(\rho x^2) - \frac{1}{5}\rho\left(5x^4 e + x^5\frac{de}{dx}\right)$$
$$-x^4\int_x^a \rho\frac{d}{dy}(e)\,dy + \frac{1}{5}x^5\rho\frac{de}{dx} = \frac{5\omega^2 x^4}{8\pi\gamma}$$

or

$$\left(x^2\frac{de}{dx} + 2xe\right)\int_0^x \rho y^2\,dy - x^4\int_x^a \rho\frac{d}{dy}(e)\,dy = \frac{5\omega^2 x^4}{8\pi\gamma} \qquad (8.109)$$

Multiplying Eq. (8.109) by x^{-4} and differentiating again with respect to x, we obtain

$$\left(x^{-2}\frac{d^2 e}{dx^2} - 2x^{-3}\frac{de}{dx} + 2x^{-3}\frac{de}{dx} - 6x^{-4}e\right)\int_0^x \rho y^2\,dy$$
$$+ \left(x^{-2}\frac{de}{dx} + 2x^{-3}e\right)(\rho x^2) + \rho\frac{de}{dx} = 0$$

or

$$\left(\frac{d^2 e}{dx^2} - \frac{6e}{x^2}\right)\int_0^x \rho y^2\,dy + 2\rho x^2\left(\frac{de}{dx} + \frac{e}{x}\right) = 0 \qquad (8.110)$$

and substituting from Eq. (8.107),

$$\rho_0\left(\frac{d^2 e}{dx^2} - \frac{6e}{x^2}\right) + \frac{6\rho}{x}\left(\frac{de}{dx} + \frac{e}{x}\right) = 0 \qquad (8.111)$$

Equation (8.111) is known as *Clairaut's differential equation*. It is an equation relating the density ρ and the ellipticity e as a function of distance x from the center of the earth. Since we know, from gravity measurements, the value of the ellipticity at the earth's outer surface, $e = e(a)$, it might be hoped that by choosing an appropriate density function, $\rho = \rho(x)$, we could find a solution to Clairaut's equation, which gave the proper value of $e(a)$ at the outer surface. This, unfortunately, does not yield any particularly useful results, for it is found that several widely different laws of density will give nearly the same value of $e(a)$. This is not an unexpected result, since we already know from Section 8.2 that the zero-order term (sphere) can be met by any $\rho = \rho(r)$ meeting the single condition that the total mass M of the earth remain constant.

Let us, however, see what we can determine about the density distribution in the earth and about the validity of the assumption of a hydrostatic earth. For $x = a$, we shall have from Eq. (8.109)

$$\left[a^2 \left(\frac{de}{dx} \right)_a + 2ae_a \right] \frac{\bar{\rho} a^3}{3} = \frac{5\omega^2 a^4}{8\pi\gamma} \tag{8.112}$$

where we have also substituted for the integral from Eq. (8.106). We have from the results of Problem 8.4(a) that our coefficient m is given by

$$m = \frac{\omega^2 a}{\gamma \dfrac{M}{a^2}} = \frac{3\omega^2}{4\pi\gamma\bar{\rho}} \tag{8.113}$$

so that Eq. (8.112) becomes

$$a \left(\frac{de}{dx} \right)_a + 2e_a = \frac{5}{2} m \tag{8.114}$$

If we were to assume that the earth were of constant density, $\partial e/\partial x$ would be zero, and Eq. (8.114) would become

$$e_a = \tfrac{5}{4} m \tag{8.115}$$

Substituting the known value of $m = 1/289$, we would obtain $e_a = 1/231$, which is considerably larger than the actual value of $e_a = 1/297$. This result implies that the density increases from the earth's outer surface toward the earth's center.

Returning to Eq. (8.114), let us introduce the variable η, defined by

$$\eta = \frac{d \log e}{d \log x} = \frac{x}{e} \frac{de}{dx} \tag{8.116}$$

Then,

$$\frac{de}{dx} = \frac{\eta e}{x}$$

and

$$\frac{d^2 e}{dx^2} = \left(\frac{1}{x} \frac{d\eta}{dx} + \frac{\eta^2 - \eta}{x^2} \right) e$$

Substituting these two expressions into Eq. (8.111), we obtain

$$x\frac{d\eta}{dx} + \eta^2 - \eta - 6 + \frac{6\rho}{\rho_0}(\eta+1) = 0 \tag{8.117}$$

Now, we can express ρ in terms of ρ_0 by taking the derivative of Eq. (8.107) with respect to x

$$\rho x^2 = \frac{1}{3}\frac{d}{dx}(\rho_0 x^3)$$

or

$$\frac{\rho}{\rho_0} = 1 + \frac{1}{3}\frac{x}{\rho_0}\frac{d\rho_0}{dx} \tag{8.118}$$

which gives for Eq. (8.117)

$$x\frac{d\eta}{dx} + \eta^2 + 5\eta + 2\frac{x}{\rho_0}\frac{d\rho_0}{dx}(1+\eta) = 0 \tag{8.119}$$

To eliminate $d\eta/dx$ and $d\rho_0/dx$ from Eq. (8.119), we shall want to introduce the seemingly cumbersome expression

$$\frac{\frac{d}{dx}(\rho_0 x^5 \sqrt{1+\eta})}{\rho_0 x^5 \sqrt{1+\eta}} = \frac{1}{\rho_0}\frac{d\rho_0}{dx} + \frac{5}{x} + \frac{1}{2(1+\eta)}\frac{d\eta}{dx}$$

Substituting into Eq. (8.119), we obtain

$$x\frac{d\eta}{dx} + 2\frac{x}{\rho_0}\frac{d\rho_0}{dx}(1+\eta) = -\eta^2 - 5\eta$$

$$\frac{2\sqrt{1+\eta}}{\rho_0 x^4}\frac{d}{dx}(\rho_0 x^5 \sqrt{1+\eta}) - 10(1+\eta) = -\eta^2 - 5\eta$$

$$\frac{d}{dx}(\rho_0 x^5 \sqrt{1+\eta}) = 5\rho_0 x^4 \chi(\eta) \tag{8.120}$$

where

$$\chi(\eta) = \frac{1 + \frac{1}{2}\eta - \frac{1}{10}\eta^2}{\sqrt{1+\eta}} \tag{8.121}$$

Now $\chi(\eta)$ has an interesting property, which allows us to make a useful approximation to Eq. (8.120). We can show that $\chi(\eta)$ has a minimum for $\eta = 0$ and a maximum for $\eta = \frac{1}{3}$. At the earth's surface, we have from Eqs.

(8.114) and (8.116), using the previous values of m and e_a, that $\eta_a = 0.57$. We have, then, for representative values of $\chi(\eta)$

η	0	0.33	0.57
$\chi(\eta)$	1.0000	1.0007	0.9993

As $\eta = 0$ at $r = 0$, we are safe in taking $\chi(\eta)$ equal to a constant for our problem. Equation (8.120) then becomes

$$\frac{d}{dx}(\rho_0 x^5 \sqrt{1+\eta}) = 5\rho_0 x^4 \qquad (8.122)$$

This is known as *Radau's approximation* to Clairaut's equation.

From Radau's approximation, we can develop a relation for e_a, which is *independent* of the density distribution. This result verifies to us that we cannot, from a knowledge of e_a alone, reach any meaningful conclusions as to the internal density distribution of the earth. We proceed as follows. We are first interested in deriving an expression for the external potential Φ_0 of a body in terms of its moments of inertia. We shall take the center of origin of our coordinates (Fig. 8.17) to be the center of mass of the body and the three coordinate axes to correspond with the principal axes of the body. Let P be a point external to the body and Q a point internal to the body. Then the potential at P may be written as

$$\Phi_0 = \gamma \int \frac{\rho \, d\tau}{R} \qquad (8.123)$$

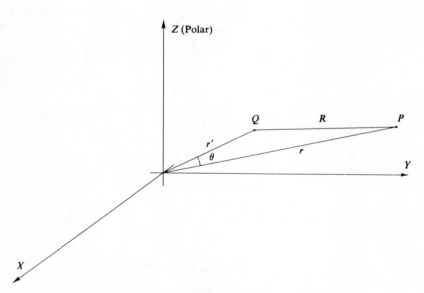

Fig. 8.17

Using the expansion for R in terms of r', r, and the Legendre polynomials of Section 1.6 and substituting for P_1 and P_2, we have for $1/R$

$$\frac{1}{R} = \frac{1}{r} + \frac{r'}{r^2}P_1 + \frac{r'^2}{r^3}P_2 + \cdots$$

$$= \frac{1}{r} + \frac{r'}{r^2}\cos\theta + \frac{r'^2}{2r^3}(3\cos^2\theta - 1) + \cdots \qquad (8.124)$$

Substituting Eq. (8.124) into Eq. (8.123), we see that the first integral is simply $\gamma M/r$, where M is the total mass of the body. The second integral vanishes, since we have chosen our origin at the center of mass. We are interested in evaluating the third integral. From Eqs. (1.1) and (1.2) we can write

$$\begin{aligned} r'^2 \cos^2\theta &= (ll' + mm' + nn')^2 r'^2 \\ &= l^2 x'^2 + m^2 y'^2 + n^2 z'^2 + 2lm x'y' + 2ln x'z' + 2mn y'z' \\ &= -l^2(y'^2 + z'^2) - m^2(x'^2 + z'^2) - n^2(x'^2 + y'^2) + r'^2 \\ &\quad + 2lm x'y' + 2ln x'z' + 2mn y'z' \end{aligned}$$

where l, m, n and l', m', n' are the direction cosines, respectively, of r and r', and x', y', z' are the Cartesian coordinates of Q. Now the integrals, [Eq. (8.123)], of the last three terms are expressions for the products of inertia; since we have chosen our coordinate axes as the principal axes of the body, these integrals vanish. The integrals of the first three terms are expressions for the principal moments of inertia A, B, C, about the three axes X, Y, Z, respectively. We know from texts on dynamics that the integral of the fourth term is an expression for one-half the sum of A, B, and C. We then have

$$\Phi_0 = \gamma\left[\frac{M}{r} + \frac{A + B + C - 3(l^2 A + m^2 B + n^2 C)}{2r^3}\right] \qquad (8.125)$$

to the first order. Expression (8.125) is known as *MacCullagh's formula*. For the earth, A is equal to B to the first order, and we can write for Eq. (8.125) after some reduction

$$\Phi_0 = \gamma\left[\frac{M}{r} - \frac{C - B}{r^3}P_2(\varphi)\right] \qquad (8.126)$$

where we have substituted for the direction cosine n its equivalent $\sin\varphi$, φ being the latitude, and where we have substituted for P_2 in terms of the latitude. Comparing Eq. (8.126) with Eq. (8.32), we see that, to the first order, the coefficient J_2 in terms of the moments of inertia is

$$\frac{C - B}{Ma^2} = \frac{2}{3}J_2 = \frac{2}{3}\left(e_a - \frac{1}{2}m\right) \qquad (8.127)$$

where we also have substituted for J_2 from Eq. (8.38).

Returning to our hydrostatic theory, we now wish to develop an expression for the moment of inertia C about the polar axis through the use of Radau's approximation. From texts on dynamics, we have, for the moment of inertia of a nearly spherical body about its axis of revolution,

$$C = \frac{8}{3}\pi \int_0^a \rho x^4 \, dx$$

$$= \frac{8}{9}\pi \int_0^a \left(3\rho_0 x^4 + x^5 \frac{d\rho_0}{dx}\right) dx$$

$$= \frac{8}{9}\pi \left[\bar{\rho} a^5 - 2\int_0^a \rho_0 x^4 \, dx\right]$$

where we have substituted from Eq. (8.118) and have integrated the second term by parts. Now from Eq. (8.122) we have directly the evaluation of the second integral, giving us,

$$C = \tfrac{8}{9}\pi\bar{\rho} a^5 [1 - \tfrac{2}{5}\sqrt{1+\eta_a}] \tag{8.128}$$

or

$$\frac{C}{Ma^2} = \tfrac{2}{3}[1 - \tfrac{2}{5}\sqrt{1+\eta_a}] \tag{8.129}$$

Combining Eqs. (8.127) and (8.129), we obtain, finally, for H the precessional constant for the precession of the equinoxes of Section 10.2, in terms of e_a and m,

$$H = \frac{C-B}{C} = \frac{e_a - \tfrac{1}{2}m}{1 - \tfrac{2}{5}\sqrt{1+\eta_a}} \tag{8.130}$$

where η_a is given in terms of e_a from Eqs. (8.114) and (8.116),

$$\eta_a = \frac{a}{e_a}\left(\frac{de}{dx}\right)_a = \frac{5}{2}\frac{m}{e_a} - 2 \tag{8.131}$$

From Eqs. (8.130) and (8.131) we have an expression for e_a in terms of H and m, independent of the distribution of density. Both H and m are known to a high degree of accuracy. Substituting their known values of $H = 1/306$ and $m = 1/289$, we obtain $e_a = 1/297$. This is in agreement with the value of $e_a = 1/297$ obtained from gravity measurements and the resultant gravity formula (8.48). To this order of accuracy, we have results that are consistent with the assumption of a hydrostatic earth.

Although we have not been able to determine very much about the internal density of the earth from gravity measurements at its outer surface, there are two useful values which we have found. These are the mass M of the earth and its moment of inertia about the polar axis C. The mass M comes directly from the gravity formula of Section 8.4. Its value is 5.98×10^{27} g.

302 ¶ *Seismology, Gravity and Magnetism*

In Eq. (8.127) the value of J_2 is obtained from the gravity formula, and in Eq. (8.130) the value of H comes from the measurement of the precession of the equinoxes. Taking the ratio of these two equations and using the observed values of J_2 and H, we find that $C = 0.334\ Ma^2$. From M we can also determine that the mean density of the earth, $\bar{\rho}$, is 5.52 g/cm^3.

Problem 8.8(a)   Derive Eq. (8.115) directly from Eq. (8.97) and the expression for the external potential of Section 8.3.

Problem 8.8(b)   Demonstrate that the value of e_a obtained from Eq. (8.115) does imply that density increases with depth.

Problem 8.8(c)   Determine the minimum and maximum of $\chi(\eta)$.

Problem 8.8(d)   Carry through the reduction for Eq. (8.126).

References

Bomford, G. 1952. *Geodesy*. Oxford: Oxford University Press.
Daly, R. A. 1940. *Strength and Structure of the Earth*. Englewood Cliffs, N.J.: Prentice-Hall.
Garland, G. D. 1971. *Introduction to Geophysics*. Philadelphia: W. B. Saunders.
Grant, F. S., and G. F. West. 1965. *Interpretation Theory in Applied Geophysics*. New York: McGraw-Hill.
Heiskanen, W. A., and F. A. Vening Meinesz. 1958. *The Earth and Its Gravity Field*. New York: McGraw-Hill.
Jeffreys, H. 1952. *The Earth*. Cambridge: Cambridge University Press.
Kaula, W. M. 1968. *An Introduction to Planetary Physics*. New York: Wiley.
Kellogg, O. D. 1929. *Foundations of Potential Theory*, New York: Dover.
LeJay, R. P. P. 1947. *Développements Modernes de la Gravimétrie*. Paris: Gauthier-Villars.
Nettleton, L. L. 1940. *Geophysical Prospecting for Oil*. New York: McGraw-Hill.
Page, L. 1935. *Introduction to Theoretical Physics*. New York: Van Nostrand.
Stacey, F. D. 1969. *Physics of the Earth*. New York: Wiley.

CHAPTER 9

GEOMAGNETISM

9.1 Electromagnetism

We shall restate here, for review purposes, some of the basic relations of electrostatics, magnetostatics, and electromagnetism that will be of use to us.

The force law in electrostatics and magnetostatics is of the same form as expression (8.3), relating in this case to unit electric charges or unit magnetic poles. Using the *electromagnetic system of units* (EMU), the constant of proportionality, corresponding to γ in Eq. (8.3), is unity for the magnetostatic force law and is c^2, the square of the velocity of propagation of electromagnetic waves in a vacuum, for the electrostatic force law. Corresponding to the gravitational attraction **g**, we have the *electric intensity* or field strength **E**, and the *magnetic intensity* or field strength **H**. The same relations with regard to an electric or magnetic potential V, Gauss' law, and the integral or differential relations of Eqs. (8.9) and (8.12) or Eqs. (8.10) and (8.11) of Sections 8.1 and 8.2 also apply. In EMU the unit of magnetic intensity is called the *gauss*; because of the numerical size of this unit, most geomagnetic measurements and calculations are quoted in *gamma*, equal to 10^{-5} gauss.

For a dielectric medium, we shall have the relation

$$\mathbf{D} = \frac{\mathbf{E}}{c^2} + 4\pi\mathbf{P} \tag{9.1}$$

where **D** is the *electric displacement* and **P** the *polarization*. The relations (8.10) and (8.11) then become

$$\nabla \cdot \mathbf{D} = 4\pi\rho \tag{9.2}$$

and

$$\nabla \times \mathbf{E} = 0 \tag{9.3}$$

where ρ is the density of free charges in the dielectric medium. Since the

304 ¶ Seismology, Gravity and Magnetism

polarization is produced by the electric field, it might be expected that **P** would be proportional to **E** for an isotropic medium, giving

$$\mathbf{P} = \frac{\epsilon}{c^2} \mathbf{E} \tag{9.4}$$

and

$$\mathbf{D} = (1+4\pi\epsilon)\frac{\mathbf{E}}{c^2} = \frac{\kappa}{c^2} \mathbf{E} \tag{9.5}$$

where ϵ is the *electric susceptibility* and κ the *dielectric constant*. For a magnetic medium, we have similar relations

$$\mathbf{B} = \mathbf{H} + 4\pi\mathbf{I} \tag{9.6}$$

where **B** is the *magnetic induction* and **I** the *intensity of magnetization*, and

$$\nabla \cdot \mathbf{B} = 0 \tag{9.7}$$

and

$$\nabla \times \mathbf{H} = 0 \tag{9.8}$$

since there are no free magnetic poles, and

$$\mathbf{I} = \epsilon\mathbf{H} \tag{9.9}$$

and

$$\mathbf{B} = (1+4\pi\epsilon)\mathbf{H} = \mu\mathbf{H} \tag{9.10}$$

where ϵ is here the *magnetic susceptibility* and μ the *permeability*. For the boundary conditions between two dielectric or magnetic media, we have that the normal component of the electric displacement or magnetic induction is continuous and that the field strength potential is continuous.

We shall have need of the concept of a *magnetic dipole*. Let us calculate the field due to a magnetic dipole, as shown in Fig. 9.1, at a distance large compared with the distance of separation of the poles. Denoting the two poles

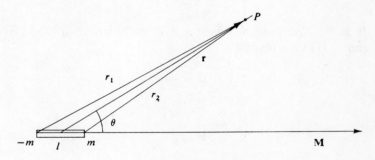

Fig. 9.1

by m and $-m$ and the distance between them by l, the potential at a point P distant from the center O of the dipole is

$$V = \frac{m}{r_1} - \frac{m}{r_2} = \frac{m}{r - \frac{l}{2}\cos\theta} - \frac{m}{r + \frac{l}{2}\cos\theta}$$

$$= \frac{m}{r}\left(1 + \frac{l}{2r}\cos\theta + \cdots - 1 + \frac{l}{2r}\cos\theta + \cdots\right)$$

$$= \frac{ml}{r^2}\cos\theta = \frac{M}{r^2}\cos\theta = \frac{\mathbf{M}\cdot\mathbf{r}}{r^3} \tag{9.11}$$

where $\mathbf{M}$ is the *magnetic moment* of the dipole. The component of magnetic intensity in the directions of increasing r and θ are, then, simply

$$H_r = -\frac{\partial V}{\partial r} = \frac{2M}{r^3}\cos\theta \tag{9.12}$$

and

$$H_\theta = -\frac{\partial V}{r\partial\theta} = \frac{M}{r^3}\sin\theta \tag{9.13}$$

In electromagnetism, we have for the magnetic field created by an electric current, *Ampere's law*, which we shall take for our purposes in the alternate form as defined by the integral relation

$$\oint \mathbf{H}\cdot d\mathbf{l} = 4\pi\int \mathbf{j}\cdot d\mathbf{s} = 4\pi i \tag{9.14}$$

or as defined by the corresponding differential relation

$$\nabla \times \mathbf{H} = 4\pi\mathbf{j} = 4\pi\rho\mathbf{v} \tag{9.15}$$

where i is the electric current, $\mathbf{j}$ the current density, ρ the charge density, and $\mathbf{v}$ the charge velocity. Equation (9.14), in the form of the line integral, is an expression for the work done by the field on a unit pole and is referred to as the *magnetomotive force* (MMF). The conventional unit of electric current, the ampere, is equal to 10^{-1} EMU. In electromagnetism, we also have the companion relation for the electric current induced in a conductor by a changing magnetic field, *Faraday's law*, which may be written in integral form as

$$\oint \mathbf{E}\cdot d\mathbf{l} = -\int \frac{\partial \mathbf{B}}{\partial t}\cdot d\mathbf{s} \tag{9.16}$$

or in differential form as

$$\nabla \times \mathbf{E} = -\frac{\partial \mathbf{B}}{\partial t} \tag{9.17}$$

where the line integral here is the work done by the electric field on a unit charge and is referred to as the *electromotive force* (EMF) $\mathscr{E}$. In addition, we also have that an EMF, referred to as a *motional electromotive force*, is induced in a conducting wire moving with velocity **v** through a magnetic field, which may be expressed by

$$\mathbf{E} = \mathbf{v} \times \mathbf{H} \tag{9.18}$$

or, corresponding to Eq. (9.16), as

$$\oint \mathbf{E} \cdot d\mathbf{l} = -\oint \mathbf{H} \times \mathbf{v} \cdot d\mathbf{l} = -\int \mathbf{B} \cdot \frac{\partial}{\partial t} d\mathbf{s} \tag{9.19}$$

We have further the fundamental relation between electric current density **j** and electric field strength **E** in a conducting medium, *Ohm's law*, that

$$\mathbf{j} = \sigma \mathbf{E} \qquad \mathbf{E} = \rho \mathbf{j} \tag{9.20}$$

where σ is the *electric conductivity* and ρ its reciprocal the *electric resistivity*. For most purposes, we may summarize the electromagnetic relations in *Maxwell's equations* from Eqs. (9.2), (9.3), (9.17), and (9.15) with Maxwell's revision of Ampere's law by the addition of the *displacement current* $(1/4\pi)$ $(\partial \mathbf{D}/\partial t)$ as

$$\nabla \cdot \mathbf{D} = 4\pi\rho \qquad \nabla \cdot \mathbf{B} = 0$$
$$\nabla \times \mathbf{E} = -\frac{\partial \mathbf{B}}{\partial t} \qquad \nabla \times \mathbf{H} = 4\pi \left(\rho \mathbf{v} + \frac{1}{4\pi} \frac{\partial \mathbf{D}}{\partial t} \right) \tag{9.21}$$

Problem 9.1(a) Calculate the magnetic intensity at a point P on the axis of a circular turn of wire, carrying a current i, at a distance x from its center.

Ans.

$$H = H_x = \frac{2\pi i a^2}{(a^2 + x^2)^{3/2}}$$

Problem 9.1(b) Find the magnetic intensity at a distance R from (1) a finite straight circuit in terms of the angles β_1 and β_2 made by lines drawn to the ends of the circuit with the perpendicular to the circuit, (2) an infinite straight circuit.

Ans.

(1) $(i/R)(\sin \beta_2 - \sin \beta_1)$, (2) $2(i/R)$

Problem 9.1(c) Determine the magnetic field (1) inside and (2) outside a toroidal solenoid of larger radius R and smaller radius a and n number of turns per unit length of the solenoid.

Ans.

(1) $H = 4\pi n i$, (2) $H = 0$

Problem 9.1(d) A magnet of moment M with its axis perpendicular to the wall of a room is moved away with velocity $\mathbf{v}$. Discuss the electric field in the plane of the wall when the magnet is at a distance h.

Ans.
$$E = \frac{3Mphv}{(h^2+p^2)^{5/2}}$$
where p is the radius of the circular line of force.

Problem 9.1(e) Show that the velocity of propagation of electromagnetic waves in a vacuum is the constant c of the electrostatic force equation. *Hint*: Eliminate $\mathbf{H}$ or $\mathbf{E}$ from Maxwell's equations.

9.2 Earth's Main Magnetic Field

To a first approximation, the measured values of the earth's main magnetic field fit the derived values for a magnetic dipole centered at the earth's center with the extension of its south pole axis intersecting the earth's surface at approximately 79°N, 70°W, and the extension of its north pole axis correspondingly intersecting the earth's surface at approximately 79°S, 100°E. The magnetic potential and the components of the magnetic field strength for a magnetic dipole are given by Eqs. (9.11), (9.12), and (9.13).

The earth's magnetic field, being a vector quantity, can be described by a number of observable magnetic quantities. These include X, the component of the horizontal intensity in the geographical meridian measured positive to the north; Y, the component of the horizontal intensity transverse to the geographical meridian measured positive to the east; Z, the vertical component of the intensity measured positive down; H, the magnitude of the horizontal intensity; F, the magnitude of the total intensity; D, the *declination* or *variation* defined as the azimuth of the horizontal component reckoned positively from geographical north; and I, the *inclination* or *dip* defined as the angle between the total intensity and the horizontal reckoned positive down and measured in a vertical plane. The interrelations among these quantities are simply

$$\begin{aligned} X &= H\cos D & Y &= H\sin D \\ H &= F\cos I & Z &= F\sin I \\ F^2 &= H^2+Z^2 = X^2+Y^2+Z^2 & & \end{aligned} \quad (9.22)$$

To specify the field at any point, three elements, of course, are needed. Those commonly used are H, D, I; H, D, Z; or X, Y, Z. It is sometimes convenient to show the distribution of the magnetic field over the earth's surface by lines of constant magnetic values. Those for declination are called *isogonic* lines, those for inclination *isoclinic* lines, and those for total intensity *isodynamic* lines.

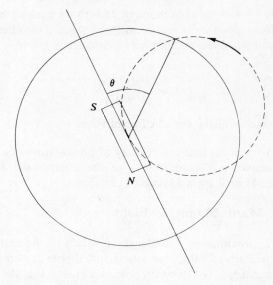

Fig. 9.2

For the specified earth dipole as shown in Fig. 9.2, the relations (9.11), (9.12), and (9.13) become

$$V = -\frac{M \cos \theta}{r^2} \tag{9.23}$$

and

$$H_r = -\frac{2M \cos \theta}{r^3}$$
$$H_\theta = -\frac{M \sin \theta}{r^3} \tag{9.24}$$

where θ is the magnetic colatitude, that is, the complement of the latitude as measured in the magnetic meridian. From the definitions for H, Z, I, and F, we have from Eqs. (9.24)

$$H = \frac{M}{r^3} \sin \theta = H_0 \sin \theta$$
$$Z = \frac{2M}{r^3} \cos \theta = 2H_0 \cos \theta \tag{9.25}$$

and

$$\tan I = \frac{Z}{H} = 2 \cot \theta$$
$$F = H_0(1 + \cos^2 \theta) \tag{9.26}$$

Problem 9.1(d) A magnet of moment M with its axis perpendicular to the wall of a room is moved away with velocity **v**. Discuss the electric field in the plane of the wall when the magnet is at a distance h.

Ans.
$$E = \frac{3Mphv}{(h^2+p^2)^{5/2}}$$

where p is the radius of the circular line of force.

Problem 9.1(e) Show that the velocity of propagation of electromagnetic waves in a vacuum is the constant c of the electrostatic force equation. *Hint*: Eliminate **H** or **E** from Maxwell's equations.

9.2 Earth's Main Magnetic Field

To a first approximation, the measured values of the earth's main magnetic field fit the derived values for a magnetic dipole centered at the earth's center with the extension of its south pole axis intersecting the earth's surface at approximately 79°N, 70°W, and the extension of its north pole axis correspondingly intersecting the earth's surface at approximately 79°S, 100°E. The magnetic potential and the components of the magnetic field strength for a magnetic dipole are given by Eqs. (9.11), (9.12), and (9.13).

The earth's magnetic field, being a vector quantity, can be described by a number of observable magnetic quantities. These include X, the component of the horizontal intensity in the geographical meridian measured positive to the north; Y, the component of the horizontal intensity transverse to the geographical meridian measured positive to the east; Z, the vertical component of the intensity measured positive down; H, the magnitude of the horizontal intensity; F, the magnitude of the total intensity; D, the *declination* or *variation* defined as the azimuth of the horizontal component reckoned positively from geographical north; and I, the *inclination* or *dip* defined as the angle between the total intensity and the horizontal reckoned positive down and measured in a vertical plane. The interrelations among these quantities are simply

$$X = H \cos D \qquad Y = H \sin D$$
$$H = F \cos I \qquad Z = F \sin I \qquad (9.22)$$
$$F^2 = H^2 + Z^2 = X^2 + Y^2 + Z^2$$

To specify the field at any point, three elements, of course, are needed. Those commonly used are H, D, I; H, D, Z; or X, Y, Z. It is sometimes convenient to show the distribution of the magnetic field over the earth's surface by lines of constant magnetic values. Those for declination are called *isogonic* lines, those for inclination *isoclinic* lines, and those for total intensity *isodynamic* lines.

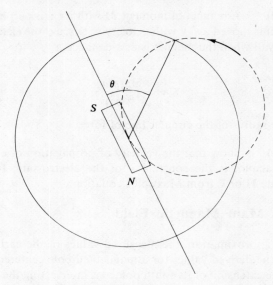

Fig. 9.2

For the specified earth dipole as shown in Fig. 9.2, the relations (9.11), (9.12), and (9.13) become

$$V = -\frac{M\cos\theta}{r^2} \tag{9.23}$$

and

$$H_r = -\frac{2M\cos\theta}{r^3}$$
$$H_\theta = -\frac{M\sin\theta}{r^3} \tag{9.24}$$

where θ is the magnetic colatitude, that is, the complement of the latitude as measured in the magnetic meridian. From the definitions for H, Z, I, and F, we have from Eqs. (9.24)

$$H = \frac{M}{r^3}\sin\theta = H_0\sin\theta$$
$$Z = \frac{2M}{r^3}\cos\theta = 2H_0\cos\theta \tag{9.25}$$

and

$$\tan I = \frac{Z}{H} = 2\cot\theta$$
$$F = H_0(1+\cos^2\theta) \tag{9.26}$$

where the constant H_0 has been substituted for M/r^3. The value of H_0 for the earth is roughly 0.3 gauss. It is of some interest to note that the magnetic intensity varies as r^3 away from the earth, whereas the gravitational attraction varies as r^2.

Implicit in the above is the assumption that the earth's main magnetic field is of an internal origin. It is of interest to establish that this is indeed the case. Consider a spherical shell in which a magnetic field is created either from magnetostatic or electromagnetic causes. Outside of this shell, the magnetic field potential will be a solution of Laplace's equation. From the discussions of Sections 1.6, 1.7, and 1.11, it is clear that the region external to the shell will have a radial dependence $r^{-(n+1)}$, and the region internal to the shell will have a radial dependence r^n, where the number n refers to the respective Legendre or associated Legendre polynomial. Restricting ourselves for illustration to the case of axial symmetry, that is, no dependence on φ, we may represent the magnetic field potential by an expression of the form of Eq. (1.128) or

$$V = a \sum_n \left[I_n \left(\frac{a}{r}\right)^{n+1} + E_n \left(\frac{r}{a}\right)^n \right] P_n(\cos \theta) \qquad (9.27)$$

where I_n are the coefficients for the internal field and E_n for the external field as measured at the earth's surface and where a is the radius of the spherical earth. The components of the magnetic field strength at the earth's surface will then be

$$X = \frac{1}{r} \frac{\partial V}{\partial \theta} = \frac{a}{r} \sum_n \left[I_n \left(\frac{a}{r}\right)^{n+1} + E_n \left(\frac{r}{a}\right)^n \right] \frac{\partial P_n}{\partial \theta}$$

$$= \sum_n [I_n + E_n] \frac{\partial P_n}{\partial \theta}$$

$$Y = 0 \qquad (9.28)$$

$$Z = \frac{\partial V}{\partial r} = \sum_n \left[-(n+1)I_n \left(\frac{a}{r}\right)^{n+2} + nE_n \left(\frac{r}{a}\right)^{n-1} \right] P_n$$

$$= \sum_n [-(n+1)I_n + nE_n] P_n$$

Upon considering only the first zonal harmonic P_1, that is, the dipole term, Eqs. (9.28) reduce to

$$X = -[I_1 + E_1] \sin \theta$$
$$Z = [-2I_1 + E_1] \cos \theta \qquad (9.29)$$

If the average values of X and Z are known at selected parallels of magnetic latitude, the values of the bracketed expressions in Eqs. (9.29) may be obtained by averaging or some other suitable means. The coefficients I_1 and E_1 are then obtained simply from the two simultaneous equations resulting from equating the brackets to these determined values. If higher-order zonal harmonic or sectorial and tesseral harmonic coefficients are desired, the process in principle remains the same. It is found that the external field amounts to the order of 1 percent of total earth's magnetic field, as measured at the earth's surface.

We should also look for the possibility of a nonpotential field, that is, in this case one caused by electric currents flowing across the earth's surface. From Eq. (9.14) we have a simple relation between the earth's magnetic field and such currents. Calculating such line integrals at various locations on the earth's surface shows that, in general, the currents determined by Eq. (9.14) would contribute a magnetic field of the order of 1 percent of the total earth's magnetic field.

Let us next consider, from magnetostatics, the problem of the magnetic field due to a uniformly magnetized sphere. Since, by definition, **I** is a constant, $\nabla \cdot \mathbf{I} = 0$, so that the magnetic potential is a solution of Laplace's equation both outside and inside the sphere. Our boundary conditions are that the normal component of **B** and V are continuous across the boundary. Taking **I** in the direction of the polar axis, we have that its radial, or normal component at the sphere boundary, is $I \cos \theta$. So only those potential functions that are linear in $\cos \theta$ will serve to satisfy the boundary conditions. Hence, from Eq. (1.128) and the results of Problem 1.6(a), we need only the two first-order Legendre polynomial solutions, the first since it is finite at the origin representing the potential inside the sphere, and the second since it vanishes at infinity representing the potential outside the sphere. Distinguishing the regions inside and outside the sphere by the subscripts i and 0, we have

$$V_i = Ar \cos \theta \qquad V_0 = C \frac{\cos \theta}{r^2} \qquad (9.30)$$

With reference to Eqs. (9.6) and (9.10) and remembering that $\mathbf{H} = -\nabla V$, we have for the boundary conditions at the surface of the sphere, $r = a$,

$$-A \cos \theta + 4\pi I \cos \theta = 2 \frac{C}{a^3} \cos \theta$$

and

$$Aa \cos \theta = \frac{C}{a^2} \cos \theta$$

where we have taken $\mu = 1$ outside the sphere. The solution of these two equations gives for C

$$C = \tfrac{4}{3}\pi a^3 I = M \qquad (9.31)$$

where from the definition of I, the quantity C is then simply the total magnetic moment M of the sphere. In conclusion, we see from Eq. (9.11) that the external potential and, consequently, the external magnetic field strength of a uniformly magnetized sphere are the same as that of a magnetic dipole.

Problem 9.2(a) Determine the intensity of magnetization necessary for a uniformly magnetized earth sphere to produce the observed magnetic field strength.

Ans. 0.07

9.3 Electromagnet Theory

One of the more interesting problems in geophysics is the origin of the earth's magnetic field. We have seen from the previous section that the observed main dipole field of the earth could be caused by a uniformly magnetized earth sphere. Unfortunately, the surface layers of the earth are, on the average, not magnetized to the extent needed. Further, below a depth of a few tens of kilometers, it is probable that the temperature exceeds the Curie point of ferromagnetic materials, the Curie point being the temperature above which ferromagnetic materials lose their properties of permanent magnetization. We must, then, look for an alternative explanation.

One such possibility is that the earth is a giant electromagnet. We suppose, then, that there are electric currents circulating within the earth, subject to free decay, and that the magnetic field produced by these currents is the observed main magnetic field of the earth. Although it is difficult to accept this hypothesis, particularly with regard to how the currents were initially started or how they are maintained, conceivably by electromagnetic induction from giant aperiodic atmospheric magnetic storms, it is of interest to follow through the theory.

We shall assume, for simplicity, that the earth is a uniform isotropic sphere, and we shall consider that there are no displacement currents or charge density. Maxwell's equations (9.21) then reduce to

$$\nabla \cdot \mathbf{D} = 0 \tag{9.32}$$

$$\nabla \cdot \mathbf{B} = 0 \tag{9.33}$$

$$\nabla \times \mathbf{E} = -\frac{\partial \mathbf{B}}{\partial t} \tag{9.34}$$

and

$$\nabla \times \mathbf{H} = 4\pi \mathbf{j} \tag{9.35}$$

We shall want to consider the magnetic induction **B** in the form of Eq.

(1.38). Since we have from Eq. (9.33) that **B** is a solenoidal vector, expression (1.38) reduces to

$$\mathbf{B} = \nabla \times \mathbf{A} \tag{9.36}$$

and

$$\nabla \cdot \mathbf{A} = 0 \tag{9.37}$$

The vector **A** is referred to as the *magnetic vector potential*. It assumes somewhat the same role for steady electric currents that the scalar potential does in electrostatics. From Eq. (9.34) we shall have, substituting for **B** from Eq. (9.36),

$$\nabla \times \mathbf{E} = -\frac{\partial}{\partial t}(\nabla \times \mathbf{A}) = -\nabla \times \frac{\partial \mathbf{A}}{\partial t}$$

or

$$\mathbf{E} = -\frac{\partial \mathbf{A}}{\partial t} + \nabla \Psi$$

where we must include, in general, $\nabla \Psi$ of an undetermined scalar since $\nabla \times \nabla \Psi$ is uniquely zero. However, in our case we have from Eq. (9.32) that **D** and **E** are solenoidal vectors, so that the second term of the above expression will be zero, and we shall have simply

$$\mathbf{E} = -\frac{\partial \mathbf{A}}{\partial t} \tag{9.38}$$

for the electric intensity in terms of the magnetic vector potential. From Eq. (9.35) we can then obtain, using Eqs. (9.10), (9.20), (9.38), (9.37), and the expansion for the triple vector product,

$$\nabla \times \mathbf{B} = 4\pi\mu\sigma\mathbf{E}$$

$$\nabla \times \nabla \times \mathbf{A} = -4\pi\mu\sigma \frac{\partial \mathbf{A}}{\partial t}$$

$$\nabla(\nabla \cdot \mathbf{A}) - \nabla \cdot \nabla \mathbf{A} = -4\pi\mu\sigma \frac{\partial \mathbf{A}}{\partial t}$$

$$\nabla^2 \mathbf{A} = 4\pi\mu\sigma \frac{\partial \mathbf{A}}{\partial t} \tag{9.39}$$

For our example, we shall be interested in the case in which there are no electric currents along the radius vector **r**, that is, the currents are confined within the earth and do not cross the earth's outer surface. A simple vector expression that will meet this requirement is

$$\mathbf{A} = \mathbf{r} \times \nabla \Phi \tag{9.40}$$

where from the definition of the vector product, **A** is normal to **r** and where from Eqs. (9.38) and (9.20), $\mathbf{A}$ is in the same direction as $\mathbf{j}$. The expression (9.40) will also satisfy our conditions [Eqs. (9.37) and (9.39)] if the scalar Φ is a solution of the differential equation

$$\nabla^2 \Phi = 4\pi\mu\sigma \frac{\partial \Phi}{\partial t} \tag{9.41}$$

From Eq. (9.37) we have

$$\nabla \cdot \mathbf{A} = \nabla \cdot (\mathbf{r} \times \nabla \Phi) = (\nabla \times \mathbf{r}) \cdot \nabla \Phi - (\nabla \times \nabla \Phi) \cdot \mathbf{r} = 0$$

since from the results of Problem 1.1(j) and from Eq. (1.35), both $\nabla \times \mathbf{r}$ and $\nabla \times \nabla \Phi$ are uniquely zero. From Eq. (9.39) we have

$$\nabla \cdot \nabla (\mathbf{r} \times \nabla \Phi) = 4\pi\mu\sigma \frac{\partial}{\partial t}(\mathbf{r} \times \nabla \Phi)$$

$$(\nabla \cdot \nabla \mathbf{r}) \times \nabla \Phi + \mathbf{r} \times \nabla \cdot \nabla (\nabla \Phi) = 4\pi\mu\sigma \frac{\partial}{\partial t}(\mathbf{r} \times \nabla \Phi)$$

$$\mathbf{r} \times \nabla^2 (\nabla \Phi) = 4\pi\mu\sigma \mathbf{r} \times \frac{\partial}{\partial t}(\nabla \Phi)$$

remembering that $\nabla \cdot \nabla \mathbf{r}$ is also uniquely zero.

The partial differential equation (9.41) is of a familiar form, which we have encountered previously in problems concerned with heat conduction and diffusion. We shall be interested here in solutions in terms of spherical coordinates for which there is azimuthal symmetry, that is, no dependence on the azimuth angle φ. Using the familiar method of solution by separation of variables, we obtain for Eq. (9.41), assuming a solution of the form,

$$\Phi = R(r)U(u)T(t) \tag{9.42}$$

where we have made the usual change of variables of $u = \cos \theta$ in spherical coordinate form of Eq. (1.121),

$$\frac{1}{r^2 R} \frac{\partial}{\partial r}(r^2 R') + \frac{1}{r^2 U} \frac{\partial}{\partial u}[(1-u^2)U'] = 4\pi\mu\sigma \frac{T'}{T} \tag{9.43}$$

Setting the right-hand side equal to the coefficient $-k^2$, we obtain for T

$$T = e^{-pt} \tag{9.44}$$

where

$$p = \frac{k^2}{4\pi\mu\sigma} \tag{9.45}$$

Multiplying Eq. (9.43) by r^2 and setting the resultant second term on the left-hand side equal to $-n(n+1)$, as before, we obtain

$$\frac{d}{du}[(1-u^2)U'] + n(n+1)U = 0 \tag{9.46}$$

the Legendre differential equation, whose solutions are the Legendre polynomials P_n, and

$$\frac{d}{dr}(r^2 R') + [k^2 r^2 - n(n+1)]R = 0 \tag{9.47}$$

This equation may be transformed into a form of Bessel's differential equation, whose solutions are the half-order Bessel functions in terms of n. However, in our case, we are only interested in the solution Φ, which will have a polar angle dependence to correspond with the observed dipole magnetic field of Eq. (9.23). We see that this can only be met for $P_n = P_1 = \cos\theta$. This reduces Eq. (9.47) to

$$\frac{d}{dr}(r^2 R') + [k^2 r^2 - 2]R = 0$$

which, upon a change of independent variable to $x = kr$ and dependent variable to $Q = xR$, gives

$$\frac{d^2 Q}{dx^2} + \left(1 - \frac{2}{x^2}\right) Q = 0 \tag{9.48}$$

This equation, with the exception of the parenthesis for the second term, is of a form satisfied by the trigonometric sine and cosine functions. We may establish by trial and error that the solutions to Eq. (9.48) are

$$Q = \frac{\sin x}{x} - \cos x$$
$$Q = \sin x + \frac{\cos x}{x} \tag{9.49}$$

To meet the condition of finite values at the origin, only the first solution will be valid for our case. Our solution will then be

$$\Phi = Af(r) \cos\theta \, e^{-pt} \tag{9.50}$$

where $f(r)$ is given by

$$f(r) = \frac{\sin(kr)}{(kr)^2} - \frac{\cos(kr)}{kr} \tag{9.51}$$

Returning to Eq. (9.40), expressing the vector product in terms of the spherical coordinate unit vectors and remembering that Φ is not a function of φ, we obtain, using Eq. (1.82),

$$\mathbf{A} = \boldsymbol{\varphi}_1 \frac{\partial \Phi}{\partial \theta} = -\boldsymbol{\varphi}_1 Af(r) \sin\theta \, e^{-pt} \tag{9.52}$$

From Eq. (9.36) and the results of Problem 1.4(a), we can then obtain for the spherical coordinate components of **B**

$$B_r = \frac{1}{r \sin \theta} \frac{\partial}{\partial \theta} (\sin \theta \, A_\varphi)$$

$$= -\frac{2A}{r} f(r) \cos \theta \, e^{-pt} \tag{9.53}$$

$$B_\theta = \frac{-1}{r} \frac{\partial}{\partial r} (r A_\varphi)$$

$$= \frac{A}{r} \cdot \frac{d}{dr} [rf(r)] \sin \theta \, e^{-pt} \tag{9.54}$$

$$B_\varphi = 0$$

Taking, for convenience, the permeability to be unity, both inside and outside the spherical earth, our boundary conditions at the earth's surface, $r = a$, are simply continuity of the tangential and vertical components of **B** or **H**. From Eqs. (9.24), (9.53), and (9.54), this will then be

$$-\frac{2A}{a} f(a) = -\frac{2M}{a^3}$$

$$\frac{A}{a} \left\{ \frac{d}{dr} [rf(r)] \right\}_{r=a} = -\frac{M}{a^3} \tag{9.55}$$

since both sets of conditions have the same polar angle, and the time dependent term will be the same outside the sphere as inside. If these equations are to have a solution for A in terms of M, the determinant of their coefficients, transposing the right-hand term to the left-hand side of the equation, must be equal to zero or

$$\left\{ \frac{d}{dr} [rf(r)] \right\}_{r=a} + f(a) = 0$$

which reduces to

$$\sin (ka) = 0$$

or

$$ka = \pi, 2\pi, \ldots \tag{9.56}$$

We can now calculate the *relaxation time*, or *time constant*, τ, of the free current system, that is, the time for the field to decrease to $1/e$ of its original value, from Eqs. (9.45) and (9.56) as

$$\tau = \frac{1}{p} = \frac{4\pi\mu\sigma}{k^2} = \frac{4\sigma a^2}{\pi} \tag{9.57}$$

for the current system with the longest time constant.

The derivation given here is for a free current in the earth. In principle, the derivation is similar to that for the problem of electromagnetic induction in a spherical earth from an external magnetic field, such external magnetic field usually being caused by atmospheric electric currents. In this latter case, the external driving field includes both the r^n and $r^{-(n+1)}$ terms in its solution, so that the boundary conditions will determine the amplitude coefficient of the internal field without a conditional relation (9.55), and the time-dependent term will no longer be a simple decay, so that the radial components of the internal field will be given in terms of hyperbolic sines and cosines rather than trigonometric functions.

Problem 9.3(a) Prove by substitution that Eqs. (9.49) are solutions of Eq. (9.48).

Problem 9.3(b) Carry through the reduction to Eq. (9.56).

Problem 9.3(c) Calculate the time constant for the earth if it had a conductivity of (1) copper and (2) iron.

Ans. (1) 10×10^6 yr, (2) 2×10^6 yr

Problem 9.3(d) Calculate the time constant for the core if it had a conductivity of $\sigma = 3 \times 10^{-6}$ EMU.

Ans. 15×10^3 yr

Problem 9.3(e) Using the concept of vector potential, determine the magnetic field external to an infinite straight wire with a constant current.

9.4 Internal and Atmospheric Dynamos

Since there is difficulty in accepting the origin of the earth's magnetic field as being caused by considering the earth as either a permanent magnet or as an electromagnet, it is necessary to look elsewhere for a possible explanation. As might be expected, several suggestions have been put forward. We shall consider one such suggestion here in part because it does provide a satisfactory explanation for the origin of the earth's field and in part because it is susceptible to a theoretical discussion. This is to consider the earth, or more particularly the core, as a *dynamo*. In electrical machinery parlance, a dynamo is a generator for the conversion of mechanical energy into electrical energy through the agency of electromagnetic induction. In this case, a conducting material moving in a magnetic field creates a motional electromotive force; the motional electromotive force causes a current to flow in the conducting material, which, in turn, creates a secondary magnetic field. If the circuit and motional relations are such, the secondary field can sustain the initial field. In our case, we are interested in creating a dynamo in the core, which will produce an external magnetic field equivalent to that measured at the earth's surface.

Assuming that $\mu = 1$ and that the displacement current can be neglected, Maxwell's equations (9.21) reduce to

$$\nabla \cdot \mathbf{E} = \frac{4\pi \rho c^2}{\kappa} \tag{9.58}$$

$$\nabla \cdot \mathbf{H} = 0 \tag{9.59}$$

$$\nabla \times \mathbf{E} = -\frac{\partial \mathbf{H}}{\partial t} \tag{9.60}$$

and

$$\nabla \times \mathbf{H} = 4\pi \mathbf{j} \tag{9.61}$$

using Eqs. (9.5) and (9.10). Since we now have an electric field strength related to the motional electromotive force as given by Eq. (9.18) in addition to the field strength from Maxwell's equations, the expression for Ohm's law [Eqs. (9.20)] for the current density and the total electric field strength will now be

$$\mathbf{j} = \sigma(\mathbf{E} + \mathbf{v} \times \mathbf{H}) \tag{9.62}$$

where $\mathbf{v}$ is the fluid velocity of the core motion and the velocity of electric charge motion. If, in addition, we assume the fluid core to be incompressible, the continuity condition for the fluid core motion will be simply

$$\nabla \cdot \mathbf{v} = 0 \tag{9.63}$$

Eliminating $\mathbf{E}$ from the above equations, we obtain, using Eqs. (1.37), (9.59), (9.61), (9.62), and (9.60),

$$\nabla \times \nabla \times \mathbf{H} = \nabla(\nabla \cdot \mathbf{H}) - \nabla \cdot \nabla \mathbf{H}$$
$$4\pi \nabla \times \mathbf{j} = -\nabla^2 \mathbf{H}$$
$$\nabla^2 \mathbf{H} = 4\pi\sigma[-\nabla \times \mathbf{E} - \nabla \times (\mathbf{v} \times \mathbf{H})]$$

$$\nabla^2 \mathbf{H} = 4\pi\sigma \left[\frac{\partial \mathbf{H}}{\partial t} - \nabla \times (\mathbf{v} \times \mathbf{H}) \right] \tag{9.64}$$

or

$$\nabla^2 \mathbf{H} = 4\pi\sigma \left[\frac{\partial \mathbf{H}}{\partial t} + (\mathbf{v} \cdot \nabla)\mathbf{H} - (\mathbf{H} \cdot \nabla)\mathbf{v} \right] \tag{9.65}$$

using the results of Problem 1.1(k) and Eqs. (9.59) and (9.63) to obtain the last equation.

For a steady state, we put $\partial \mathbf{H}/\partial t = 0$ in Eq. (9.64), obtaining

$$\nabla^2 \mathbf{H} + 4\pi\sigma \nabla \times (\mathbf{v} \times \mathbf{H}) = 0 \tag{9.66}$$

If we can find nonvanishing solutions of Eq. (9.66) for a specified **v**, we can say that a steady dynamo that maintains **H** is possible. Such a solution is also required to meet the specified boundary conditions for the magnetic and electric fields at the core boundary and to produce an external field corresponding to the observed dipole field. Several such particular solutions, which will not be derived here, do exist. The difficulty is not so much in finding a solution, but in adequately describing the energy source, such as from rotational or thermal causes, and its coupling to maintain a steady-state dynamo.

The rather complex subject described by Eq. (9.64) or (9.65) of fluid motion in a conducting medium and the resultant electromagnetic effects is known as *magnetohydrodynamics*. We can get some insight into the phenomena by looking at the individual terms on the right-hand side of Eq. (9.65). If **v** = 0, the second and third terms disappear, and Eq. (9.65) reduces to the same form as Eq. (9.39), as might be expected. It is simply the equation for the magnetic field due to electromagnetic effects in a conducting medium in the absence of motion of the conducting medium. It shows, as we obtained in Section 9.3, that in the absence of fluid motion, the magnetic field will always decay. The second term in conjunction with the first term is simply the derivative $D\mathbf{H}/Dt$ following the fluid motion. It describes the conservation of the magnetic field following the fluid motion. The third term is the magnetic field multiplied by the gradient of the velocity in the direction of the field. It represents the stretching of the magnetic lines of force.

We may gain some further insight into the electromagnetic process involved by consideration of the simple machine shown in Fig. 9.3, known as a Faraday disk generator. A conducting disk rotates about its axis. A brush attached to a circular turn of wire is in contact with the axis of the disk to complete the electrical circuit. Supposing a uniform magnetic field to be applied to the system in the direction of the axis, we see from Eq. (9.18) that a radially directed electric field will be induced in the disk by the rotational motion. The potential difference thus set up between the axis and the periphery will drive an electric current in the circular turn of wire, which will, in turn, create a magnetic field in the direction of the original inducing field. We may write down the simple equations governing this circuit. From Eq. (9.18) and the definition of the electromotive force $\mathscr{E}$, we shall have

$$\mathscr{E} = \int_0^a \mathbf{E} \cdot d\mathbf{r} = \int_0^a \omega r H_0 \, dr = \tfrac{1}{2}\omega H_0 a^2 \qquad (9.67)$$

where H_0 is the original inducing magnetic field. From Ohm's law, the current in the circular circuit will be

$$i = \frac{\mathscr{E}}{R} = \frac{\omega H_0 a^2}{2R} \qquad (9.68)$$

where R is the total resistance of the electrical circuit. From Problem 9.1(a)

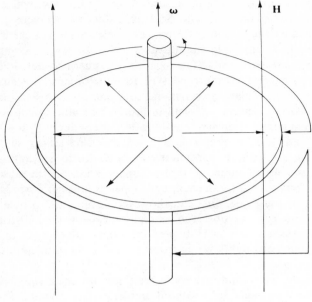

Fig. 9.3

we have that the magnetic field induced by the circular current at its center, which we shall presume to apply to the area of the disk, is

$$H = \frac{2\pi i}{a} \tag{9.69}$$

Combining Eqs. (9.68) and (9.69), we obtain

$$H = \frac{\pi \omega a H_0}{R} \tag{9.70}$$

If $\omega = R/\pi a$, the induced field will sustain the original field, and we shall have a self-exciting dynamo.

We have mentioned that of the order of 1 percent or less of the measured earth's magnetic field can be attributed to external cause. For a total field of the order of 10^4 gamma, this gives an external contribution of 10^2 gamma or less. This is an observable quantity, which is quite variable temporally and geographically and can be correlated with solar, lunar, and sun spot cycles, magnetic storm onsets, and the like. Part of this external field can be attributed to a dynamo effect in the upper atmosphere or ionosphere. In this case, the motion of the conducting, ionized, medium of the upper atmospheric regions in the presence of the earth's magnetic field creates a motional electromotive force according to Eq. (9.18). This motional electromotive force will drive electric currents according to Ohm's law, which, in turn, will create a secondary magnetic field according to Ampere's law. This secondary magnetic field is the external field measured at the earth's surface. In principle, it is simple. In application, there are several complications. Considering the upper conducting region as a spherical shell, the dynamo field is not, in general, of the right type to impel a steady electric current system without accumulating electric charge at some points. This necessitates the inclusion of an electrostatic field in order to be able to consider a continuous current circuit. Further, the ionization or conductivity σ is a function of the solar radiation, so that σ in Ohm's law can no longer be considered a constant. Further still, the motion itself can be due to gravitational effects of the sun and moon motions, that is, upper atmospheric tides, thermal expansion effects caused by the sun, and upper atmospheric winds, or some combination of the three. One can, however, reverse the problem, as has been done, and use the observed external magnetic field distribution to ascertain the electric current system in the upper atmosphere that created the field through Ampere's law and from that ascertain some notion as to the motions involved.

To conclude, we should give some brief description of the variability of the earth's main magnetic field without, unfortunately, giving the theoretical explanation or derivation of such variations as do occur. The earth's magnetic field has been one of the earliest and easiest geophysical quantities to measure, and it has had an importance in ocean navigation for a long time. Consequently, there are historical records of earth magnetic measurements extending over centuries. Such measurements show that the earth's field has undergone long-term changes in magnitude and direction measured over periods of years. Such changes are known as the *secular variation* of the earth's field. The simplest measurements, those of declination and inclination, extending over four centuries show a movement of the trace of the dipole axis on the earth's surface of the order of 0.02° latitude per year, which may be in the form of a rotation with a period longer than that of the measurements and perhaps indicates a rotation toward the geographic axis. The magnitude of the magnetic field strength shows a regular variation, which can be mapped and contoured of the order of up to 100 gamma per year, in some areas positive and in others negative, but in the aggregate negative. This contoured rate of change of magnetic field strength, as well as the nondipole portion of the field, shows a *westward drift* from one epoch to the next of the order of 0.2° latitude per year. The magnetic moment M of the dipole field itself shows an historical decrease in magnitude over the past century of from about 8.5×10^{25} EMU to about 8.0×10^{25} EMU. On a longer time scale of the order of 10^5 yr from archeomagnetic and paleomagnetic measurements, the dipole magnetic field has shown reversals in polarity from north to south and south to north. These time-calibrated measurements of *polarity reversals* have provided a useful method for geochronology. On a still longer time scale, the dipole axis has shown a *polar wandering* from the earliest Paleozoic to the present over an arcuate distance of the order of 90°. The actual path traced by the measurements from geologic samples over this time period from one area may differ from those in another continental area, depending on how the relative positions of these continental masses have changed throughout geologic time.

9.5 Electromagnetic Induction in the Earth

We have shown in Section 9.2 how the observed earth's field could be separated into portions of internal or external origin. The main portion of internal origin shows no fluctuations on a time scale measured in days. The fluctuating or transient portion is found to be of external origin and can be separated into portions having a period the same as the sun's rotation, solar daily magnetic variation S; a lesser field having a period the same as the moon's rotation, lunar daily magnetic variation L; and a transient field referred to as magnetic storms D. These external fields will, in turn, induce an electromagnetic field in the earth, which will, of course, contribute to the observed fluctuating magnetic field at the earth's surface. Using the same method as before, this field can be separated into portions of external and internal origin. It is found for the spherical harmonic components of the field that the ratio of the amplitudes of the external to internal parts is about 2.4 with a phase lag of about 20°. We have discussed briefly in the previous section the atmospheric dynamo origin of the S field. We should like to examine here the field induced in the earth by this external field.

The observations of the amplitude ratios of the external-to-internal induced fields and their phase difference provide information on the electrical conductivity properties of the earth. The magnitude of the internal induced field implies that the electrical conductivity of the earth at depth greatly exceeds the electrical conductivity of the surface rocks. A complete examination must, then, divide the earth into spherical shells of constant conductivity, or assume a conductivity function, such as an exponential, increasing with depth. We shall, however, restrict ourselves here to the derivation for an earth of constant conductivity.

Let us examine first the general relations for the depth dependence of the induced field in terms of the conductivity and the frequency of the inducing field. For electrical currents flowing parallel to the earth's surface, Eqs. (9.40) and (9.41) apply. In one-dimensional form for a periodic variation of the inducing field at the earth's surface, the determining equation and solution will be the same as that for the periodic flow of heat of Section 2.5. From Eqs. (2.67), (2.57), and (9.41), our solution will then be of the form

$$\Phi = I_0 e^{-z\sqrt{2\pi\mu\sigma\omega}} \sin(\omega t - z\sqrt{2\pi\mu\sigma\omega}) \qquad (9.71)$$

The depth at which the amplitude is reduced to $1/e$ of its surface value, referred to as the *skin depth* z_0, is then

$$z_0 = \frac{1}{\sqrt{2\pi\mu\sigma\omega}} = \frac{1}{2\pi}\sqrt{\frac{T}{\sigma}} \qquad (9.72)$$

where, in the last expression, we have substituted the period T, being given as usual as $T = 2\pi/\omega$, and have taken the permeability μ to be unity. If information is to be obtained on a certain region of the earth from measurements at the surface, it is apparent that the skin depth for the inducing field must be sufficiently great for the field disturbance to penetrate to the depth in question. In other words, a spectrum of variations of increasing period will provide information on the conductivity at increasing depths.

Returning to a spherical earth, we shall want to derive the induced field for a simple harmonic inducing field, such as S. Taking Φ in the form

$$\Phi = \Phi(r, \theta, \varphi)e^{i\omega t} \qquad (9.73)$$

Eq. (9.41) becomes in spherical coordinate form from Eq. (1.83)

$$\frac{1}{r^2}\frac{\partial}{\partial r}\left(r^2\frac{\partial \Phi}{\partial r}\right) + \frac{1}{r^2 \sin\theta}\frac{\partial}{\partial \theta}\left(\sin\theta\frac{\partial \Phi}{\partial \theta}\right) + \frac{1}{r^2 \sin^2\theta}\frac{\partial^2 \Phi}{\partial \varphi^2} - k^2\Phi = 0 \qquad (9.74)$$

where we have made the substitution

$$k^2 = 4\pi\mu\sigma\omega i \qquad (9.75)$$

Using the method of separation of variables, the solutions for the φ an the dependent functions are the same as those given in Section 1.11, and θd r dependent equation becomes

$$\frac{1}{R}\frac{d}{dr}\left(r^2\frac{dR}{dr}\right) - k^2 r^2 - n(n+1) = 0 \qquad (9.76)$$

Making a change of independent and dependent variables to

$$x = kr \qquad R = r^{-1/2}w$$

Eq. (9.76) reduces to

$$x^2\frac{d^2w}{dx^2} + x\frac{dw}{dx} - \left[\left(n+\frac{1}{2}\right)^2 + x^2\right]w = 0 \qquad (9.77)$$

which is simply the modified Bessel equation of Eq. (1.233), whose solution in terms of R and r is

$$R = Ar^{-1/2}I_{n+1/2}(kr) + Br^{-1/2}I_{n-1/2}(kr) \qquad (9.78)$$

In order that the solution remain finite at the origin, we take only the first solution. Our combined solution may be written for each spherical harmonic component n, m as

$$\Phi = ac_n^m \rho^{-1/2} I_{n+1/2}(k\rho a) S_n^m(\theta, \varphi) \tag{9.79}$$

where from Section 1.11, S_n^m is given by

$$S_n^m(\theta, \varphi) = P_n^m(\cos\theta) e^{\pm im\varphi} \tag{9.80}$$

where we have made the substitution

$$\rho = \frac{r}{a} \tag{9.81}$$

a being the earth radius, for convenience in scaling since the measurements are referred to the earth's surface, and where we have included the constant a outside the summation so that the coefficients c_n^m may be directly compared with the corresponding coefficients for the external field.

From Eqs. (9.40) and (1.82) we shall have for **A**

$$\mathbf{A} = -\mathbf{\theta}_1 \frac{1}{\sin\theta} \frac{\partial \Phi}{\partial \varphi} + \mathbf{\varphi}_1 \frac{\partial \Phi}{\partial \theta} \tag{9.82}$$

From Eq. (9.36), using the results of Problem 1.4(a), the components of **B** become

$$B_r = \frac{1}{r\sin\theta} \left[\frac{\partial}{\partial \theta}\left(\sin\theta \frac{\partial \Phi}{\partial \theta}\right) + \frac{1}{\sin\theta} \frac{\partial^2 \Phi}{\partial \varphi^2} \right]$$

$$= -\frac{1}{r}\left[\frac{\partial}{\partial r}\left(r^2 \frac{\partial \Phi}{\partial r}\right) - k^2 r^2 \Phi\right]$$

$$= -\frac{1}{r} n(n+1) \Phi$$

$$= -\rho^{-1} n(n+1) f_n S_n^m \tag{9.83}$$

$$B_\theta = \frac{1}{r\sin\theta}\left[-\sin\theta \frac{\partial}{\partial r}\left(r\frac{\partial \Phi}{\partial \theta}\right)\right]$$

$$= -\rho^{-1} \frac{d}{d\rho}(\rho f_n) \frac{\partial S_n^m}{\partial \theta} \tag{9.84}$$

$$B_\varphi = -\frac{1}{r}\left[\frac{\partial}{\partial r}\left(\frac{r}{\sin\theta}\frac{\partial \Phi}{\partial \varphi}\right)\right]$$

$$= -\rho^{-1} \frac{d}{d\rho}(\rho f_n) \frac{\partial S_n^m}{\sin\theta \, \partial \varphi} \tag{9.85}$$

where we have used Eqs. (9.74) and (9.76) in obtaining Eq. (9.83) and where we have made the substitution

$$f_n = c_n^m \rho^{-1/2} I_{n+1/2}(k\rho a) \tag{9.86}$$

for convenience in all three expressions.

For the nonconducting medium exterior to the earth, we have from Eq. (9.27) for the scalar potential

$$V = a(e_n^m \rho^n + i_n^m \rho^{-n-1})S_n^m(\theta, \varphi) \qquad (9.87)$$

Taking $\mu = 1$ we have from Eq. (1.82) for the components of **B** or **H**

$$B_r = -\frac{\partial V}{\partial r} = -[ne_n^m \rho^{n-1} - (n+1)i_n^m \rho^{-n-2}]S_n^m \qquad (9.88)$$

$$B_\theta = -\frac{1}{r}\frac{\partial V}{\partial \theta} = -[e_n^m \rho^{n-1} + i_n^m \rho^{-n-2}]\frac{\partial S_n^m}{\partial \theta} \qquad (9.89)$$

$$B_\varphi = -\frac{1}{r\sin\theta}\frac{\partial V}{\partial \varphi} = -[e_n^m \rho^{n-1} + i_n^m \rho^{-n-2}]\frac{\partial S_n^m}{\sin\theta\, \partial \varphi} \qquad (9.90)$$

The coefficient e_n^m is the nm spherical harmonic component for the inducing field of external origin, and the coefficient i_n^m is the corresponding component for the induced field of internal origin.

For the boundary conditions at the earth's surface, $\rho = 1$, for continuity of the normal component of **B** and the tangential components of **H**, we then have, assuming $\mu = 1$ in the earth,

$$ne_n^m - (n+1)i_n^m = n(n+1)f_n \qquad (9.91)$$

and

$$e_n^m + i_n^m = \left[\frac{d}{d\rho}(\rho f_n)\right]_{\rho=1} \qquad (9.92)$$

Substituting from Eq. (9.86) and solving for the ratio i_n^m/e_n^m, we obtain

$$\frac{i_n^m}{e_n^m} = \frac{n}{n+1}\left[1 - \frac{(2n+1)I_{n+1/2}(ka)}{ka I_{n-1/2}(ka)}\right] \qquad (9.93)$$

Since k^2 is imaginary, the modified Bessel functions of Eq. (9.93) will be complex. The amplitude ratio and phase difference between the internal and external components are, then, given simply by the modulus and argument of Eq. (9.93).

Problem 9.5(a)   Obtain the φ and θ solutions and the r dependent equation (9.76) from Eq. (9.74).

Problem 9.5(b)   Reduce Eq. (9.76) to Eq. (9.77).

Problem 9.5(c)   Solve Eqs. (9.91) and (9.92) for Eq. (9.93).

References

Chapman, S. 1951. *The Earth's Magnetism*. New York: Wiley.
———, and J. Bartels. 1940. *Geomagnetism*, vols. I and II. Oxford: Oxford University Press.
Fleming, J. A., ed. 1939. *Terrestrial Magnetism and Electricity*. New York: McGraw-Hill.
Garland, G. D. 1971. *Introduction to Geophysics*. Philadelphia: W. B. Saunders.
Grant, F. S., and G. F. West. 1965. *Interpretation Theory in Applied Geophysics*. New York: McGraw-Hill.
Harnwell, G. P. 1949. *Principles of Electricity and Electromagnetism*. New York: McGraw-Hill.
Lamb, H. 1883. "On electrical motions in a spherical conductor," *Phil. Trans. London*, vol. 174, pp. 519–549.
Jeffreys, H. 1952. *The Earth*. Cambridge: Cambridge University Press.
Nagata, T. 1953. *Rock Magnetism*. Tokyo: Maruzen.
Page, L. 1935. *Introduction to Theoretical Physics*. New York: Van Nostrand.
Rikitake, T. 1966. *Electromagnetism and the Earth's Interior*. New York: Elsevier.
Stacey, F. D. 1969. *Physics of the Earth*. New York: Wiley.
Stratton, J. A. 1941. *Electromagnetic Theory*. New York: McGraw-Hill.
Takeuchi, H. 1966. *Theory of the Earth's Interior*. Waltham: Blaisdell.
Vacquier, V., N. C. Steenland, R. G. Henderson, and I. Zeitz. 1951. "Interpretation of aeromagnetic maps," *Geol. Soc. Am.*, Memoir 47.

Part Four

Dynamics of the Earth

CHAPTER 10

EARTH MOTION, ROTATION, AND DEFORMATION

10.1 Motion under an Inverse Square Force of Attraction

We shall restate here, for review purposes, the basic relations for motion under an inverse square force law of attraction. This, of course, applies to the motion of the earth about the sun.

In polar coordinate form, the scalar equations of motion will be from Eq. (6.4)

$$m\left[\frac{d^2r}{dt^2} - r\left(\frac{d\theta}{dt}\right)^2\right] = -\gamma \frac{mM}{r^2} \tag{10.1}$$

and

$$m\left[2\frac{dr}{dt}\frac{d\theta}{dt} + r\frac{d^2\theta}{dt^2}\right] = 0 \tag{10.2}$$

where M is the mass of the sun and m the mass of the earth, where r is the radius vector measured from M to m and θ is the polar angle measured in the plane of orbit of m, and where we have also assumed here at the outset that the mass of the sun is so much greater than the mass of the earth that the motion of the former under the attraction of the latter can be neglected.

To find the equation of the path, it is necessary to eliminate t from Eqs. (10.1) and (10.2), obtaining

$$\frac{1}{r} = \frac{\gamma M}{h^2} + A\cos\theta \tag{10.3}$$

where h and A are constants of integration, h being given by

$$r^2 \frac{d\theta}{dt} = h \tag{10.4}$$

which is simply the integral of Eq. (10.2). Equation (10.3) is the equation in polar coordinate form of a conic section; in the case of the earth, the motion is that of an ellipse, which is referred to as Kepler's first law of planetary motion. Equation (10.4) simply states that the radius vector describes equal areas in equal times, which is referred to as Kepler's second law of planetary motion. From Eq. (10.4) we can also obtain the period P of the elliptical motion in terms of its semimajor axis a as

$$P = \frac{2\pi a^{3/2}}{\sqrt{\gamma M}} \tag{10.5}$$

which is referred to as Kepler's third law of planetary motion.

To return to the effect of the motion of the sun due to the attraction of the earth, we find that the same equations of motion (10.1) and (10.2) for the motion of the earth relative to the sun would apply if m of the left-hand side of these equations were replaced by μ, defined by

$$\frac{1}{\mu} = \frac{1}{m} + \frac{1}{M} \tag{10.6}$$

or what is the same by changing γ to $\gamma m/\mu$ wherever it appears in our results, Kepler's third law then becoming

$$P = \frac{2\pi a^{3/2}}{\sqrt{\gamma(M+m)}} \tag{10.7}$$

Problem 10.1(a) Derive the solution [Eq. (10.3)] from Eqs. (10.1) and (10.2).

Problem 10.1(b) Derive Kepler's third law of planetary motion [Eq. (10.5)].

Problem 10.1(c) Derive the result expressed by Eq. (10.6).

10.2 Precession of the Equinoxes

Some of the more interesting problems in geophysics have to do with the motion, rotation, and deformation of the earth as a whole. Most of the solutions, as given in their basic, simplified form here, have been a long-standing part of the theoretical literature in classical physics and geophysics. They provide a thoughtful application of the principles of dynamics to earth problems.

Let us consider, for the moment, the *celestial sphere*. This is a spherical shell of arbitrarily large radius with the earth as center on which celestial objects appear projected. The path described by the sun over this sphere is referred to as the *ecliptic*, and the two points of intersection of the ecliptic with the projection of the earth's equator on this sphere are known as the *equinoxes*.

We shall consider the earth here as a rigid body, and we shall be interested in the effect on the earth's motion caused by the torque of the sun, or the moon, created by the unequal mass distribution related to the earth's equatorial bulge. We must then return to Euler's equation (6.19) for the motion of a rigid body.

If a rotating body possesses symmetry of revolution about an axis of spin, such as the earth, we may use, in place of axes attached to the body, a set of axes with origin at the center of mass and the Z axis parallel to the axis of spin, but with X and Y axes that rotate with an angular velocity different from that of the spinning body. If $\boldsymbol{\omega}$ is the angular velocity of the axes XYZ and s the angular velocity of spin, the total angular velocity of the body will be $\boldsymbol{\omega}+\mathbf{k}s$, and Eq. (6.14) will become

$$\mathbf{H} = \Phi \cdot (\boldsymbol{\omega}+\mathbf{k}s) = \mathbf{i}B\omega_x + \mathbf{j}B\omega_y + \mathbf{k}C(\omega_z+s) \tag{10.8}$$

since the products of inertia vanish on account of symmetry, and the two moments of inertia A and B are equal. Now, since the moments of inertia about all axes perpendicular to the axis of symmetry are the same, B does not change with time even though it is not referred to an axis attached to the body, and Euler's equation (6.19) becomes

$$\mathbf{L} = \mathbf{i}[B\dot{\omega}_x + (C-B)\omega_y\omega_z + Cs\omega_y]$$
$$+ \mathbf{j}[B\dot{\omega}_y + (B-C)\omega_z\omega_x - Cs\omega_x] + \mathbf{k}C(\dot{\omega}_z+\dot{s}) \tag{10.9}$$

Let us now refer to Fig. 10.1 where the plane X_0Y_0 is the plane of the ecliptic, and XY is the earth's equatorial plane. The axes $X_0Y_0Z_0$ are the inertial axis system, and the axes XYZ are those attached to the earth in the manner proscribed by Eq. (10.9). We have taken the X axis to be the line of intersection OX of the plane through O perpendicular to the Z axis with the fixed X_0Y_0 plane, that is, the line defining the equinoxes. We shall denote by θ the angle between the axis OZ and the axis OZ_0, that is, the angle between the plane of the ecliptic and the equatorial plane. We shall also denote by ψ the precession angle that the intersection line OX makes with the fixed axis OX_0. The angular rate $d\psi/dt$ will, then, be the rate of precession of the equinoxes, which is the quantity we wish to find.

From Fig. 10.1 we have for $\boldsymbol{\omega}$ in terms of θ and ψ,

$$\omega_x = \frac{d\theta}{dt} \quad \omega_y = \frac{d\psi}{dt}\sin\theta \quad \omega_z = \frac{d\psi}{dt}\cos\theta \tag{10.10}$$

If in Eq. (10.9) there is no component of the torque $\mathbf{L}$ about the Z axis, which will turn out to be the case in our example, we shall have, then, in scalar form

$$B\dot{\omega}_x + (C-B)\omega_y\omega_z + Cs\omega_y = L_x$$
$$B\dot{\omega}_y + (B-C)\omega_z\omega_x - Cs\omega_x = L_y \tag{10.11}$$
$$C(\dot{\omega}_z+\dot{s}) = 0$$

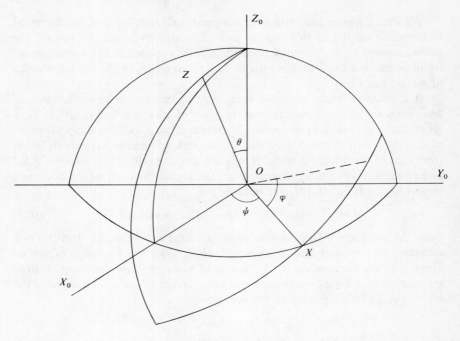

Fig. 10.1

The last equation shows that the angular velocity $\omega_z + s$ of spin about the Z axis remains constant during the motion. Putting Ω for this constant angular velocity and eliminating s from the first two equations, we get

$$B\dot{\omega}_x - B\omega_y\omega_z + C\Omega\omega_y = L_x$$
$$B\dot{\omega}_y + B\omega_z\omega_x - C\Omega\omega_x = L_y \qquad (10.12)$$

Equations (10.12) in conjunction with Eqs. (10.10) are the equations that we wish to solve. It should be clear that if the earth were a sphere of constant density or a function of radial distance only, there would be no torque and no precession.

To calculate the torque due to the sun, let S in Fig. 10.2 be the center of the sun and E the center of the earth. We shall consider the earth to be rigid and the path of the sun to be circular. Take origin at the center of the earth and orient the axes to conform with Fig. 10.1 so that the Z axis is parallel to the axis of rotation of the earth and $\mathbf{R}$ lies in the YZ plane. Let P be a point in the interior of the earth with coordinates x, y, z relative to E. The position vector of P is then

$$\boldsymbol{\rho} = \mathbf{i}x + \mathbf{j}y + \mathbf{k}z \qquad (10.13)$$

and the force with which the sun attracts a mass dm situated at P is

$$dF = -\gamma \frac{Mdm}{r^3} [\mathbf{i}x + \mathbf{j}(R\sin\alpha + y) + \mathbf{k}(-R\cos\alpha + z)] \quad (10.14)$$

The torque $\mathbf{L}$ about E due to this force is $\boldsymbol{\rho} \times d\mathbf{F}$. However, it is clear from symmetry about the XY plane that the resultant torque on the entire earth is about the X axis. Therefore, we need calculate only the X component of $\mathbf{L}$, that is

$$\begin{aligned} dL_x &= (\boldsymbol{\rho} \times d\mathbf{F})_x \\ &= -\gamma \frac{Mdm}{r^3} [y(-R\cos\alpha + z) - z(R\sin\alpha + y)] \\ &= \gamma \frac{MR}{r^3} [y\cos\alpha + z\sin\alpha]\, dm \end{aligned} \quad (10.15)$$

Now,

$$\begin{aligned} r^2 &= R^2 + 2\mathbf{R}\cdot\boldsymbol{\rho} + \rho^2 \\ &= R^2 + 2R(y\sin\alpha - z\cos\alpha) + \rho^2 \end{aligned}$$

and

$$\frac{1}{r^3} = \frac{1}{R^3}\left[1 - \frac{3}{R}(y\sin\alpha - z\cos\alpha) + \cdots\right] \quad (10.16)$$

to first-order terms in ρ. Substituting Eq. (10.16) into Eq. (10.15), we have

$$L_x = \gamma \frac{M}{R^2} \int \left[1 - \frac{3}{R}(y\sin\alpha - z\cos\alpha)\right][y\cos\alpha + z\sin\alpha]\, dm$$

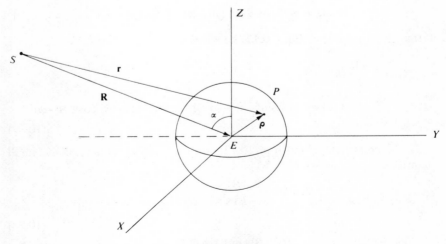

Fig. 10.2

Dynamics of the Earth

On account of symmetry, integrals containing the first power of y and z or the product yz vanish, giving

$$L_x = -3\frac{\gamma M}{R^3}\sin\alpha\cos\alpha\left[\int y^2\,dm - \int z^2\,dm\right]$$

$$= -3\frac{\gamma M}{R^3}\sin\alpha\cos\alpha\left[\int(x^2+y^2)\,dm - \int(z^2+x^2)\,dm\right]$$

The first of these integrals is the moment of inertia C about the Z axis and the second the moment of inertia B about the Y axis so that, in vector form, we have

$$\mathbf{L} = 3\frac{\gamma M}{R^5}(C-B)\mathbf{R}\cdot\mathbf{k}\,\mathbf{R}\times\mathbf{k} \tag{10.17}$$

Now let us return to Fig. 10.1, if p is the angular velocity of the earth's revolution about the sun, we have

$$\mathbf{R} = R(\mathbf{i}_0\cos pt + \mathbf{j}_0\sin pt) \tag{10.18}$$

$\mathbf{i}_0$ and $\mathbf{j}_0$ being unit vectors along the fixed X_0 and Y_0 axes. In terms of the unit vectors $\mathbf{i}$, $\mathbf{j}$, $\mathbf{k}$ along the moving axes X, Y, Z, we have the following vector relations

$$\mathbf{i}_0 = \mathbf{i}\cos\psi - \mathbf{j}\cos\theta\sin\psi + \mathbf{k}\sin\theta\sin\psi$$

$$\mathbf{j}_0 = \mathbf{i}\sin\psi + \mathbf{j}\cos\theta\cos\psi - \mathbf{k}\sin\theta\cos\psi$$

and

$$\mathbf{R}\cdot\mathbf{k} = -R\sin\theta\sin(pt-\psi)$$

$$\mathbf{R}\times\mathbf{k} = R[\mathbf{i}\cos\theta\sin(pt-\psi) - \mathbf{j}\cos(pt-\psi)]$$

so that the torque $\mathbf{L}$ of Eq. (10.17) becomes

$$\mathbf{L} = 3\frac{\gamma M}{R^3}(C-B)$$

$$[-\mathbf{i}\sin\theta\cos\theta\sin^2(pt-\psi) + \mathbf{j}\sin\theta\sin(pt-\psi)\cos(pt-\psi)] \tag{10.19}$$

In the equations of motion (10.12), we may omit products of the small quantities ω_x, ω_y, ω_z, giving

$$B\dot\omega_x + C\Omega\omega_y = -\frac{3}{2}\frac{\gamma M}{R^3}(C-B)\sin\theta\cos\theta\,[1-\cos 2(pt-\psi)]$$

$$B\dot\omega_y - C\Omega\omega_x = \frac{3}{2}\frac{\gamma M}{R^3}(C-B)\sin\theta\sin 2(pt-\psi) \tag{10.20}$$

Earth Motion, Rotation, and Deformation ¶ 335

These equations show that ω_x and ω_y have the angular frequency $2p$. Therefore, $\dot{\omega}_x$ and $\dot{\omega}_y$ are of the order of magnitude of $2p\omega_x$ and $2p\omega_y$, which are negligible as compared with $\Omega\omega_x$ and $\Omega\omega_y$. So neglecting the terms involving $\dot{\omega}_x$ and $\dot{\omega}_y$ in Eqs. (10.20), we have

$$\frac{d\theta}{dt} = \omega_x = -\frac{3}{2}\frac{\gamma M}{R^3\Omega}\frac{C-B}{C}\sin\theta\sin 2(pt-\psi) \tag{10.21}$$

and

$$\frac{d\psi}{dt} = \frac{\omega_y}{\sin\theta} = -\frac{3}{2}\frac{\gamma M}{R^3\Omega}\frac{C-B}{C}\cos\theta[1-\cos 2(pt-\psi)] \tag{10.22}$$

The first equation represents the nutation, and the second, the precession of the earth's rotational axis about the normal to the plane of the ecliptic, the mean precession being given by the first term on the right-hand side of Eq. (10.22). The negative sign indicates that the precession has the opposite sense to the velocity Ω of the earth's rotation, that is, the equinoxes precess from east to west. The precession is very small, having a period of 25,800 yr, and the greater part of it is due to the moon whose proximity more than compensates for its smaller mass. The development of the equations for the lunar precession is exactly the same as that given here for the solar precession, and the two rates will add vectorially or to a first approximation, since the plane of the lunar orbit is close to the plane of ecliptic, add algebraically. From the observed value of the precession, the precessional constant $H = (C-B)/C$, referred to in Section 8.8, can be calculated.

Problem 10.2(a) Obtain the expression for the torque, using MacCullagh's formula.

10.3 Gyroscopic Compass

The gyroscopic compass consists of a rapidly rotating disk D, as shown in Fig. 10.3, suspended at its center of mass. The axis of the disk is kept horizontal by means of a pendulum bob B or other suitable counterbalance attached to the axle ab on which the disk is turning. Take the Z axis along the axis of rotation of the disk and denote the velocity of spin by s. As the disk is symmetrical about the axis, it is not necessary for the X and Y axes to rotate with the disk in order that the equal moments of inertia A and B about these axes may remain constant. Therefore, we shall take the Y axis vertical and the X axis in the horizontal plane perpendicular to Z and use Eq. (10.9) in our investigation of the motion. Denote by Ω the angular velocity of the earth's rotation, by γ the latitude, and by φ the angle measured positive to the west that Z makes with the meridian. The vertical component

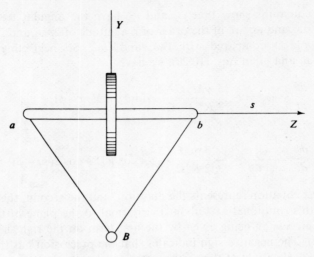

Fig. 10.3

of the angular velocity of the earth is $\Omega \sin \gamma$ and the northward horizontal component $\Omega \cos \gamma$. Therefore, the components of angular velocity of the axes XYZ resolved along these axes are

$$\begin{aligned} \omega_x &= -\Omega \cos \gamma \sin \varphi \\ \omega_y &= \Omega \sin \gamma + \dot{\varphi} \\ \omega_z &= \Omega \cos \gamma \cos \varphi \end{aligned} \qquad (10.23)$$

We are concerned only with rotation about the Y axis. As there is no applied torque about this axis, Eq. (10.9) gives with Eqs. (10.23)

$$A\ddot{\varphi} - (A-C)\Omega^2 \cos^2 \gamma \sin \varphi \cos \varphi + Cs\Omega \cos \gamma \sin \varphi = 0$$

As Ω is very small compared to s, we may neglect the term in Ω^2, giving

$$\ddot{\varphi} + \left(\frac{C}{A} s\Omega \cos \gamma\right) \sin \varphi = 0 \qquad (10.24)$$

This equation has the same form as that for the motion of a pendulum, or for φ small that of simple harmonic motion in general. Therefore, our solution states that the axis of spin oscillates about the meridian, the effective torque of restitution being maximum at the equator and falling off with the cosine of the latitude to zero at the poles. The period for small amplitudes is simply

$$P = 2\pi \sqrt{\frac{A}{Cs\Omega \cos \gamma}} \qquad (10.25)$$

which is greater the higher the latitude and less the more rapid the spin. If there is some damping, the axis of the gyroscope ultimately comes to rest pointing toward the true geographical north.

10.4 Eulerian Free Motion

We have discussed in the second section of this chapter precession of the earth's rotational axis around the normal to the plane of the ecliptic. We shall be interested here in local variation of the instantaneous pole of rotation about a mean position of the rotation axis; such variation is referred to as *wobble*. It is clear that the earth is in stable rotation about its axis of greatest moment of inertia C, where the moments of inertia about the other two principal axes are equal and smaller than C. It should also be apparent that if the earth were a sphere of constant density, there would be no preferred orientation for the rotation axis.

Similar to our previous discussions of free and forced oscillations for a simple harmonic motion system under a restoring force or torque, we shall be interested here in determining the period of the free oscillations of the earth about its mean axis of rotation. The governing equation for a rigid earth is Euler's equation, (6.19), where we set the torque **L** equal to zero and where $A = B$, or

$$B\frac{d\omega_x}{dt} + (C-B)\omega_y\omega_z = 0$$

$$B\frac{d\omega_y}{dt} - (C-B)\omega_z\omega_x = 0 \qquad (10.26)$$

$$C\frac{d\omega_z}{dt} = 0$$

From the third of Eqs. (10.26), we have immediately

$$\omega_z = \text{constant} \qquad (10.27)$$

Substituting

$$n = \frac{C-B}{B}\omega_z \qquad (10.28)$$

into the first two equations, we have

$$\frac{d\omega_x}{dt} + n\omega_y = 0$$

$$\frac{d\omega_y}{dt} - n\omega_x = 0 \qquad (10.29)$$

338 ¶ Dynamics of the Earth

Differentiating the first and using the second of Eqs. (10.29) to eliminate ω_y, we obtain

$$\frac{d^2\omega_x}{dt^2} + n^2\omega_x = 0$$

and, similarly, (10.30)

$$\frac{d^2\omega_y}{dt^2} + n^2\omega_y = 0$$

If we choose

$$\omega_x = a \cos nt \qquad (10.31)$$

for the solution of the first of Eqs. (10.30), the solution of the second will be

$$\omega_y = a \sin nt \qquad (10.32)$$

from Eqs. (10.29). The period of this *Eulerian free oscillation* or *nutation* will be simply

$$\tau_0 = \frac{2\pi}{n} = \frac{B}{C-B} \text{ days} \qquad (10.33)$$

where $\omega_z = 2\pi$ as measured in days. From the measurement of the precessional constant H, the value of τ_0 is known and is about 305 days.

A consequence of such a wobble in the earth's rotational axis will be a variation in the latitude for any point on the earth's surface. The colatitude,

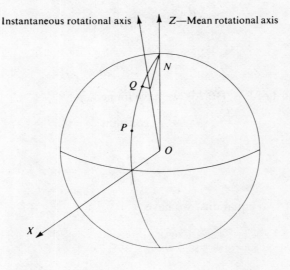

Fig. 10.4

as determined from celestial observations, is the angle between the instantaneous axis of rotation and the observing point. Consider the point P in Fig. 10.4 with origin at the center of the earth and Z axis along the mean axis of rotation. The angular velocity of the earth will be

$$\boldsymbol{\omega} = \mathbf{i}\omega_x + \mathbf{j}\omega_y + \mathbf{k}\omega_z \tag{10.34}$$

so that the direction numbers of the instantaneous axis of rotation will be ω_x, ω_y, ω_z. Since ω_z is much larger than ω_x and ω_y, the angle NQ may be approximated by its tangent, or from Eqs. (10.31) and (10.34),

$$NQ = \frac{a}{\omega_z} \cos nt \tag{10.35}$$

The colatitude of the point P will then be

$$PQ = PN - \frac{a}{\omega_z} \cos nt \tag{10.36}$$

The measured latitude should then show a variation with the same period τ_0.

In addition to this free oscillation, there can be forced oscillations. A forced oscillation, which one might expect, would be that due to periodic changes in the moments of inertia related to seasonal changes in the weather or other similar effects with a yearly cycle. In actuality, the empirical formula for latitude variation deduced from celestial observations has two periodic terms, one with a period of a year and the other with a period of about 430 days. This latter variation, named after its discoverer, is referred to as the *Chandler wobble*. We shall see in a later section that the Chandler wobble can be ascribed to a free oscillation of an elastic earth as contrasted with the 305-day period deduced here for a rigid earth.

10.5 Marine Tides

We have examined in Section 5.9 the equilibrium theory of the tides and, in Section 5.10, have included the motion of the ocean itself in the dynamical theory of the tides. We should like to examine here the tides, or tidal components, related to the variation of the moon motion with respect to the earth considered as a rigid sphere covered with water. We shall restrict ourselves to consideration of the tide-raising potential function.

From Eq. (5.151) we have for the tide-raising potential due to the moon, using the direction notation of Section 8.1 rather than that of Section 5.9,

$$W_2 = \frac{1}{2} \frac{\gamma M a^2}{R^3} (3 \cos^2 \theta - 1) \tag{10.37}$$

340 ¶ *Dynamics of the Earth*

and from Eq. (5.153) for the theoretical height ξ of the equilibrium tide for a rigid earth,

$$\xi = \frac{W_2}{g} = \frac{1}{2} \frac{\gamma M a^2}{g R^3} (3\cos^2\theta - 1)$$
$$= \frac{1}{2} \frac{M}{E} \left(\frac{a}{R}\right)^3 a(3\cos^2\theta - 1) \qquad (10.38)$$

where M is the mass of the moon, E the mass of the earth, a the radius of the earth, R the distance from the center of the earth to the center of the moon, and θ the local angle as shown in Fig. 5.9. We further note that the parenthesis of Eq. (10.37) is of the form of the second-degree Legendre

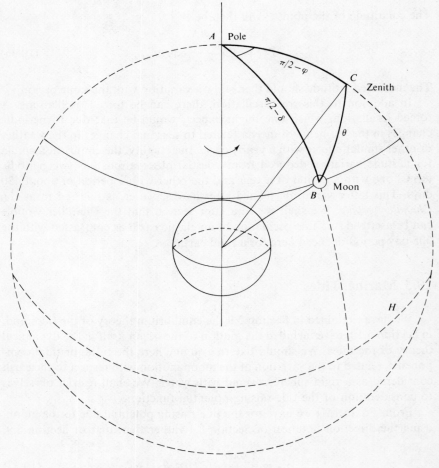

Fig. 10.5

polynomial of Problem 1.6(a), so that we may alternatively write for Eq. (10.37)

$$W_2 = \frac{\gamma M a^2}{R^3} P_2 \qquad (10.39)$$

That the tide-raising potential can be written in this form in terms of Legendre polynomials greatly simplifies some of the later considerations we shall make with regard to the effects of the tide-raising potential on the earth as a whole.

The local coordinate angle θ is inconvenient for consideration of tidal effects with respect to any point on the earth's surface. Further, the function P_2 is asymmetrical with respect to θ. We should rather have an expression for W_2 in terms of the latitude φ of any point on the earth's surface, the moon's declination angle δ, equivalent to latitude, and the local hour angle H of the moon, equal to the difference in longitude of the observing point and the moon. Consider the spherical triangle of Fig. 10.5 formed on the celestial sphere by the pole of rotation of the earth, the zenith of the observing point, and the moon. The moon will trace the indicated circular path with respect to the earth, considered stationary, as shown, and the sides and angles of the spherical triangle will also be as shown. From the relations between the sides a, b, c and angles A, B, C of a spherical triangle, we have the formula

$$\cos a = \cos b \cos c + \sin b \sin c \cos A$$

which, in terms of our triangle, will be

$$\cos \theta = \sin \varphi \sin \delta + \cos \varphi \cos \delta \cos H \qquad (10.40)$$

For the expression in the parenthesis of Eq. (10.37), we shall then have

$$\begin{aligned}
3\cos^2 \theta - 1 &= 3\sin^2 \varphi \sin^2 \delta + 3\cos^2 \varphi \cos^2 \delta \cos^2 H \\
&\quad + 6 \sin \varphi \cos \varphi \sin \delta \cos \delta \cos H - 1 \\
&= 3\sin^2 \varphi \sin^2 \delta - 1 + \tfrac{3}{2} \cos^2 \varphi \cos^2 \delta [1+\cos 2H] \\
&\quad + \tfrac{3}{2} \sin 2\varphi \sin 2\delta \cos H \\
&= \tfrac{3}{2} \cos^2 \varphi \cos^2 \delta \cos 2H + \tfrac{3}{2} \sin 2\varphi \sin 2\delta \cos H \\
&\quad + 3\sin^2 \varphi \sin^2 \delta - 1 + \tfrac{3}{2}(1-\sin^2 \varphi)(1-\sin^2 \delta) \\
&= \tfrac{3}{2} \cos^2 \varphi \cos^2 \delta \cos 2H + \tfrac{3}{2} \sin 2\varphi \sin 2\delta \cos H \\
&\quad + \tfrac{3}{2}(3\sin^2 \varphi \sin^2 \delta - \sin^2 \varphi - \sin^2 \delta + \tfrac{1}{3})
\end{aligned}$$

or for W_2

$$W_2 = \frac{3}{4} \frac{\gamma M a^2}{R^3}$$
$$[\cos^2 \varphi \cos^2 \delta \cos 2H + \sin 2\varphi \sin 2\delta \cos H$$
$$+ 3(\sin^2 \varphi - \tfrac{1}{3})(\sin^2 \delta - \tfrac{1}{3})] \qquad (10.41)$$

342 ¶ Dynamics of the Earth

The potential W_2 has now been divided into three terms, which we may interpret directly. The first function has nodal lines located at 45° on either side of the meridian beneath the moon. These lines divide the earth sphere into four sections, where the function is alternatively positive and negative, corresponding on the equilibrium theory to regions of high tide, $\xi > 0$, and low tide, $\xi < 0$. The period of this tide is semidiurnal, and its amplitude is maximum at the equator when the moon's declination is zero and zero at the poles. The second function has a nodal line at 90° from the meridian beneath the moon and a nodal line along the equator. The four regions into which it divides the earth sphere also change sign with the declination of the moon. It is always zero at the equator and at the poles. The third function is dependent only on latitude and declination of the moon. Its nodal lines are the parallels 35°N and 35°S. Since it is also only a function of the square of the sine of the declination, its period will be 14 days in the case of the moon and 6 months in the case of the sun. This tide is referred to as the long-period component of the tides. The relation of these various tides is shown diagrammatically in Fig. 10.6.

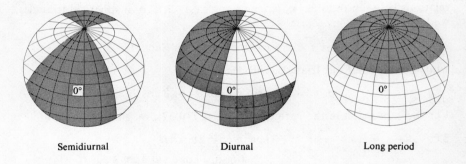

Semidiurnal   Diurnal   Long period

Fig. 10.6 (from Melchior, 1966, p. 24)

To carry an investigation of the marine tides further, we would, of course, have to include the solar tide-raising potential to the lunar potential. Since the plane of the moon's orbit is nearly the same as that of the plane of the ecliptic, the moon and sun will be in relative positions to give a reinforcing potential twice during a lunar month, or a period of approximately 14 days. The high tides during this 14-day period are known as *spring tides* and the low tides as *neap tides*. To carry the investigation still further, we would have to take account of the actual moon orbit and its variability and the ellipticity of both the lunar and solar orbits among other items. The net effect is to divide the three tidal components of Eq. (10.41) into various subcomponents related to these lunar and solar effects.

Problem 10.5(a) Graph the excitation function of Eq. (10.41) as a function of φ and discuss the tidal motion that might be observed at various locations on the earth.

10.6 Earth Tides

In all our previous discussions of tidal effects, we have implicitly assumed that the earth itself was rigid. This, of course, is clearly not the case. As the moon and the sun move relative to the earth, their gravitational force will deform the earth, considering the earth now as an elastic body. This deformation of the earth as a whole will cause, among other things, a change in the theoretical amplitude of the equilibrium marine tide, a variation in the intensity of gravity measured at the earth's surface, and a change in the vertical. This deformation, as one might expect, is referred to as the *earth tide*.

For a study of the earth tides, we may restrict ourselves to the equilibrium theory. There are no ocean currents or hydrodynamic wave motion. The displacements are small, and equilibrium is quickly established at the rate of the seismic propagation velocities. The tidal period for the moon and sun motions around the earth are long compared with the seismic periods for establishing equilibrium. We may treat the problem as a static rather than a dynamic problem.

We shall now want to use the fact that the tide-raising potential of Eq. (10.37) or (10.39) is of the form of the second-degree Legendre polynomial to give us a substantial simplification at the outset. Considering the earth as an elastic sphere, the governing conditions are Eqs. (6.34), where we set the left-hand side of the equations equal to zero. As before in Section 6.2, we may separate the displacement vector into an irrotational and a solenoidal part, obtaining equations similar to Eqs. (6.36) and (6.38), where the left-hand side is again set equal to zero. If we assume that the earth is spherical and that the density and elastic constants are a function of the radial distance only, then we shall be interested only in the portion of the solution for **s** for which $\nabla \times \mathbf{s} = 0$, for a rotation alone will produce no change in the shape of a spherical earth or the density distribution within it. For $\nabla \times \mathbf{s} = 0$, **s** may be expressed as the gradient of a scalar displacement potential, and Eq. (6.36) reduces to Laplace's equation. Under these conditions, the boundary constraint of an outer spherical surface will lead to a solution of Laplace's equation in terms of the spherical harmonic functions of Section 1.11, similar in principle to the solution in terms of trigonometric functions for the problem of a vibrating string constrained at both ends. Finally, since the driving potential W_2 is of the form of the second-degree Legendre polynomial of this solution and remembering the orthogonal relations among Legendre polynomials, the final solution will contain only $P_2(\theta)$ for the polar and azimuthal angle dependent term, similar again to the solution for the vibrating string problem driven by a sinusoidal force of the same form as one of the series expansion terms in the general solution. Then the solution for the displacement potential will be given by some function of the radial distance r multiplied by $P_2(\theta)$.

From this form for the displacement potential, we may, then, write for the radial displacement ζ at any point in the earth, using Eq. (10.39),

$$\zeta(r,\theta) = h(r)\frac{W_2(r,\theta)}{g} \qquad (10.42)$$

or for the displacement at the earth's outer surface, $r = a$, simply

$$\zeta = h\frac{W_2}{g} \qquad (10.43)$$

where $h = h(a)$ is a constant. Similarly, we may write for the horizontal displacements u_φ and u_λ at the earth's outer surface, l being a constant,

$$u_\varphi = \frac{l}{g}\frac{\partial W_2}{\partial \theta}$$

$$u_\lambda = \frac{l}{g \cos \varphi}\frac{\partial W_2}{\partial \lambda} \qquad (10.44)$$

where φ and λ are latitude and longitude, remembering also that θ is the polar angle measured from the line between the center of the earth and the center of the moon. Further, the dilatation Δ, which is here simply the Laplacian of the displacement potential, may also be expressed directly in terms of W_2, so that we have for Δ at any point in the earth

$$\Delta(r, \theta) = f(r)\frac{W_2(r, \theta)}{g} \qquad (10.45)$$

Following the arguments at the beginning of Section 8.8, we may obtain one final relation. We see that whatever these radial functions $h(r)$ and $f(r)$, relating ζ and Δ to W_2, may be, the gravitational potential due to the deformation given in terms of the variation in density accompanying the dilatation and the surface displacement can be expressed in the form

$$U = k(r)W_2$$

or again at the earth's outer surface for the total potential

$$U = kW_2 \qquad (10.46)$$

where $k = k(a)$ is a constant.

The coefficients h and k are referred to as *Love's numbers* and the coefficient l as *Shida's number*. From Eq. (10.38) we see that h is the ratio of the height of the earth tide to the height of the equilibrium marine tide. The coefficient k is the ratio of the additional potential produced by the deformation to the deforming potential, and the coefficient l may be thought of as the ratio of the horizontal displacement of the crust to that of the equilibrium marine tide.

Following the definitions of Love's numbers, the total disturbing potential is now obviously the original tide-producing potential plus the potential of the deformed earth, or $(1+k)W_2$, and the static marine tide will have an effective height of $(1+k)W_2/g$ rather than simply W_2/g as given by Eq. (10.38). However, the tide observed with tide gauges attached to the earth is the difference between the static marine tide and that of the crust, namely, the earth tide hW_2/g. The amplitude observed for the equilibrium marine tide will, therefore, be

$$\xi = (1+k)\frac{W_2}{g} - h\frac{W_2}{g} = (1+k-h)\frac{W_2}{g} \tag{10.47}$$

where
$$\Gamma = 1+k-h \tag{10.48}$$

is a quantity that can be determined from observations of the tides. The quantity Γ, of course, is simply the ratio of the observed equilibrium marine tide on an elastic earth to the theoretical equilibrium marine tide on a rigid earth.

We can obtain another combination of Love's numbers related to other observable phenomena as follows. If V_0 is the initial potential of the earth, the potential of an earth deformed by the tide-producing potential W_2 will be

$$V_1 = V_0 + W_2 + U + \zeta\frac{\partial V_0}{\partial r} \tag{10.49}$$

where W_2 is the tide-producing potential of the moon or sun; U is the additional gravitational potential due to the deformation; and the final term is simply the elevation, or free air, correction of Section 8.7. The change in gravity related to this change in potential will then be simply, to first-order terms,

$$\Delta g = g_1 - g_0 = -\frac{\partial W_2}{\partial r} - \frac{\partial U}{\partial r} - \zeta\frac{\partial^2 V_0}{\partial r^2} \tag{10.50}$$

From Eq. (8.72) we have for the zero-order term

$$\frac{\partial^2 V_0}{\partial r^2} = -\frac{\partial g}{\partial r} = \frac{2g}{r}$$

or, using Eq. (10.43),

$$\zeta\frac{\partial^2 V_0}{\partial r^2} = 2h\frac{W_2}{r} \tag{10.51}$$

From Eq. (10.39) we have, since a in that expression corresponds to r in Eq. (10.50),

$$\frac{\partial W_2}{\partial r} = 2\frac{W_2}{r} \tag{10.52}$$

346 ¶ Dynamics of the Earth

Referring to Section 8.8 and, in particular, to expression (8.97), we see that the zero-order r dependence for U will be r^{-3} so that we shall have, using Eq. (10.46),

$$\frac{\partial U}{\partial r} = -3\frac{U}{r} = -3k\frac{W_2}{r} \tag{10.53}$$

The difference between the derivatives of Eqs. (10.52) and (10.53) is simply that for a second-order spherical harmonic function of external and internal origin, respectively. Substituting Eqs. (10.51), (10.52), and (10.53) into Eq. (10.50), we have

$$\Delta g = -(1 - \tfrac{3}{2}k + h)\, 2\frac{W_2}{r} \tag{10.54}$$

where

$$\delta = 1 - \tfrac{3}{2}k + h \tag{10.55}$$

is a quantity that can be determined directly from observations of the gravity anomalies produced by the tide-producing potential of the moon or sun. The quantity δ is simply the ratio of the observed gravity anomaly on an elastic earth to the theoretical anomaly on a rigid earth.

We can obtain still another combination of Love's and Shida's numbers related to observable phenomena. If we have a disturbing potential W_2, the horizontal components of gravity related to it will be at the earth's surface

$$g_\varphi = -\frac{1}{a}\frac{\partial W_2}{\partial \varphi}$$

$$g_\lambda = -\frac{1}{a\cos\varphi}\frac{\partial W_2}{\partial \lambda}$$

and the deflection of the vertical related to it, to zero-order terms,

$$\begin{aligned} n_\varphi = \tan n_\varphi &= -\frac{1}{ag}\frac{\partial W_2}{\partial \varphi} \\ n_\lambda = \tan n_\lambda &= -\frac{1}{ag\cos\varphi}\frac{\partial W_2}{\partial \lambda} \end{aligned} \tag{10.56}$$

As before, the total disturbing potential for an elastic earth is not W_2 but $(1+k)W_2$. As before, with the consideration of the measurement base for the marine tides, we have a horizontal deformation of the earth's crust producing a deflection given by Eqs. (10.44) to zero-order terms

$$\begin{aligned} m_\varphi &= \frac{l}{ag}\frac{\partial W_2}{\partial \varphi} \\ m_\lambda &= \frac{l}{ag\cos\varphi}\frac{\partial W_2}{\partial \lambda} \end{aligned} \tag{10.57}$$

The total measured deflection of the vertical with reference to the mean position of the earth's axis of rotation will then be simply the difference of these two expressions

$$p_\varphi = -(1+k-l)\frac{1}{ag}\frac{\partial W_2}{\partial \varphi}$$

$$p_\lambda = -(1+k-l)\frac{1}{ag\cos\varphi}\frac{\partial W_2}{\partial \lambda} \qquad (10.58)$$

where

$$\Lambda = 1+k-l \qquad (10.59)$$

is again a quantity that can be determined directly from astronomical observations of the deflection of the vertical determined by the tide-producing potential of the moon or sun. As before, the quantity Λ is simply the ratio of the observed deflection of the vertical on an elastic earth to the theoretical deflection of the vertical on a rigid earth.

Observations of Γ, δ, and Λ will then provide values for h, k, and l. In addition, we can obtain a separate estimate of k, as will be discussed in the next section, from the Chandler wobble period. Characteristic values that result from these various determinations are $h = 0.58$, $k = 0.29$, and $l = 0.05$.

Problem 10.6(a) Calculate the periodic variation of gravity due to the moon motion on a rigid earth.

Ans. 0.2 mg.

Problem 10.6(b) Explain how the water level in a well may be affected by earth tides.

Problem 10.6(c) With reference to Section 8.8, derive an expression for k in terms of $h(r)$ and $f(r)$.

Ans.

$$k = \frac{3}{5\rho a^6}\int_0^a \rho_0 \left\{\frac{d}{dr}[r^6 h(r)] - r^6 f(r)\right\} dr$$

where ρ is the average density of the earth, and $\rho_0 = \rho_0(r)$ is the density as a function of depth in the undisturbed state.

10.7 Chandler Wobble

In Section 10.4 we determined the period of the free nutation of the earth, considering it as a rigid body. During the nutation, the instantaneous axis of rotation will, of course, differ from the mean axis of rotation. This will cause a variation in the distribution of the centrifugal force of rotation throughout for the earth. For a rigid earth, such a change in centrifugal force will not create any change in the mass distribution within the earth and, consequently, no change in the period of free nutation. For an elastic earth, this will not be the case. The centrifugal force will be decreased in the equatorial plane and increased in all other directions, producing therefore a slight decrease in the earth's ellipticity, a decrease in the difference $C-B$ of the moments of inertia, and an increase in the period τ_0 of Eq. (10.33) of the free nutation. Through this mechanism, we shall therefore look for an explanation for the increase in the period of free nutation for the earth to explain the observed Chandler wobble period.

We could proceed by considering an instantaneous axis of rotation displaced from the mean axis with direction cosines l, m as measured with relation to X and Y axes in the equatorial plane of the mean axis. The driving potential for the elastic deformation of the earth will be the difference in the centrifugal potential of the earth's rotation from the mean axis to the instantaneous axis. As the centrifugal potential [Eq. (8.102)] is of the form of a second-degree Legendre polynomial, the resultant potential of the deformed earth will be related to it by the Love number coefficient of Eq. (10.46). In turn, the equation of motion, Euler's equation, could be written for free motion about the instantaneous axis of rotation, including in this case the small products of inertia and their time derivatives. The products of inertia could then be found from the potential change in the deformed earth through MacCullagh's formula, each in turn being a linear function of the direction cosines l, m, and the equations of motion solved for the resultant free period.

Instead, we shall proceed here from a somewhat more general consideration, which produces the same result. We have from Eq. (8.126) that the gravitational potential of the earth is given by MacCullagh's formula as

$$V = \gamma \left[\frac{M}{r} - \frac{C-B}{r^3} P_2 \right] \qquad (10.60)$$

and to first-order terms that the same potential is given from the solution of Laplace's equation by Eq. (8.32) as

$$V = \frac{\gamma M}{a} \left[\frac{a}{r} - \frac{2}{3} J_2 \left(\frac{a}{r} \right)^3 P_2 \right] \qquad (10.61)$$

from which we obtain as before the relation (8.127) or

$$\frac{C-B}{Ma^2} = \frac{2}{3} J_2 \qquad (10.62)$$

From Eq. (8.15) we have for the zero-order gravity term

$$g = \gamma \frac{M}{a^2} \qquad (10.63)$$

from Eq. (8.38) for the first-order expression for J_2 in terms of the ellipticity e and the coefficient m,

$$J_2 = e - \tfrac{1}{2}m \qquad (10.64)$$

and from Eq. (8.34) for the first-order expression for m being as before to the first order the ratio of the centrifugal acceleration at the equator to the zero-order gravity term,

$$m = \frac{\omega^2 a}{\gamma \dfrac{M}{a^2}} = \frac{\omega^2 a^3}{\gamma M} = \frac{\omega^2 a}{g} \qquad (10.65)$$

Substituting Eqs. (10.63), (10.64), and (10.65) into Eq. (10.62), we obtain for $C-B$

$$C - B = \frac{2}{3} Ma^2 \left(e - \frac{1}{2}m\right) = \frac{2}{3}\frac{g}{\gamma} a^4 \left(e - \frac{\omega^2 a}{2g}\right) \qquad (10.66)$$

Further, we have from Eq. (8.102) that the kinetic potential due to the rotation of the earth is given by

$$W = \tfrac{1}{2}\omega^2 r^2 \sin^2 \theta = \tfrac{1}{3}\omega^2 r^2 (1 - P_2) \qquad (10.67)$$

where θ is the polar angle measured from the mean axis of rotation. Considered as a deforming potential, the first term in Eq. (10.67) will produce a uniform expansion, which will have a negligible effect on $C-B$. The second term will produce an asymmetrical deformation, which will affect the quantity $C-B$. Since the second term is proportional to P_2, we may write for the deformed potential, following Eq. (10.46),

$$U = -\tfrac{1}{3}k\omega^2 r^2 P_2 \qquad (10.68)$$

Now consider the time average condition for the earth over several nutation periods around the earth. The time average value of the difference of the moments of inertia will have changed from $C-B$ to a new value $C'-B'$. In order for the geopotential to remain constant, the difference in the earth's gravitational potential as given by Eq. (10.60) must equal that of the deformed earth potential [Eq. (10.68)], or

$$-\frac{\gamma}{r^3}[(C-B)-(C'-B')]P_2 = -\tfrac{1}{3}k\omega^2 r^2 P_2 \qquad (10.69)$$

or from Eq. (10.66)

$$\frac{C'-B'}{C-B} = 1 - \frac{\tfrac{1}{3}k\dfrac{\omega^2 a^5}{\gamma}}{C-B}$$

$$= 1 - \frac{k\dfrac{\omega^2 a}{2g}}{e - \dfrac{\omega^2 a}{2g}} \qquad (10.70)$$

We may then write for the period of the free nutation of an elastic earth τ in terms of the free period of a rigid earth τ_0 from Eq. (10.33)

$$1 - \frac{\tau_0}{\tau} = 1 - \frac{C'-B'}{C-B} = k\frac{\dfrac{\omega^2 a}{2g}}{e - \dfrac{\omega^2 a}{2g}} \qquad (10.71)$$

since B is approximately equal to B'. From Eq. (10.65) we then have for k

$$k = \left(\frac{2e}{m} - 1\right)\left(1 - \frac{\tau_0}{\tau}\right) \qquad (10.72)$$

Taking the observed value of the Chandler wobble period τ, a value of $k = 0.28$ is found. It is to be noted that, as in the discussion of Section 8.8 of the relation of the ellipticity to the internal density distribution of the earth, this relation is independent of any hypothesis as to the internal structure of the earth except in the definition and the value of k.

Problem 10.7(a) Calculate the value of k from Eq. (10.72).

Problem 10.7(b) Carry through the derivation of paragraph two of Section 10.7.

10.8 Tidal Friction

We have examined the marine tides in the ocean. We have also examined the earth tides in an elastic earth. We could also have looked at the tides in

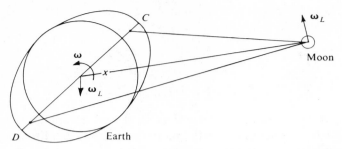

Fig. 10.7

the earth core considered as a fluid. In each of these three cases, there will be some frictional dissipation of energy, in the case of the oceans and the core from turbulence, viscosity, and the like, and in the case of the earth from departures from perfect elasticity. This energy dissipation takes place slowly, but is irreversible. Its effects take place over long periods of time and may have produced great changes in the rotation of the earth and the orbit of the moon.

Let us look at this phenomenon briefly here and at the equations defining these long-term effects. Considering the equilibrium theory of the tides if there were no tidal friction, the tidal bulges would be in line with the line through the center of the earth and moon and symmetrical about this line; there would be no torque on the moon from the tidal bulges on the earth. However, if there is a frictional loss, the tides on the earth will lag behind the moon. As shown in Fig. 10.7, the earth rotation with an angular velocity ω is counterclockwise looking down from the north pole, and the revolution of the earth-moon system with an orbital angular velocity ω_L about their common center of gravity is also as shown. The relative motion of the moon with respect to the earth will, of course, be clockwise, and the lag of the tidal bulge will be as shown in the figure. Now consider the attraction of the tides, that is, tidal bulges, on the moon. We may consider the mass of the near high tide to be concentrated at C and that of the far high tide to be concentrated at D. Neither force acts along the line joining the center of the earth and moon, and both will, therefore, have components at right angles to it. The component due to the mass at C will be greater than that due to the mass at D, because the total distance is shorter, and the angle to the line of centers is greater. Thus, there will be a net force on the moon in the direction of its revolution. We see, then, that one effect of tidal friction is that over a long period of time, the moon will swing out on increasingly wider orbits. From the principle of conservation of angular momentum for the earth-moon system, we also see that as the angular momentum of revolution is increased, the angular momentum of the earth's rotation must decrease, and consequently the angular velocity of the earth's rotation itself.

Dynamics of the Earth

Let us examine some of the simpler equations defining these motions. From Kepler's third law, Eq. (10.7), we have

$$\frac{4\pi^2}{\omega_L^2} = P^2 = \frac{4\pi^2 R^3}{\gamma(M+m)}$$

$$\omega_L^2 R^3 = \gamma(M+m) \tag{10.73}$$

where R is the distance between centers for the nearly circular orbit, and M and m are the masses of the earth and moon, respectively. From Eq. (10.73) we obtain by differentiation

$$\frac{dR}{dt} = -\frac{2}{3}\frac{R}{\omega_L}\frac{d\omega_L}{dt} \tag{10.74}$$

We see that as the orbital distance R increases, the angular velocity of revolution decreases.

From the conservation of angular momentum, we shall have, since the moon maintains a constant face toward the earth, for the sum of the angular momenta of rotation and revolution,

$$C\omega + \frac{Mm}{M+m} R^2 \omega_L = \text{constant} \tag{10.75}$$

where, as before, C is the moment of inertia about the axis of rotation and where the second term is obtained from Eq. (10.6) and the definition of angular momentum. Differentiating Eq. (10.75) and substituting from Eq. (10.74), we obtain

$$C\frac{d\omega}{dt} + \frac{Mm}{M+m}\left(2R\omega_L \frac{dR}{dt} + R^2 \frac{d\omega_L}{dt}\right) = 0$$

$$C\frac{d\omega}{dt} + \frac{Mm}{M+m}\left(-\frac{4}{3}R^2 \frac{d\omega_L}{dt} + R^2 \frac{d\omega_L}{dt}\right) = 0$$

$$\frac{d\omega}{dt} = \frac{1}{3}\frac{m}{M+m}\frac{MR^2}{C}\frac{d\omega_L}{dt} \tag{10.76}$$

The total energy of the system, being comprised of the kinetic energy of rotation, the kinetic energy of revolution, and the moon-earth gravitational potential energy, will be

$$E = \frac{1}{2}C\omega^2 + \frac{1}{2}R^2\omega_L^2 \frac{Mm}{M+m} - \gamma\frac{Mm}{R}$$

$$= \frac{1}{2}C\omega^2 - \frac{1}{2}\gamma\frac{Mm}{R} \tag{10.77}$$

using Eq. (10.73). Differentiating, we obtain for the energy dissipation term

$$\frac{dE}{dt} = C\omega \frac{d\omega}{dt} + \frac{1}{2}\gamma \frac{Mm}{R^2}\frac{dR}{dt}$$

$$= C\omega \frac{d\omega}{dt} - \gamma C \frac{M+m}{\omega_L R^3}\frac{d\omega}{dt}$$

$$= C(\omega - \omega_L)\frac{d\omega}{dt} \tag{10.78}$$

using Eqs. (10.74), (10.76), and (10.73). As $d\omega_L/dt$ is a physically measurable quantity, we can obtain estimates of dR/dt, $d\omega/dt$, and dE/dt from it through Eqs. (10.74), (10.76), and (10.78).

From these simple relations, we may obtain one further expression. Let us rewrite Eq. (10.75) as

$$C\omega + \frac{Mm}{M+m} R^2 \omega_L = (1+K)C\omega_0 \tag{10.79}$$

where K is the present ratio of orbital angular momentum to rotational angular momentum, and ω_0 is the present rotational angular velocity. Substituting for R from Eq. (10.73), we get

$$[(1+K)C\omega_0 - C\omega]^3 = \gamma^2 \frac{M^3 m^3}{\omega_L(M+m)} \tag{10.80}$$

Now, for $\omega = \omega_L$, the moon will revolve around the earth synchronously with the rotation of the earth. There will be no tidal motions and no tidal dissipation. Imposing this condition on Eq. (10.80), we get

$$\frac{\omega}{\omega_0}\left[1 + K - \frac{\omega}{\omega_0}\right]^3 = \frac{\gamma^2 M^3 m^3}{C^3 \omega_0^4 (M+m)} \tag{10.81}$$

whose real solutions for ω/ω_0 after inserting the proper numerical values for the constants are $\omega/\omega_0 = 4.95, 0.0211$. This corresponds to rotational periods of 4.85 hr and 47.4 days and from Eq. (10.73) orbital distances at 14,000 km and 645,000 km. Apparently, the moon is moving away from the close synchronous orbit toward the distant synchronous orbit.

Problem 10.8(a) Show that the same result will be obtained as to the direction of force on the moon from the earth tidal bulges if the dynamical instead of the equilibrium theory of the tides were used.

Problem 10.8(b) From a value $d\omega_L/dt = -1.09 \times 10^{-23}$ rad/sec^2, calculate dR/dt, $d\omega/dt$, and dE/dt.

Ans. 3.3 cm/yr, -4.85×10^{-22} rad/sec^2, -2.74×10^{19} erg/sec.

10.9 Free Oscillations of the Earth

Considering the earth as an elastic body of spherical shape, we shall be interested here in examining its free elastic vibrations, or *free oscillations* as they are referred to in seismology. Understandably, these free oscillations, being defined in large part simply by the dimensions of the earth, will have long periods, much longer than those associated with the usual earth seismic wave propagation phenomena. However, with sufficiently sensitive, long-period seismic instruments, these free oscillations can be recorded and form an important part of seismological investigations. Both the period and the time decay of the amplitudes of these oscillations are subjects of investigation. In general, then, the appropriate theory must include both the radial variation of the elastic constants and the density and a dissipation function. Here we shall examine the simpler example of a uniform sphere with no dissipation.

The governing equation is the vector elastic wave propagation equation of Eq. (6.35), and the boundary condition is that the stress vanish at the earth's outer spherical surface. We shall make the same reduction as before for the displacement $\mathbf{s}$ to an irrotational component $\mathbf{s}_1$ and a solenoidal component $\mathbf{s}_2$, so that the governing vector wave equations for these two quantities become Eqs. (6.36) and (6.38), respectively. Making the usual substitution for the time-dependent term in the form

$$\mathbf{s} = \mathbf{s}(x, y, z)e^{i\omega t} \tag{10.82}$$

Eqs. (6.36) and (6.38) reduce to

$$(\nabla^2 + h^2)\mathbf{s}_1 = 0 \tag{10.83}$$

and

$$(\nabla^2 + k^2)\mathbf{s}_2 = 0 \tag{10.84}$$

where

$$h^2 = \frac{\rho\omega^2}{\lambda + 2\mu} \tag{10.85}$$

$$k^2 = \frac{\rho\omega^2}{\mu} \tag{10.86}$$

and

$$\nabla \times \mathbf{s}_1 = 0 \tag{10.87}$$

$$\nabla \cdot \mathbf{s}_2 = 0 \tag{10.88}$$

Since $\mathbf{s}_1$ is irrotational, we can from Eq. (10.87) represent it by

$$\mathbf{s}_1 = \nabla \Phi \tag{10.89}$$

where Φ is a scalar function, reducing Eq. (10.83) to

$$(\nabla^2 + h^2)\Phi = 0 \tag{10.90}$$

It will be convenient to divide $\mathbf{s}_2$ into a translational or rotational component with respect to the spherical boundary and into a complementary distortional component, much the same as we did in Section 6.3 in dividing this component into a portion parallel and perpendicular to a plane boundary for the discussion of reflection and refraction of plane elastic waves from a plane boundary. From Section 9.3 and Eq. (9.40) we have that the translational component $\mathbf{s}_2'$ can be represented by

$$\mathbf{s}_2' = -\mathbf{r} \times \nabla \Psi \qquad (10.91)$$

where Ψ is a scalar function. $\mathbf{s}_2'$ satisfies the condition Eq. (10.88) and also satisfies Eq. (10.84) if Ψ is a solution of

$$(\nabla^2 + k^2)\Psi = 0 \qquad (10.92)$$

We note that $\nabla \times \mathbf{s}_2'$ will also be a solution of Eq. (10.84) so that the complementary solution $\mathbf{s}_2''$ to $\mathbf{s}_2'$ can be represented by

$$\mathbf{s}_2'' = \nabla \times \mathbf{s}_2' \qquad (10.93)$$

Our problem is then reduced to finding solutions of Eqs. (10.90) and (10.92) satisfying the boundary conditions. We see that these equations are the same as Eq. (9.74) with the substitution of h^2 and k^2, respectively, for $-k^2$. The solution will then be in terms of the half-order Bessel functions rather than the modified half-order Bessel functions. In order to meet the condition of finite motion at the origin, we shall want only the positive argument Bessel functions. From Eqs. (9.79), (9.80), and Section 1.12, our solutions will then be

$$\Phi = A_{nm} r^{-1/2} J_{n+1/2}(hr) P_n^m(\theta) e^{im\varphi} \qquad (10.94)$$

and

$$\Psi = B_{nm} r^{-1/2} J_{n+1/2}(kr) P_n^m(\theta) e^{im\varphi} \qquad (10.95)$$

From Eqs. (10.89), (10.91), and (10.93), using Eq. (1.82) and the results of Problem 1.4(a), the displacement components in the r, θ, φ directions will be, in terms of Φ and Ψ,

$$\begin{aligned} s_{1r} &= \frac{\partial \Phi}{\partial r} \\ s_{1\theta} &= \frac{1}{r} \frac{\partial \Phi}{\partial \theta} \\ s_{1\varphi} &= \frac{1}{r \sin \theta} \frac{\partial \Phi}{\partial \varphi} \end{aligned} \qquad (10.96)$$

and

$$s'_{2r} = 0$$
$$s'_{2\theta} = \frac{1}{\sin\theta}\frac{\partial\Psi'}{\partial\varphi} \quad (10.97)$$
$$s'_{2\varphi} = -\frac{\partial\Psi'}{\partial\theta}$$

and

$$s''_{2r} = \frac{-1}{r\sin\theta}\left[\frac{\partial}{\partial\theta}\left(\sin\theta\,\frac{\partial\Psi''}{\partial\theta}\right) + \frac{1}{\sin\theta}\frac{\partial^2\Psi''}{\partial\varphi^2}\right]$$
$$s''_{2\theta} = \frac{1}{r}\frac{\partial^2(r\Psi'')}{\partial r\,\partial\theta} \quad (10.98)$$
$$s''_{2\varphi} = \frac{1}{r\sin\theta}\frac{\partial^2(r\Psi'')}{\partial r\,\partial\varphi}$$

We can develop expressions for the stress components in the spherical coordinate directions referred to the spherical coordinate displacements comparable to the expressions (6.26) for rectangular coordinates. For the stress components referred to the outer spherical surface of the earth, $r = a$, over which they must vanish, we would then have

$$T_{rr} = \lambda\Delta + 2\mu\frac{\partial s_r}{\partial r} = 0$$
$$T_{r\theta} = \mu\left(\frac{\partial s_\theta}{\partial r} - \frac{s_\theta}{r} + \frac{1}{r}\frac{\partial s_r}{\partial\theta}\right) = 0 \quad (10.99)$$
$$T_{r\varphi} = \mu\left(\frac{1}{r\sin\theta}\frac{\partial s_r}{\partial\varphi} + \frac{\partial s_\varphi}{\partial r} - \frac{s_\varphi}{r}\right) = 0$$

where, as before, Δ is the dilatation, and the components of **s** in Eqs. (10.99) are the components of the sum of $\mathbf{s}_1$, $\mathbf{s}'_2$, and $\mathbf{s}''_2$.

It can be shown, which we shall not do here, that these boundary conditions can be satisfied independently by $\mathbf{s}'_2$ or by a combination of $\mathbf{s}_1$ and $\mathbf{s}''_2$, similar to the two types of elastic surface waves, Love and Rayleigh, respectively, of Sections 7.5 and 7.4. The free oscillations involving $\mathbf{s}'_2$ only are referred to as *torsional* oscillations and for their lowest mode, $n = 1$, *rotational* oscillations. The free oscillations involving $\mathbf{s}_1$ and $\mathbf{s}''_2$ are referred to as *spheroidal* oscillations and for their lowest mode, $n = 0$, *radial* oscillations. We shall derive here, from the boundary conditions, the period equation defining the free oscillations for the lowest mode in each of these cases.

For $n = 1$ and $m = 0$, we have from Eq. (10.95) for Ψ for the rotational oscillations

$$\Psi = Br^{-1/2}J_{3/2}(kr)\cos\theta$$

and from Eqs. (10.97), for $s'_{2\varphi}$, s'_{2r}, and $s'_{2\theta}$ being zero,

$$s'_{2\varphi} = Br^{-1/2}J_{3/2}(kr)\sin\theta$$

showing that the motion in this mode is rotational. For the boundary condition [Eq. (10.99)] that $T_{r\varphi}$ vanish, we shall then have

$$B\frac{d}{dr}[r^{-1/2}J_{3/2}(kr)]\sin\theta - Br^{-3/2}J_{3/2}(kr)\sin\theta = 0$$

or letting $x = kr$, and using Eq. (1.225),

$$\frac{d}{dx}[x^{-1/2}J_{3/2}(x)] - x^{-3/2}J_{3/2}(x) = 0$$

$$-\frac{1}{2}x^{-3/2}J_{3/2}(x) + x^{-1/2}\left[\frac{3}{2}\frac{1}{x}J_{3/2}(x) - J_{5/2}(x)\right] - x^{-3/2}J_{3/2}(x) = 0$$

$$J_{5/2}(x) = 0$$

which, from the results of Problem 1.12(c), gives

$$\tan x = \frac{3x}{3-x^2}$$

whose solutions are

$$\frac{ka}{\pi} = 1.84, 2.90, \ldots \tag{10.100}$$

For $n = 0$ and $m = 0$ we have, from Eq. (10.94) for the radial oscillations,

$$\Phi = Ar^{-1/2}J_{1/2}(ha)$$

and from Eqs. (10.96) for s_{1r}, all other components being zero,

$$s_{1r} = A\frac{d}{dr}[r^{-1/2}J_{1/2}(hr]$$

showing that the motion in this mode is radial. From the results of Problem 1.4(a), we have for the dilatation

$$\Delta = \nabla \cdot \mathbf{s} = \frac{ds_{1r}}{dr} + \frac{2}{r}s_{1r}$$

For the boundary condition [Eqs. (10.99)] that T_{rr} vanish, we shall then have

$$\lambda A\frac{d^2}{dr^2}[r^{-1/2}J_{1/2}(hr)] + \frac{2\lambda}{r}A\frac{d}{dr}[r^{-1/2}J_{1/2}(hr)] + 2\mu A\frac{d^2}{dr^2}[r^{-1/2}J_{1/2}(hr)] = 0$$

358 ¶ Dynamics of the Earth

or letting $x = hr$, and using first Eq. (1.226) and then (1.227),

$$\lambda \frac{d^2}{dx^2}[x^{-1/2}J_{1/2}(x)] + \frac{2\lambda}{x}\frac{d}{dx}[x^{-1/2}J_{1/2}(x)] + 2\mu\frac{d^2}{dx^2}[x^{-1/2}J_{1/2}(x)] = 0$$

$$-\lambda \frac{d}{dx}[x^{-1/2}J_{3/2}(x)] - 2\lambda x^{-3/2}J_{3/2}(x) - 2\mu\frac{d}{dx}[x^{-1/2}J_{3/2}(x)] = 0$$

$$\lambda[2x^{-3/2}J_{3/2}(x) - x^{-1/2}J_{1/2}(x)] - 2\lambda x^{-3/2}J_{3/2}(x)$$
$$+ 2\mu[2x^{-3/2}J_{3/2}(x) - x^{-1/2}J_{1/2}(x)] = 0$$

$$-\frac{1}{h^2}x^{-1/2}J_{1/2}(x) + \frac{4}{k^2}x^{-3/2}J_{3/2}(x) = 0$$

which, from Eq. (1.231) and the results of Problem 1.12(c), gives

$$\frac{J_{3/2}(x)}{J_{1/2}(x)} = -\cot x + \frac{1}{x} = \frac{k^2 x}{4h^2}$$

whose solutions for $\lambda = \mu$, or $k^2/h^2 = 3$, are

$$\frac{ha}{\pi} = 0.82, 1.93, \ldots \qquad (10.101)$$

Problem 10.9(a) Derive the six stress components in spherical coordinates in terms of the spherical coordinate displacements.

References

Danby, J. M. A. 1962. *Fundamentals of Celestial Mechanics*. New York: Macmillan.
Garland, G. D. 1971. *Introduction to Geophysics*. Philadelphia: W. B. Saunders.
Jeffreys, H. 1952. *The Earth*. Cambridge: Cambridge University Press.
Kuala, W. M. 1968. *An Introduction to Planetary Physics*. New York: Wiley.
Larmor, J. 1909. "The relation of the earth's free precessional nutation to its resistance against tidal deformation," *Proc. Royal Soc. London*, Ser. A, vol. 82, pp. 89–96.
Love, A. E. H. 1909. "The yielding of the earth to disturbing forces," *Proc. Royal Soc. London*, Ser. A, vol. 82, pp. 73–88.
———. 1911. *Some Problems of Geodynamics*. New York: Dover.
Melchior, P. J. 1957. *Latitude Variation, Physics and Chemistry of the Earth*, vol. 2. New York: Pergamon.
———. 1966. *The Earth Tides*. New York: Pergamon.
Munk, W. H., and G. J. F. MacDonald. 1960. *The Rotation of the Earth*. Cambridge: Cambridge University Press.
Page, L. 1935. *An Introduction to Theoretical Physics*. New York: Van Nostrand.
Stacey, F. D. 1969. *Physics of the Earth*. New York: Wiley.
Takeuchi, H. 1966. *Theory of the Earth's Interior*. Waltham: Blaisdell.

Chapter 11

EARTH CRUSTAL AND MANTLE DEFORMATION

11.1 Introduction

One of the more interesting aspects of geophysics, and also unfortunately one of the least susceptible to theoretical description, has to do with the deformation of the earth's crust and upper mantle through geologic time. The surficial features of these deformations and related processes are readily apparent in continents and ocean basins, island arcs and deep-sea trenches, mountain ranges, volcanoes, midocean ridges, and the like. Field geological and geophysical information adds a great deal more as to the structure, composition, and geologic history of these features. The abundance of this information and the rapidity with which new information is added encourage many to hypothesize as to the origin of these features and to criticize or corroborate hypotheses of others as each new piece of significant information is added. Further, sometimes a given new piece of information in itself suggests a process that theretofore had not been considered. It is sort of like a gigantic, intellectual guessing game, and any number can play. There is no question that it is one of the more fascinating aspects of earth science.

Unfortunately, however, these deformations are not particularly susceptible to explanation by the present methods of theoretical physics. The reasons for this are fairly obvious. First, we have considered materials as being either rigid, elastic, viscous, or fluid. For the purposes of this chapter, we might consider the materials as being in the limit either elastic or viscous. These considerations, nevertheless, would be limiting cases, and we should rather consider the non-elastic, dislocation and fracture, and plastic and viscoelastic properties of the materials. These descriptions can be given in part, but are beyond the scope of what has been covered in this book. Further, the deformations at depth in the crust and in the upper mantle take place at high pressures and high temperatures. Small samples of earth materials can be studied in the laboratory at such pressures and temperatures, and some of their properties can be defined empirically. Again, it is difficult to write down any simple or complete relations defining their properties. Second, earth crustal and mantle deformations are often accompanied or controlled by thermodynamic and chemical phase change effects. Again, these effects are only partially understood, and the theoretical equations defining them are difficult to write down or incorporate in the nonelastic deformational effects. Third, and perhaps of greatest importance is the fact that these deformations take place over a long time period measured in terms of tens of thousands to tens of millions of years. All the laboratory physical experience is in terms of effects measured in much shorter time periods. The extrapolation from such laboratory measurements to the geologic time scale is exceedingly difficult to make. It does not take much astuteness to guess that a study of the earth deformational processes will remain for many years to come, as it has been in the past, paramount in earth science.

We shall consider here only some of the simpler aspects, which can be defined in terms of elastic or viscous materials.

11.2 Loading of an Elastic Crust

Let us consider first the bending, or deformation, of an elastic crust under a vertical load. Following the notation of Chapter 6, we shall designate the displacement components in the X, Y, Z directions by u, v, w, and we shall take the Z axis as vertical downward. We shall restrict ourselves to the two-dimensional case, so that $v = 0$ and all derivatives $\partial/\partial y = 0$. From Eq. (6.31) we then have, for the static equilibrium equations to be met everywhere in the elastic crust,

$$\frac{\partial \sigma_{xx}}{\partial x} + \frac{\partial \sigma_{xz}}{\partial z} = 0$$

$$\frac{\partial \sigma_{zx}}{\partial x} + \frac{\partial \sigma_{zz}}{\partial z} = 0$$

(11.1)

where the stresses σ are those in excess of gravity. From Eq. (6.26), using the relations (6.23) and (6.22), the relations between these stresses and the resultant strains will be

$$\sigma_{xx} = (\lambda+2\mu)\frac{\partial u}{\partial x} + \lambda\frac{\partial w}{\partial z}$$

$$\sigma_{xz} = \sigma_{zx} = \mu\left(\frac{\partial u}{\partial z} + \frac{\partial w}{\partial x}\right)$$

(11.2)

$$\sigma_{zz} = \lambda\frac{\partial u}{\partial x} + (\lambda+2\mu)\frac{\partial w}{\partial z}$$

We see that the relations (11.1) for the stresses σ will be met for the stresses defined in terms of a function X by

$$\sigma_{xx} = \frac{\partial^2 X}{\partial z^2}$$

$$\sigma_{xz} = \sigma_{zx} = -\frac{\partial^2 X}{\partial x \partial z} \quad (11.3)$$

$$\sigma_{zz} = \frac{\partial^2 X}{\partial x^2}$$

From the first and the third of Eqs. (11.2), we have, directly solving for $\partial u/\partial x$ and $\partial w/\partial z$,

$$\frac{\partial u}{\partial x} = \frac{1}{4\mu(\lambda+\mu)}[(\lambda+2\mu)\sigma_{xx} - \lambda\sigma_{zz}] \quad (11.4)$$

and

$$\frac{\partial w}{\partial z} = \frac{1}{4\mu(\lambda+\mu)}[(\lambda+2\mu)\sigma_{zz} - \lambda\sigma_{xx}] \quad (11.5)$$

We may then derive from Eqs. (11.3) and (11.2), using Eqs. (11.4) and (11.5),

$$-\frac{\partial^4 X}{\partial x^2 \partial z^2} = \frac{\partial^2 \sigma_{xz}}{\partial x \partial z} = \mu\left(\frac{\partial^3 u}{\partial x \partial z^2} + \frac{\partial^3 w}{\partial x^2 \partial z}\right)$$

$$= \frac{1}{4(\lambda+\mu)}\left\{\frac{\partial^2}{\partial z^2}\left[(\lambda+2\mu)\frac{\partial^2 X}{\partial z^2} - \lambda\frac{\partial^2 X}{\partial x^2}\right]\right.$$

$$\left. + \frac{\partial^2}{\partial x^2}\left[(\lambda+2\mu)\frac{\partial^2 X}{\partial x^2} - \lambda\frac{\partial^2 X}{\partial z^2}\right]\right\}$$

$$= \frac{1}{4(\lambda+\mu)}\left[(\lambda+2\mu)\frac{\partial^4 X}{\partial x^4} - 2\lambda\frac{\partial^4 X}{\partial x^2 \partial z^2} + (\lambda+2\mu)\frac{\partial^4 X}{\partial z^4}\right]$$

or

$$\frac{\partial^4 X}{\partial x^4} + 2\frac{\partial^4 X}{\partial x^2 \partial z^2} + \frac{\partial^4 X}{\partial z^4} = 0$$

$$\left(\frac{\partial^2}{\partial x^2} + \frac{\partial^2}{\partial z^2}\right)^2 X = 0 \quad (11.6)$$

We see that the parenthesis of Eq. (11.6) is the two-dimensional form of ∇^4, referred to as the *biharmonic operator*, and Eq. (11.6) is the two-dimensional form of the *biharmonic equation*.

We shall be interested here only in the case of a vertical load at the surface of simple harmonic form, given by

$$\sigma_{zz} = \rho g h \cos kx$$
$$\sigma_{xz} = \sigma_{zx} = 0 \qquad (z = 0) \qquad (11.7)$$

where h is the height, or amplitude, of the harmonic load and ρ its density. These load conditions will be met if X is given by

$$X = -\frac{\rho g h}{k^2} \cos kx$$
$$\frac{\partial X}{\partial z} = 0 \qquad (z = 0) \qquad (11.8)$$

at the surface. We may then take X in the elastic crust to be of the form

$$X = -\frac{\rho g h}{k^2} X(z) \cos kx \qquad (11.9)$$

which reduces Eq. (11.6) to the ordinary differential equation

$$\left(\frac{d^2}{dz^2} - k^2\right)^2 X(z) = 0 \qquad (11.10)$$

whose solutions are simply

$$X(z) = Ae^{-kz} + Be^{kz} + Cze^{-kz} + Dze^{kz} \qquad (11.11)$$

In order to meet the conditions of vanishing stresses at great depths, $B = D = 0$. Then to satisfy the second of the conditions [Eqs. (11.8)] at the surface, we get $C = kA$; then for the first of the conditions (11.8), $A = 1$. The final solution for X of Eq. (11.9) is then

$$X = -\frac{\rho g h}{k^2}(1+kz)e^{-kz} \cos kx \qquad (11.12)$$

For the stresses in the elastic crust, we shall have from Eqs. (11.12), (11.3), and (6.26)

$$\sigma_{xx} = \rho g h(1-kz)e^{-kz} \cos kx$$
$$\sigma_{xz} = \sigma_{zx} = -\rho g h k z e^{-kz} \sin kx$$
$$\sigma_{zz} = \rho g h(1+kz)e^{-kz} \cos kx$$
$$\sigma_{yy} = \lambda \Delta = \frac{\lambda}{2(\lambda+\mu)}(\sigma_{xx} + \sigma_{zz}) \qquad (11.13)$$
$$= \frac{\lambda}{\lambda+\mu} \rho g h e^{-kz} \cos kx$$
$$\sigma_{xy} = \sigma_{yx} = \sigma_{yz} = \sigma_{zy} = 0$$

We see that σ_{yy} is a principal stress. The other two principal stresses may be determined, using the vector transformation relations, and are found to be

$$\rho g h (\cos kx \pm kz) e^{-kz} \qquad (11.14)$$

At shallow depths, σ_{yy} will be an extremum stress. At moderate to great depths, σ_{yy} will be the intermediate stress, and the difference for the extremum stresses will be from expression (11.14)

$$2\rho g h k z e^{-kz} \qquad (11.15)$$

We see that in this latter case, the stress difference at any given depth is uniform, independent of x. The expression (11.15) will have a maximum at a depth given by

$$\frac{d}{du}(ue^{-u}) = e^{-u} - u^2 e^{-u} = 0$$

or

$$u = kz_m = 1$$

so that

$$z_m = \frac{1}{k} = \frac{\lambda}{2\pi} \qquad (11.16)$$

We may then expect failure to occur if at all at a depth of $1/2\pi$ times the wavelength of the harmonic load, that the failure will be uniform, and that it will tend to flatten out the inequalities of the surface load.

For the vertical displacement, we shall have from Eqs. (11.5) and (11.13)

$$\begin{aligned} w &= \frac{1}{4\mu(\lambda+\mu)} \int_z^\infty [(\lambda+2\mu)\sigma_{zz} - \lambda\sigma_{xx}]\, dz \\ &= \frac{\rho g h \cos kx}{4\mu(\lambda+\mu)} \int_z^\infty [(\lambda+2\mu)(1+kz)e^{-kz} - \lambda(1-kz)e^{-kz}]\, dz \\ &= \frac{\rho g h \cos kx}{4\mu(\lambda+\mu)k} [(\lambda+2\mu)(2+kz) + \lambda kz] e^{-kz} \\ &= \frac{\rho g h}{4\mu(\lambda+\mu)k} [2\lambda(1+kz) + 2\mu(2+kz)] e^{-kz} \cos kx \qquad (11.17) \end{aligned}$$

For $z = 0$, the surface displacement is then

$$w_0 = \frac{\rho g h}{2\mu k} \frac{\lambda+2\mu}{\lambda+\mu} \cos kx \qquad (11.18)$$

Problem 11.2(a)   Show that Eq. (11.11) are solutions of Eq. (11.10).

Problem 11.2(b)   Determine the principal stresses [expression (11.14)] from Eqs. (11.13).

11.3 Loading of a Floating Lithosphere

Referring back to the general results of Chapter 8 having to do with isostasy, we should next like to consider an elastic lithosphere resting on a substratum, or asthenosphere, of negligible strength. Let us designate by H the thickness of the lithosphere, its density and the density of a superimposed load by ρ, and the density of the asthenosphere by ρ'.

In the previous section, we found for an elastic crust that the greatest stress differences occurred at depths comparable to the horizontal width of the load. We may properly consider it as a valid approximation to this problem for loads whose horizontal dimensions are small compared with H. We shall examine here only the other extreme in which the horizontal dimensions of the vertical load are large compared with H, so that the lithosphere may be considered as a thin elastic plate. The problem, then, reduces to that of the bending of a thin plate.

We must first examine the elastic properties associated with the bending of a thin plate. We shall do so as follows in a somewhat crude and approximate manner. First, let us rewrite the stress-strain relations of Eq. (6.26) as solutions in terms of the strains and in terms of the elastic constants E and σ of Eqs. (6.27) and (6.28), Young's modulus and Poisson's ratio, respectively, which give

$$\frac{\partial u}{\partial x} = \frac{1}{E}[\sigma_{xx} - \sigma(\sigma_{yy} + \sigma_{zz})]$$

$$\frac{\partial v}{\partial y} = \frac{1}{E}[\sigma_{yy} - \sigma(\sigma_{xx} + \sigma_{zz})]$$

$$\frac{\partial w}{\partial z} = \frac{1}{E}[\sigma_{zz} - \sigma(\sigma_{xx} + \sigma_{yy})] \qquad (11.19)$$

$$\frac{1}{2}\left(\frac{\partial w}{\partial y} + \frac{\partial v}{\partial z}\right) = \frac{1+\sigma}{E}\sigma_{yz}$$

$$\frac{1}{2}\left(\frac{\partial u}{\partial z} + \frac{\partial w}{\partial x}\right) = \frac{1+\sigma}{E}\sigma_{zx}$$

$$\frac{1}{2}\left(\frac{\partial v}{\partial x} + \frac{\partial u}{\partial y}\right) = \frac{1+\sigma}{E}\sigma_{xy}$$

For a bent plate, as shown in Fig. 11.1, the upper portion, above a median plane corresponding with the coordinate axes, will be under compression and the lower portion under tension. Let us assume for a thin plate that these compressions and tensions may be given by a linear function of the z coordinate

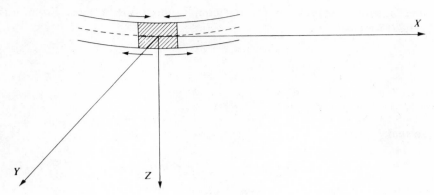

Fig. 11.1

distance away from the median plane so that the σ_{xx} and σ_{yy} stresses may be represented by

$$\sigma_{xx} = E\alpha z \qquad \sigma_{yy} = E\beta z \tag{11.20}$$

where α and β are constants, and all other stresses are zero. The stress-strain relations (11.19) then reduce to

$$\frac{\partial u}{\partial x} = (\alpha - \sigma\beta)z$$

$$\frac{\partial v}{\partial y} = (\beta - \sigma\alpha)z$$

$$\frac{\partial w}{\partial z} = -\sigma(\alpha + \beta)z \tag{11.21}$$

$$\frac{\partial w}{\partial y} + \frac{\partial v}{\partial z} = \frac{\partial u}{\partial z} + \frac{\partial w}{\partial x} = \frac{\partial v}{\partial x} + \frac{\partial u}{\partial y} = 0$$

These relations are satisfied for displacements u, v, w given by

$$\begin{aligned} u &= (\alpha - \sigma\beta)xz \\ v &= (\beta - \sigma\alpha)yz \\ w &= -\tfrac{1}{2}(\alpha - \sigma\beta)x^2 - \tfrac{1}{2}(\beta - \sigma\alpha)y^2 - \tfrac{1}{2}\sigma(\alpha + \beta)z^2 \end{aligned} \tag{11.22}$$

Now the internal couple about the Y axis associated with the bending in the XZ plane will be

$$M = \int_{-H/2}^{H/2} z\sigma_{xx}\, dz = E\alpha \int_{-H/2}^{H/2} z^2\, dz = \tfrac{1}{12} E\alpha H^3 \tag{11.23}$$

and a similar expression for the couple about the X axis. From Eqs. (11.22) we have

$$\frac{\partial^2 w}{\partial x^2} = \sigma\beta - \alpha$$

$$\frac{\partial^2 w}{\partial y^2} = \sigma\alpha - \beta$$

so that

$$\alpha = -\frac{1}{1-\sigma^2}\left(\frac{\partial^2 w}{\partial x^2} + \sigma\frac{\partial^2 w}{\partial y^2}\right) \tag{11.24}$$

and a similar expression for β. Substituting Eq. (11.24) into Eq. (11.23), we get

$$M = -\frac{1}{12}\frac{EH^3}{1-\sigma^2}\left(\frac{\partial^2 w}{\partial x^2} + \sigma\frac{\partial^2 w}{\partial y^2}\right) \tag{11.25}$$

or since we shall only consider bending in the XZ plane,

$$M = -D\frac{d^2 w}{dx^2} \tag{11.26}$$

where D, referred to as the *flexural rigidity* of the plate, is given by

$$D = \frac{1}{12}\frac{EH^3}{1-\sigma^2} \tag{11.27}$$

For many problems in elastic plate or beam deformation, the bending is sufficiently slight that $\partial w/\partial z$ is small compared with unity, so that the curvature of the bending may be approximated by $\partial^2 w/\partial z^2$. Equation (11.26) then states that the internal bending moment, or couple, is directly proportional to the curvature of the bending; this relation is sometimes referred to as the *theorem of the bending moment*.

We shall now develop the equilibrium equations for our example of a vertically loaded, floating, elastic plate. Let us designate by T the total vertical force over the thickness H of the plate acting on an XZ surface of unit dimensions in the Y direction and by Z the net vertical external force per unit length. Then, as shown in Fig. 11.2, the equation for the equilibrium of the vertical forces will be

$$(T+\Delta T) - T - Z\Delta x = 0$$

or

$$\frac{dT}{dx} = Z \tag{11.28}$$

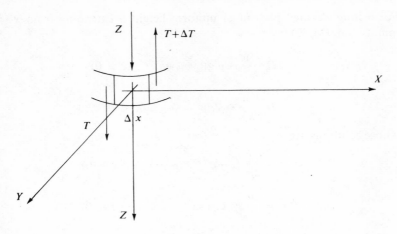

Fig. 11.2

And the equation for the equilibrium of the moments about the Y axis will be

$$T\Delta x + \Delta M = 0$$

or

$$\frac{dM}{dx} = -T \qquad (11.29)$$

Combining Eqs. (11.29) and (11.28) with Eq. (11.27), we have then finally as our equilibrium equation

$$D\frac{d^4 w}{dx^4} = Z \qquad (11.30)$$

For our example, the net vertical external force will be the weight of the load at the surface P minus the buoyancy force of the vertically displaced floating plate, or

$$Z = P - \rho' g w \qquad (11.31)$$

For harmonic loading, corresponding to Eq. (11.7), we shall have

$$P = \rho g h \cos kx \qquad (11.32)$$

so that Eq. (11.30) becomes

$$D\frac{d^4 w}{dx^4} + \rho' g w = \rho g h \cos kx \qquad (11.33)$$

whose solution is

$$w = \frac{\rho g h}{D k^4 + \rho' g} \cos kx \qquad (11.34)$$

For a long elevated plateau of uniform height h extending from $x = 0$ to infinity, Eq. (11.30) becomes

$$D\frac{d^4w}{dx^4} + \rho'gw = \rho gh \qquad (x > 0)$$
$$D\frac{d^4w}{dx^4} + \rho'gw = 0 \qquad (x < 0) \qquad (11.35)$$

whose solutions are

$$w = \frac{\rho h}{2\rho'}(2 - e^{-\gamma x}\cos\gamma x) \qquad (x > 0)$$
$$w = \frac{\rho h}{\rho'}e^{\gamma x}\cos\gamma x \qquad (x < 0) \qquad (11.36)$$

where γ is given by

$$\gamma^4 = \frac{\rho' g}{4D} \qquad (11.37)$$

We see that in this latter example that $1/\gamma$ is a measure of the horizontal extent of the regional isostatic compensation. We could, if desired, develop a system for regional isostatic compensation of gravity anomalies for a given distribution of surficial mass excesses and deficiencies based on calculations such as those given in this section.

Problem 11.3(a)   Derive the expressions (11.19).

Problem 11.3(b)   Show that Eq. (11.34) is the solution of Eq. (11.33) and that Eqs. (11.36) are the solutions of Eqs. (11.35).

Problem 11.3(c)   Show that in the limit as $x \to \infty$, the first of expressions (11.36) approaches an Airy local isostatic compensation.

Problem 11.3(d)   Taking $\lambda = \mu = 3.3 \times 10^{11}$ dyne/cm^2, $H = 50$ km, $\rho' = 3.3$ g/cm^3, $\rho = 2.5$ g/cm^3, calculate D. Then for $Dk^4 = \rho'g$, calculate $2\pi/k$, w, and the resultant height referred to sea level of a 1-km harmonic load. Calculate the same displacement and resultant height on the assumption of an Airy local isostatic compensation.

Ans. $D = 9 \times 10^{30}$ CGS, $2\pi/k = 460$ km, 0.38 km, 0.62 km, 0.76 km, 0.24 km

Problem 11.3(e)   For these same values of the constants, calculate $1/\gamma$ of Eq. (11.37) and the distance to the left of the origin at which w becomes zero.

Ans. $1/\gamma = 100$ km, 160 km

11.4 Postglacial Uplift

There has been one particular experiment by nature over the time scale of the phenomena being investigated here that can provide some knowledge of the viscous, or plastic, properties of the earth's crust and upper mantle. During the glacial period, the two large continental areas of Fennoscandia and a part of North America were covered with ice. At the end of the last glacial period, about 16,000 yr ago, the ice disappeared, and these two areas have since shown clear evidence of rising, that is, a long period rebound following removal of the ice load. This rising can be traced in systems of uplifted shorelines on an archeological time scale and by precise leveling on today's time scale, which also shows that the rebound is still proceeding.

On this time scale, we shall presume that the elastic lithosphere does not materially affect the phenomenon, but that the earth's surface follows more or less accurately the movement of the underlying asthenosphere. We shall presume again on this time scale that the asthenosphere acts as a simple, viscous liquid, and we shall be interested in deriving an expression for its viscosity in terms of the measured uplift.

If we presume that the original ice load can be approximated by a two-dimensional harmonic load, we see that in effect we have already derived our needed expressions in Section 11.2. Referring to Section 3.6, we see that the derivations for a viscous fluid can be obtained directly from those for an elastic solid by replacing μ by η, λ by $-2\eta/3$, w by dw/dt, and σ_{xx} and σ_{zz} by $\sigma_{xx}-\sigma_H$ and $\sigma_{zz}-\sigma_H$, where η is the coefficient of viscosity, and σ_H is the mean hydrostatic stress. From Eqs. (11.13) we then have

$$\sigma_{xx}-\sigma_H = -\rho g h k z e^{-kz} \cos kx$$
$$\sigma_{zz}-\sigma_H = \rho g h k z e^{-kz} \cos kx \tag{11.38}$$

so that Eqs. (11.17) and (11.18) become

$$\frac{dw_0}{dt} = \frac{\rho g h}{4\eta} \int_0^\infty (4kze^{-kz} - 2kze^{-kz}) \cos kx \, dz$$

$$= \frac{\rho g h \cos kx}{2\eta} \int_0^\infty kze^{-kz} \, dz$$

$$= \frac{\rho g h}{2\eta k} \cos kx \tag{11.39}$$

In our example, the vertical load stress is not Eqs. (11.7), but that of the upward buoyant force of the displaced asthenosphere or

$$\sigma_{zz} = -\rho' g \zeta(t) \cos kx \tag{11.40}$$

where $\zeta(t)$ is the vertical displacement of the asthenosphere from its equilibrium position at some time t. The expression (11.39) then becomes

$$\frac{dw_0}{dt} = \frac{d\zeta}{dt}\cos kx = -\frac{\rho' g\zeta}{2\eta k}\cos kx \qquad (11.41)$$

or

$$\zeta = \zeta_0 e^{-\rho' gt/2\eta k} \qquad (11.42)$$

where ζ_0 is the value of ζ at some initial time. Further, since we are considering a two-dimensional harmonic load, we shall want to replace $\cos kx$ of Eq. (11 40) by $\cos lx \cos my$. From the form of Eq. (11.10) we then see that k in Eq. (11.42) will be replaced by a value given by

$$k^2 = l^2 + m^2 \qquad (11.43)$$

From Eqs. (11.41) and (11.42) we then have that the time constant T of the viscous rebound is given by

$$T = \frac{\zeta \Delta t}{-\Delta \zeta} = \frac{2\eta k}{\rho' g} \qquad (11.44)$$

from which the value of the viscosity can be calculated.

Problem 11.4(a)   For the central area of Fennoscandia, the surface level has been estimated to have risen 130 m from 5000 B.C. referred to the 1950 level. From the remaining gravity anomalies in this area, it is estimated that there is 150 m left to rise. The orthogonal surface dimensions of the original ice sheet are estimated to be 1200 km and 1800 km. Using a value of $\rho' = 3.3$, calculate the time constant and the viscosity.

Ans. $T = 15,000$ yr, $\eta = 2.4 \times 10^{22}$ poise.

11.5 Thermal Contraction of the Earth

One of the more obvious features of the earth's surface is the mountain ranges. Geological investigations of various ranges indicate that they have almost invariably been formed as the result of horizontal compression in the crust with the geologic strata showing varying degrees of thrust faulting and folding and that the original geologic strata forming the mountain ranges show a crustal shortening associated with the mountain-building process. It can be argued from other geological information as to whether the crustal shortening associated with mountain building represents an overall crustal shortening throughout geologic time or whether there has been a compensating crustal lengthening elsewhere in the crust.

We shall examine here an hypothesis, which accounts for the crustal shortening on a worldwide basis and which provides for the horizontal compressive stresses to produce the mountains. This hypothesis is simply that the earth has been cooling throughout geologic time, which in itself is an assumption, and that the thermal contraction associated with this cooling has been sufficient to provide the crustal shortening observed in the world mountain ranges.

We shall assume, as stated above, that the earth has been cooling through geologic time, that is, that the radioactive heat generation within the earth has been sufficiently small to give an overall cooling. We shall assume, in the derivation given here, that the cooling has been by conduction only, that is, that there are no convection currents. With these assumptions in mind, let us discuss first, in a qualitative manner, what may have occurred. We know from the discussion of Section 2.7 that the cooling by conduction has only extended down to a depth of 300 km or so since the time of solidification of the mantle and crust. Initially, starting at the time of final solidification, the cooling would have been most rapid near the earth's surface. Consider a shell near the earth's outer surface as shown in Fig. 11.3. Its inner radius may be considered to be bounded, and the thermal contraction will have to take place by a thinning of the shell or, if it is sufficiently solidified, by fracturing and filling by molten material from below. The shell will be in a state of tensional stress. At some later time, the cooling will be most rapid at some depth in the earth. For a spherical shell below this depth, as shown in Fig. 11.3(b), the cooling conditions will be the same as that shown in Fig. 11.3(a) with a decrease in its outer radius and a tensional stress in the shell. For a shell at the earth's outer surface, which may now be considered to be solidified, its inner surface will be too small to fit the contracting earth, and the shell will be under a state of horizontal compression stress. At some intermediate depth, presumably at present between 0 and 300 km, there will be a level of no strain.

Let us now examine this phenomenon quantitatively. Consider a shell of inner radius r and thickness dr. Let its thermal coefficient of linear expansion

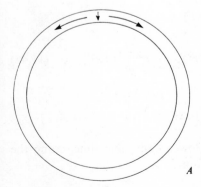
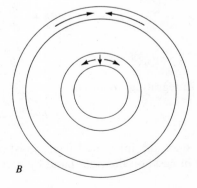

Fig. 11.3

Dynamics of the Earth

be n, similar to the coefficient of volume expansion of Eq. (2.21) and where n for any given material will also, in general, be a function of the temperature T. Let its initial density be ρ and the rise of temperature of the shell be ΔT, where in our case ΔT will be negative. The density of the shell following the rise in temperature then becomes $\rho(1-3n\Delta T)$. Now let the inner radius of the shell become $r(1+\alpha)$ following the expansion associated with ΔT. The external radius then becomes

$$r(1+\alpha)+dr\left[1+\frac{\partial}{\partial r}(r\alpha)\right] \qquad (11.45)$$

The original mass of the shell is $4\pi\rho r^2 dr$. The mass of the shell after the change in temperature will then be, neglecting second-order terms in α and ΔT,

$$4\pi r^2(1+\alpha)^2 dr\left[1+\frac{\partial}{\partial r}(r\alpha)\right]\rho(1-3n\Delta T)$$
$$= 4\pi\rho r^2 dr\left[1+2\alpha+\frac{\partial}{\partial r}(r\alpha)-3n\Delta T\right] \qquad (11.46)$$

The equation of mass continuity will then be simply

$$2\alpha+\frac{\partial}{\partial r}(r\alpha)-3n\Delta T = 0 \qquad (11.47)$$

Now if the shell were allowed to expand without stretching, its radius would be increased by $rn\Delta T$ instead of $r\alpha$. The stretching required to make it continue to fit the interior is then $r(\alpha-n\Delta T)$. Let us denote by k the quantity $\alpha-n\Delta T$. This is the quantity that we wish to determine. Substituting k for α in Eq. (11.47), we obtain

$$2(k+n\Delta T)+\frac{\partial}{\partial r}[r(k+n\Delta T)]-3n\Delta T = 0$$

$$3k+r\frac{dk}{dr}+r\frac{d}{dr}(n\Delta T) = 0$$

$$\frac{d}{dr}(kr^3) = -r^3\frac{d}{dr}(n\Delta T) \qquad (11.48)$$

Integrating, we obtain

$$kr^3 = -\int_0^r r^3\frac{d}{dr}(n\Delta T)\,dr$$

since the integrated part vanishes at the lower limit; then integrating the integral on the right hand by parts, we obtain

$$k = -n\Delta T+\frac{1}{r^3}\int_0^r 3r^2 n\Delta T\,dr \qquad (11.49)$$

since the integrated term again vanishes at the lower limit. Finally, as k is the stretching that is associated with a given temperature change ΔT occurring over some time interval Δt during the cooling of the earth, the integral stretching K, defined by

$$k = \frac{\partial K}{\partial t}\Delta t \qquad \Delta T = \frac{\partial T}{\partial t}\Delta t \qquad (11.50)$$

will be from Eq. (11.49)

$$\frac{\partial K}{\partial t} = -n\frac{\partial T}{\partial t} + \frac{1}{r^3}\int_0^r 3r^2 n \frac{\partial T}{\partial t} dr \qquad (11.51)$$

The total contraction at the earth's surface is then found from Eq. (11.51) by taking the integral of the second term on the right-hand side for the earth's radius and by then integrating with respect to time from an assumed time of original solidification of the earth's crust. The level of no strain is simply found by setting Eq. (11.51) equal to zero. Such evaluations can be made using a given, assumed thermal history for the earth, such as that of expression (2.121). For such evaluations, there are several relatively undetermined quantities including, among others, the quantity and distribution of radioactive heat sources in the earth, initial values of T and m in Eq. (2.121), variation of the coefficient n with temperature and depth, and the initial time from which the second integral should be taken.

Problem 11.5(a) Using expressions (2.121) and (11.51), a total value of $K = -4.3 \times 10^{-3}$ has been determined for the earth. Calculate the shortening on a great circle at the earth's surface. Discuss.

Ans. 170 km

Problem 11.5(b) Assuming a breaking stress for the earth's crust of 8×10^8 dyne/cm², a modulus of rigidity of 5×10^{11} dyne/cm², and a Young's modulus of 12×10^{11} dyne/cm² and assuming that Hooke's law held up to the breaking point, what would the crustal extension be. Using the value of K from Problem 11.5(a), how many times would the breaking stress be reached during the earth's geologic history. Discuss.

Ans. -0.7×10^{-3}, 6

11.6 Thermal Convection Currents in the Mantle

Up to this point, we have considered that heat dissipation in the earth's crust and mantle has been by conduction alone. We shall be interested in examining here what the conditions are that would permit thermal convection currents to exist in the upper mantle, particularly those conditions having to do with the dimensions of the upper mantle, the apparent viscosity of the asthenosphere, and the anticipated vertical temperature gradient. If convection currents can exist in the mantle, which does appear to be the case, the thermal history of the earth, as given by Section 2.7, and the temperature of the earth as a function of time and depth, as given by expressions such as Eq. (2.121), will have to be altered. Further, the convection cells in the asthenosphere will produce a horizontal, viscous shear stress on the lithosphere, which could, in turn, move sections of the earth's crust relative to one another throughout geologic time and be an important deformational force in the development of the observed earth crustal features.

There are five differential equations that govern this type of motion: the three equations of motion for a viscous fluid, the equation of continuity, and the heat conduction equation; and five dependent variables: the three components of velocity, pressure, and temperature. We shall examine here only the simplest case, that of a harmonic cell distribution in Cartesian coordinates and that for which the coefficients entering the differential equations are taken as constant. We shall impose a constant vertical temperature gradient β and shall be interested in determining what the conditions are for a steady-state convection cell to exist.

For convenience, we shall take our pressure p and temperature T to be those in excess of the hydrostatic condition of a constant downward increase in pressure due to gravity and in excess of the thermostatic condition of the imposed constant vertical temperature gradient. We shall then have that the density at any location will be

$$\rho = \rho_0(1 - \gamma T) \qquad (11.52)$$

where γ is the volume coefficient of expression (2.21); ρ_0 is the hydrostatic, thermostatic density; and where we have ignored the pressure effects on the change in density in comparison with the temperature effects. For steady state in an incompressible fluid for which the velocity of the fluid elements is small, the viscous equations of motions [Eq. (3.51)] become, using Eq. (11.52) where ρ is now used for ρ_0,

$$-\frac{\partial p}{\partial x} + \eta \nabla^2 v_x = 0 \qquad (11.53)$$

$$-\frac{\partial p}{\partial y} + \eta \nabla^2 v_y = 0 \qquad (11.54)$$

$$-\frac{\partial p}{\partial z} + \eta \nabla^2 v_z - g\rho\gamma T = 0 \qquad (11.55)$$

where η is the coefficient of viscosity, and v_x, v_y, v_z are the velocity components.

Repeating Eq. (3.18), we see that the equation of continuity for an incompressible medium is simply

$$\frac{\partial v_x}{\partial x} + \frac{\partial v_y}{\partial y} + \frac{\partial v_z}{\partial z} = 0 \tag{11.56}$$

From the derivation of the heat conduction equation in Section 2.4, it is apparent that the heat flow in our case will be simply to first order, the imposed vertical temperature gradient β multiplied by the vertical particle velocity v_z, so that Eq. (2.54) becomes

$$\alpha \nabla^2 T = \frac{\partial T}{\partial t} = \beta v_z \tag{11.57}$$

where α is the coefficient of thermal diffusivity.

As stated, we shall assume that the small quantities of particle velocity and excess pressure and temperature can be represented in the XY plane by a simple harmonic distribution so that v_x can be given by

$$v_x = v_x(z)\, e^{ilx} e^{imy} \tag{11.58}$$

and similarly for v_y, v_z, p, and T. Substituting expression (11.58) into Eqs. (11.53), (11.54), (11.55), (11.56), and (11.57), we obtain

$$-ilp + \eta\left(-l^2 v_x - m^2 v_x + \frac{d^2 v_x}{dz^2}\right) = 0 \tag{11.59}$$

$$-imp + \eta\left(-l^2 v_y - m^2 v_y + \frac{d^2 v_y}{dz^2}\right) = 0 \tag{11.60}$$

$$-\frac{dp}{dz} + \eta\left(-l^2 v_z - m^2 v_z + \frac{d^2 v_z}{dz^2}\right) - g\rho\gamma T = 0 \tag{11.61}$$

$$ilv_x + imv_y + \frac{dv_z}{dz} = 0 \tag{11.62}$$

$$\alpha\left(-l^2 T - m^2 T + \frac{d^2 T}{dz^2}\right) = \beta v_z \tag{11.63}$$

To eliminate v_x, v_y, and p from the first four equations, we take d/dz of the first, second, and fourth equations, multiply Eq. (11.59) by il and Eq. (11.60) by im and add, and substitute from Eq. (11.62), obtaining

$$(l^2 + m^2)\frac{dp}{dz} + \eta\left(-l^2 - m^2 + \frac{d^2}{dz^2}\right)\left(-\frac{d^2 v_z}{dz^2}\right) = 0$$

and then substitute this for dp/dz in Eq. (11.61), obtaining

$$-\frac{\eta}{l^2 + m^2}\left(-l^2 - m^2 + \frac{d^2}{dz^2}\right)\frac{d^2 v_z}{dz^2} + \eta\left(-l^2 - m^2 + \frac{d^2}{dz^2}\right) v_z - g\rho\gamma T = 0 \tag{11.64}$$

The other equation involving v_z and T only is simply Eq. (11.63).

At the upper and lower boundaries of the convection cell, the motion is horizontal, and there are no horizontal temperature gradients so that $v_z = T = 0$ at both boundaries. Taking, as before, a simple harmonic function satisfying these conditions, we shall assume that

$$v_z = v_{z0} \sin sz \qquad T = T_0 \sin sz \qquad (11.65)$$

where

$$D = \frac{\pi}{s} \qquad (11.66)$$

is the thickness of the convection cell. Substituting Eqs. (11.65) into Eqs. (11.64) and (11.63), we have

$$-\eta(l^2+m^2+s^2)s^2 v_{z0} - \eta(l^2+m^2+s^2)(l^2+m^2)v_{z0} - g\rho\gamma(l^2+m^2)T_0 = 0$$

$$\eta(l^2+m^2+s^2)^2 v_{z0} + g\rho\gamma(l^2+m^2)T_0 = 0 \qquad (11.67)$$

and

$$\beta v_{z0} + \alpha(l^2+m^2+s^2)T_0 = 0 \qquad (11.68)$$

If Eqs. (11.67) and (11.68) are to have a solution, the determinant of their coefficients must vanish, or

$$(l^2+m^2+s^2)^3 - \frac{g\rho\beta\gamma}{\eta\alpha}(l^2+m^2) = 0 \qquad (11.69)$$

so that

$$s^2 = \sqrt[3]{\frac{g\rho\beta\gamma}{\eta\alpha}(l^2+m^2)} - (l^2+m^2) \qquad (11.70)$$

Now the expression (11.70) is an equation of the form

$$y = ax^{1/3} - x \qquad (11.71)$$

which will have a graph as shown in Fig. 11.4. The minimum thickness D, or maximum s, for Eq. (11.71) to have a real-value solution is simply

$$\frac{dy}{dx} = \frac{1}{3} ax_m^{-2/3} - 1 = 0$$

or

$$27 x_m^2 = a^3 \qquad (11.72)$$

so that

$$y_m = 3x_m - x_m = 2x_m = \sqrt{\frac{4}{27} a^3} \qquad (11.73)$$

From Eqs. (11.66), (11.70), and (11.73), the minimum thickness D_m of the cell is then simply

$$D_m = \frac{\pi}{s_m} = \pi \sqrt[4]{\frac{27\eta\alpha}{4g\rho\beta\gamma}} \qquad (11.74)$$

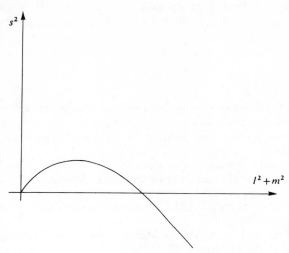

Fig. 11.4

We also note from Eqs. (11.73) and (11.70) that for a convection cell in which $l = m$, the rectangular horizontal dimensions will be twice the minimum thickness dimension.

Equation (11.74) and any other results that might be developed from it as to values of the five dependent variables or the stresses associated with the viscous motion are of limited direct application because, among other things, we have assumed that all the coefficients are constant and that the cell motion can be represented by a simple harmonic solution. Further, we have assumed that the upper mantle is a simple, viscous liquid, which would imply that there would be a cell with complex solution, that is, nonsteady state, exponential time decreasing solution, down to zero temperature gradient. We would anticipate, however, that this would not be the case and that some minimum stress must be overcome to start any motion. For the earth we should also keep in mind that any vertical, or horizontal, temperature gradient that might produce, or might have in the past produced, convection would, in itself, tend to be eliminated by the convective motion, so that any given earth convection cell might continue only for one-quarter to one-half turn before stopping, unless the temperature gradient producing the convection were maintained or rejuvenated by radioactive heating.

Problem 11.6(a) Taking a value of $\eta = 2 \times 10^{22}$ poise and using a value of $\alpha = 0.01$ CGS and $\gamma = 2 \times 10^{-5}/°C$ and using a typical upper mantle density $\rho = 4.0$ and a modest temperature gradient $\beta = 1°C/km$, calculate the minimum thickness for a steady-state convection cell. Discuss.

Ans. 200 km

Problem 11.6(b) Taking an average cell velocity of 1 cm/yr, calculate the total time for one circuit of the cell of Problem 11.6(a). Discuss.

Ans. 120×10^6 yr

References

Anderson, E. M. 1942. *The Dynamics of Faulting.* Edinburgh: Oliver and Boyd.
Biot, M. A. 1965. *Mechanics of Incremental Deformations.* New York: Wiley.
Daly, R. A. 1940. *Strength and Structure of the Earth.* Englewood Cliffs, N.J.: Prentice-Hall.
Garland, G. D. 1971. *Introduction to Geophysics.* Philadelphia: W. B. Saunders.
Hart, P. J., ed. 1969. "The earth's crust and upper mantle," *Am. Geophys. Un.*, Geophys. Monograph 13.
Haskell, N. A. 1935. "The motion of a viscous fluid under a surface load," *Physics*, vol. 6, pp. 265–269.
Heiskanen, W. A., and F. A. Vening Meinesz. 1958. *The Earth and Its Gravity Field.* New York: McGraw-Hill.
Jacobs, J. A., R. D. Russell, and J. T. Wilson. 1959. *Physics and Geology.* New York: McGraw-Hill.
Jeffreys, H. 1952. *The Earth.* Cambridge: Cambridge University Press.
Knopoff, L. 1964. "The convection current hypothesis," *Reviews of Geophysics*, vol. 2, pp. 89–122.
Pekeris, C. L. 1935. "Thermal convection in the interior of the earth," *Royal Astron. Soc., Monthly Notices, Geophys. Suppl.*, vol. 3, pp. 343–367.
Poldervaart, A., ed. 1955. "Crust of the earth," *Geol. Soc. Am.*, Special Paper 62.
Rayleigh, L. 1916. "On convection currents in a horizontal layer of fluid," *Phil. Mag.*, vol. 32, pp. 529–546.
Scheidegger, A. E. 1958. *Principles of Geodynamics.* Berlin: Springer-Verlag.
Stacey, F. D. 1969. *Physics of the Earth.* New York: Wiley.
Takeuchi, H. 1966. *Theory of the Earth's Interior.* Waltham: Blaisdell.
———, S. Uyeda, and H. Kanamori, 1970. *Debate About the Earth.* San Francisco: Freeman.
Vening Meinesz, F. A. 1964. *The Earth's Crust and Mantle.* New York: Elsevier.

Index

Index

Acceleration, Coriolis, 58
   geostrophic, 126
   normal, 182
   referred to moving axes, 56–59
   tangential, 182
   transport, 58
Adiabatic equilibrium, gravitational, 70–72
Adiabatic transformation, 64
Advection, 109
Airy hypothesis, 287
Airy phase, 252
Ampere's law, 305
Amphidromic region, 158
Amplitude, 40
Anomaly, Bouguer, 285
   free air, 285
   isostatic, 285
Associated Legendre polynomials, 49–51
Asthenosphere, 284

Bending moment theorem, 366
Bernoulli's equation, 99
Bessel differential equation, modified, 55
   order n, 53
   zero order, 31
Bessel function, half order, 54
   modified, 55
   order n, 52
   zero order, 32
Biharmonic equation, 361
Bulk modulus, 185

Calorimetric coefficients, 67
Cauchy-Poisson problem, 169
Cauchy's relation, 186
Celestial sphere, 330
Chandler wobble, 347–350
Characteristic equation, 242
Circulation, Bjerknes theorem, 125
   definition, 100
   Kelvin theorem, 99
   oceanographic, 139–142
Clairaut's differential equation, 296
Clairaut's formula, 275

Clausius-Clapeyron equation, 69
Coefficient, bulk modulus, 185
   calorimetric, 67
   conductivity, 306
   diffusion, 110
   electric susceptibility, 304
   exchange, 111
   flexural rigidity, 366
   incompressibility, 185
   magnetic susceptibility, 304
   Poisson's ratio, 185
   Rayleigh reflection, 199
   reflection, 194
   refraction, 194
   resistivity, 306
   rigidity, 186
   shear modulus, 186
   thermal conductivity, 72
   thermal diffusivity, 74
   thermoelastic, 66
   thermometric, 66
   viscosity, 102
   Young's modulus, 185
Compensation, Airy hypothesis, 287
   isostatic, 284
   Pratt hypothesis, 286
Conjugate functions, 93
Continuity, equation of, 8, 94–96
Contractions, 185
Coordinates, curvilinear, 13
   cylindrical, 20–21
   orthogonal, 15
   spherical, 18–20
Core, 230
Coriolis acceleration, 58
Correction, Bouguer, 283
   elevation, 283
   free air, 283
   isostatic, 284
   topographic, 283
Critical angle, 196
Crust, 230, 284
Curl, 9
Current, bottom, 138
   deep, 138

drift, 133
geostrophic, 126
gradient, 133
inertia, 131
surface, 138

Declination, 307
Deflection of the vertical, 281
Dielectric constant, 304
Diffusion, definition, 109
eddy, 109
molecular, 109
Dilatation, 185
Dip, 307
Dispersion, definition, 242
general, 241–248
geometrical, 247
material, 247
Displacement current, 306
Divergence, 8
Doublet, 79
Dyadic, conjugate, 12
idemfactor, 12
momental, 183
pure strain, 185
rate of strain, 101
rotation, 185
skew-symmetric, 12
strain, 184
stress, 184
symmetric, 12
Dynamo, atmospheric, 319–321
internal, 316–319

Ecliptic, 330
Eikonal equation, 203
Ekman spiral, 135
Elastic constants, 185
Electric conductivity, 306
Electric displacement, 303
Electric resistivity, 306
Electric susceptibility, 304
Electromagnetic induction, 321–324
Electromotive force, 306
Ellipsoid of revolution, 269
Energy, conservation, 182
free, 64
intrinsic, 63
kinetic, 182
potential, 182
Enthalpy, 64
Entropy, 63

Equilibrium, gravitational adiabatic, 70–72
indifferent, 71
Equinoxes, 330
Equipotential surface, 92, 265
Error function, 78
Eulerian free motion, 338
Euler's equation, 184
Exchange coefficients, 111
Expansion, 94
Extensions, 185

Faraday's law, 305
Fermat's principle, 207
Flexural rigidity, 366
Fluid, perfect, 89
viscous, 101
Flushing time, 108
Fourier integral theorem, 39
Fourier series, even function, 35
odd function, 35
Fourier transforms, 40
Free energy, 64
Free motion, 338
Free oscillations, 354
Frequency, 41
Frictional depth, 136

Gal, 261
Gamma, 303
Gauss, 303
Gauss' law, 263
Gauss' theorem, 11
Geocentric latitude, 274
Geographic latitude, 274
Geoid, fictitious, 284
real, 284
Geopotential, 269
Geostrophic acceleration, 126
Geostrophic current, 126
Geothermal flux, 82
Gibbs' functions, 65
Gradient operator, 7
Gravitational attraction, 261
Gravitational flux, 262
Gravity, anomaly, 276
definition, 261
Green's equivalent layer, 266
Green's theorem, 11
Gyroscopic compass, 335–337

Harmonics, sectorial, 51

spherical, 50
tesseral, 51
zonal, 51
Heat conduction equation, 72–74
Heat flux, 72
Heat, periodic flow, 74–76
Helland-Hansen formula, 128
Hooke's law, 185
Huygens' principle, 209
Hydrodynamic equations, Eulerian form, 89
  Lagrangian form, 89
Hydrostatic pressure, 184

Idemfactor, 12
Image source, 239
Inclination, 307
Incompressibility, 185
Index of refraction, 203
Indirect effect, 284
Inertia, moments of, 183
  principal moments of, 183
  products of, 183
Infinitesimal strain theory, 184
Intensity, electric, 303
  magnetic, 303
Intensity of magnetization, 304
Intramission angle, 200
Irrotational motion, 91
Isobaric surfaces, 124
Isoclinic lines, 307
Isodynamic lines, 307
Isogonic lines, 307
Isopycnal surfaces, 124
Isoteric surfaces, 124
Isothermal transformation, 63

Kelvin's circulation theorem, 99
Kelvin waves, 157
Kepler's laws, 329

Lamé constants, 185
Laplace's equation, 27–34
Laplacian operator, 9
Latent heat, 69
Latitude, geocentric, 274
  geographic, 274
Legendre differential equation, 25
Legendre polynomials, associated, 49–51
  definition, 22
  P_0 to P_4, 27
  Rodrigues' formula, 25

Lithosphere, 284
Lloyd mirror dependence, 240
Love waves, 256
Love's numbers, 344

MacCullagh's formula, 300
Magnetic dipole, 304
Magnetic induction, 304
Magnetic moment, 305
Magnetic susceptibility, 304
Magnetomotive force, 305
Mantle, 230, 284
Margules' equation, 130
Mass, center of, 182
Maxwell's equations, 306
Maxwell's relations, 65
Mean sea level, 269
Merian's formula, 153
Milligal, 261
Mixing, estuary, 117–122
  ocean, 114–117
Mohorovicic discontinuity, 230
Momentum, angular, 182
  linear, 182
Motion, equation of, hydrodynamic, 96–99
  internal friction, 131
  viscous fluid, 103
Motion, tidal, amphidromic region, 158
  cotidal lines, 157
  direct, 152
  inverted, 152

Navier-Stokes equation, 103
Nodes, 234
Normal modes, definition, 202, 234
  dispersive, 248–253
Nutation, 338

Ohm's law, 306

Period, 40
Permeability, 304
Poisson's equation, 264
Poisson's ratio, 185
Poisson's relation, 186
Polar wandering, 321
Polarity reversals, 320
Polarization, 303
Postglacial uplift, 369–370
Potential, displacement, 192
  magnetic vector, 312

scalar, 10
thermodynamic, 64
vector, 10
velocity, 91
Potential temperature, 71
Pratt hypothesis, 286
Precession of the equinoxes, 330–335
Pressure, hydrostatic, 184
stress, 184

Radau's approximation, 299
Rayleigh method, 146
Rayleigh reflection coefficient, 199
Rayleigh waves, 256
Rays, definition, 202
flat earth, 210–225
spherical earth, 225–233
Reflections, 190
Refractions, 190
Relaxation time, 315
Rigidity, 186
Rodrigues' formula, 25
Rotation, 9

Salinity, 95
Secular variation, 320
Seiches, 152
Shadow zone, 224
Shear modulus, 186
Shears, 185
Shida's number, 344
Simple harmonic motion, amplitude, 40
frequency, 41
period, 40
phase, 41
wave number, 45
wavelength, 44
Skin depth, 322
Snell's law, 207, 212
Specific volume, 63
Spectrum, 41
Spherical harmonics, 50–51
State, equation of, 65
Stefan-Boltzmann law, 65
Stokes' integral, 280
Stokes' theorem, 11
Stress, compressive, 184
shearing, 184
tensional, 184
Surface waves, 253

Temperature, potential, 71

Tension, 184
Thermal conductivity, 72
Thermal contraction, 370–373
Thermal convection currents, 373–378
Thermal diffusivity, 74
Thermodynamic potential, 64
Thermodynamics, implicit functions, 65–70
Thermoelastic coefficients, 66
Thermometric coefficients, 66
Tidal friction, 350–353
Tidal motion, 152
Tides, amphidromic region, 158
co-oscillation, 153
cotidal lines, 158
diurnal, 342
dynamical theory, 176–178
earth, 343–347
equilibrium theory, 173–176
long period, 342
marine, 339–342
neap, 342
semidiurnal, 342
spring, 342
Torque, 182
Total reflection, 196
Transformation, adiabatic, 64
isothermal, 63
Trochoid, 168

Variation, 307
Vectors, cross product, 5
curl, 9
differential operator ∇, 7
divergence, 8
dot product, 5
gradient, 7
irrotational, 9
rectangular components, 4
rotation, 9
scalar product, 5
solenoidal, 9
triple scalar product, 6
triple vector product, 6
unit, 3
vector product, 5
vortex, 142
Velocity, dilatational, 189
distortional, 189
equivoluminal, 189
group, 244
irrotational, 189

periodic heat flow, 75
permanent waves, 168
phase, 242
propagation, 42
referred to moving axes, 56–59
rotational, 189
surface waves, 165
tidal waves, 146
tidal waves, second order, 147
trace, 217
Vibration, longitudinal, 189
   transverse, 189
Viscosity, coefficient, 102
   eddy, 113, 132
   molecular, 113, 132
Vorticity, equation, 142
   planetary, 142

Wave equation, 42–49
Wave fronts, 206
Wave guide, 234
Wave number, 45
Wave surfaces, 202, 206
Wavelength, 44
Waves, dilatational, 189
   distortional, 189
   driven tidal, 150–152

elastic, 186–190
equivoluminal, 189
forced, 150
free, 150
gravity, 144
internal, 162
internal, tidal, 159–163
irrotational, 189
Kelvin, 157
long, 144
Love, 256
permanent, 166–169
Rayleigh, 256
reflection, 190
refraction, 190
rotational, 189
seiches, 152
short, 144
standing, 234
surface, 163–166, 253
tidal, 144
tidal, geostrophic, 154
Westward drift, 320
Wobble, Chandler, 347–350
   definition, 337

Young's modulus, 185